Catastrophe Theory

Catastrophe
Theory
Second Edition

Domenico P. L. Castrigiano
Sandra A. Hayes

Technische Universität München
Zentrum Mathematik
Garching, Germany

40 Computer-generated Graphics by Frank Hofmaier
93 Solved Exercises

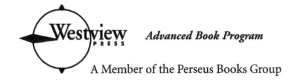

Westview PRESS   *Advanced Book Program*

A Member of the Perseus Books Group

Copyright © 2004 by Westview Press, A Member of the Perseus Books Group
First edition published by Addison-Wesley Publishing Company, 1993.

Westview Press books are available at special discounts for bulk purchases in the United States by corporations, institutions, and other organizations. For more information, please contact the Special Markets Department at the Perseus Books Group, 11 Cambridge Center, Cambridge MA 02142, or call (617) 252-5298 or (800) 255-1514 or email j.mccrary@perseusbooks.com.

Published in the United States of America by Westview Press, 5500 Central Avenue, Boulder, Colorado 80301–2877 and in the United Kingdom by Westview Press, 12 Hid's Copse Road, Cumnor Hill,Oxford OX2 9JJ.

Find us on the World Wide Web at www.westviewpress.com

A Cataloging-in-Publication data record for this book is available from the Library of Congress.

ISBN 0-8133-4126-4 (hc)   ISBN 0-8133-4125-6 (pbk)

The paper used in this publication meets the requirements of the American National Standard for Permanence of Paper for Printed Library Materials Z39.48–1984.

10   9   8   7   6   5   4   3   2   1

To the future ...

Contents

Foreword

Domenico P. L. Castrigiano and Sandra A. Hayes have written what is very probably the best book on the market for an introduction to catastrophe theory. I am very much obliged to the authors for putting so much emphasis on what most contemporary scientists consider, if they are strictly mathematicians, as just a part of the theory of local singularities of smooth morphisms, or, if they are interested in the wider ambitions of this theory, as a dubious methodology concerning the stability (or instability) of natural systems. The former category can but find cause for satisfaction in the authors' presentation. Two steps leading to the pure theory are very clearly marked. The first, relatively formal, concerns local algebra and is dominated by Nakayama's Lemma. The second, functional in aspect, appears when one introduces the germs of the diffeomorphisms generated by integration of local vector fields. Difficulties arise here which cannot be dodged, all of them centered on the Malgrange–Mather preparation theorem; the real jets have to be complexified, which requires an extension theorem like Nirenberg's. It is striking to realize that it took two or three centuries for the following question to be asked: When is a germ of a function C^∞ equivalent (via local diffeomorphisms of source and target) to its development limited to a finite order k? It will remain one of the essential historical achievements of catastrophe theory to have set this progress in motion.

On the controversial question of applications, the authors very wisely keep to examples that are indisputable because drawn from exact natural sciences— "rational" mechanics, geometrical optics, ... But if we remain in the scientific realm of what the physicist Wigner so happily called the "unreasonable efficiency of mathematics in natural sciences," then catastrophe theory holds little interest per se, since every result can be numerically computed by means of equations furnished by some existing theory. The situation is a little less simple if we consider systems depending on parameters where there reigns an opposition between slow and fast time, and where the precision of an asymptotic evolution is to be controlled. The theory of structural stability of differential systems on a manifold comes into play here, a theory in which very few general results are available. My general idea could be formulated as follows: If we consider an unfolding, we can obtain a "qualitative" intelligence about the behavior of a system in the neighborhood of an unstable equilibrium point. This idea has not enjoyed much credit with orthodox scientists, no doubt because for them only numerical exactness allows prediction, and therefore efficient action. We must observe then, that the domain governed by the exactness of laws dear to Wigner rests, in the last analysis, on analytic continuation, which alone permits reliable extrapolation of a numerical function. Now we should see that the theory of singularity unfolding is a particular case of a general theory of the analytic category, namely, the theory of "flat deformations of an analytic set." Algebraists now tell us that in the theory of flat local deformations of an analytic set, only the hypersurface case has a smooth unfolding of finite dimension. So that if we wanted to continue the scientific domain of calculable exact laws, we would be justified in considering the instance where an analytic process leads to a singularity of codimension one (in internal variables). Might we not then expect that the process (defined for example as the propagation of an action) be diffused (in external variables) and subsequently propagated in the unfolding according to a mode that is to be defined? Such an argument allows one to think that the Wignerian domain of exact laws can be extended into a region where physical processes are no longer calculable but where analytic continuation remains "qualitatively" valid.

I shall take no further these venturesome considerations. But I am sure that readers will find, in this book by D. Castrigiano and S. Hayes, a way into recently charted mathematical territory that can, without exaggeration, be called fundamental. The whole of qualitative dynamics, all the "chaos" theories talked about so much today, depend more or less on it. The book *Catastrophe Theory* seems the best way of acquiring this indispensable culture.

René Thom

The inventor of Catastrophe Theory, Professor René Thom, passed away on October 25, 2002, at the age of 79 in Bures-sur-Yvette.

Preface to the First Edition

An international workshop on "Modeling Processes of Structural Change in Social Systems" organized by the Max-Planck-Gesellschaft in Starnberg, Germany, in 1986 was the original impetus for our interest in Catastrophe Theory, a branch of analysis used to model such phenomena. It was a novelty for us to hear social scientists dealing with this theory; there are only very few branches of pure mathematics that have ever been of direct service to the humanities. Another striking aspect of the workshop was that many of the social scientists were former physicists.

As mathematicians, we wanted afterward to see the proofs of the theorems applied. According to the maxim "What better way to learn, than to teach," we decided to hold a course on Catastrophe Theory at the Technical University of Munich. Fortunately, at that time we were able to invite René Thom to come to Munich to speak on "Some Aspects of Bifurcation Theory or the Way to Chaos" and "Aristotelian Physics and Catastrophe Theory." During that same year, we also had the pleasure of having E. C. Zeeman as our guest, who gave brilliant talks on "Stability of Dynamical Systems" and "Applications of Catastrophe Theory to the Social Sciences," as well as I. Stewart, who spoke on "Whatever Happened to Catastrophe Theory?," "Catastrophes, Singularities and Bifurcations," and "Dynamics with Symmetry," bringing us up to date on recent applications in the natural sciences. An interdisciplinary seminar on applications to physics, medicine, economics, and sociology accompanied our course and was a challenging new experience for everyone involved.

A surprising feature of Catastrophe Theory is its elementary nature—an aspect emphasized in this book by presupposing only a good background in calculus of several real variables together with a little algebra and linear algebra; the theorems needed from calculus are even included in the Appendix (Implicit Function Theorem, Existence of Solutions of a Differential Equation, Derivative of a Limiting Function, Rank Theorem, Green's Theorem). It is our impression that there is no textbook aimed at advanced undergraduate students with the primary intention of giving a direct, complete proof of Thom's classification theorems for the so-called *elementary catastrophes* and for versal unfoldings, and we have attempted to fill this gap. This entailed a lot of cross-referencing, a new development of part of the material to avoid using manifolds and Lie groups, and a clarification of some ambiguous points in the standard proofs.

Catastrophe Theory is an intriguing, beautiful field of pure mathematics, which deserves to be more accessible to beginning students as well as to non-specialists who want to understand the mathematics involved. It is a natural introduction to bifurcation theory and to the rapidly growing and very popular field of dynamical systems.

There is no question that most of the material in this book can be found in some form elsewhere. After studying the known proofs of Thom's theorems, we felt our task was to minimize technicalities and maximize rigor without sacrificing insight. To increase comprehension, the book contains many examples, a large number of exercises and their solutions, an abundance of figures, and computer-generated graphics of the elementary catastrophes. The figures and graphics were produced by Stephan Haas, a student who participated in both the course and the seminar.

In everyday language, a catastrophe refers to an unexpected disastrous event, e.g., an earthquake, stock market crash, or heart attack. The unexpectedness is because the conditions surrounding the event change slowly and continuously. Nevertheless, a sudden discontinuity—a radical jump—occurs, overthrowing the existing order. This is also the essence of the mathematical theory, where not only are unstable critical points of smooth functions classified but also their embedding in stable families.

Thom's celebrated theorem on the **7** elementary catastrophes classifies degenerate critical points of smooth functions up to codimension at most 4. Insiders know that the classification can be extended to include codimension at most 6, but this classification is not easily accessible. Here the classification for codimensions 5 and 6 is treated in detail in the exercises to Chapter 6 and their solutions along with a classification for the case of codimension 7 with corank 2. It is also shown in the exercises to Chapter 6 that no further finite classification is possible by giving examples of unimodal catastrophes in the codimension 7, corank 3 case and in the codimension 8, corank 2 case.

The proof of Thom's classification theory for degenerate critical points uses Morse's Lemma and the Reduction Lemma. Here both of these lemmas follow from a local diagonalization of symmetric matrices of smooth functions (Lemma (36) of Chapter 1), and we think the resulting conciseness is an improvement over the usual proofs.

Another new feature concerns Mather's necessary condition for determinacy, which is shown to follow from a Linearization Lemma (37) proved in Chapter 4. This lemma addresses a general property of curves in transformation groups and clarifies an ambiguous point in Mather's original proof as well as in all other known proofs (see the remark in Chapter 4 after (32)). An interesting consequence of this lemma is a general lifting property for curves in orbits of transformation groups.

Two other aspects of this book that we consider improvements are a sharper estimate for the codimension than the usual bound (see Theorem (19) and Exercise 10 of Chapter 5) and the use of a criterion for the equivalence of germs to calculate determinacy (Lemma (22), Chapter 4).

Regarding the organization of this book, in Chapter 1 the first step is taken toward classifying critical points of smooth functions. First the functions of one variable are completely treated, then the nondegenerate critical points of arbitrary smooth functions. The methods employed demonstrate the crucial role of the Taylor expansion for such classifications. In Chapter 2 the notions "catastrophe" and "stability" are explained in order to give an idea what the entire theory is about at an early stage. The first two elementary catastrophes, the fold and the cusp, which motivate the further development, are introduced. A basic reason for being able to achieve a finite classification is the Reduction Lemma, which is proved in Chapter 3.

Chapter 4 is devoted to the question of when a smooth function f is determined locally by one of its Taylor polynomials, meaning that every other smooth function with the same Taylor polynomial coincides with f locally up to a smooth coordinate change. In this chapter, Mather's sufficient condition for determinacy as well as his necessary condition are proved (Theorems (24) and (38)). These conditions are algebraic, and the bit of ideal theory needed is developed from scratch. The examples included in the text illustrate the practicality of these conditions. A characterization of finite determinacy (Theorem (39)) also narrows the determinacy down to just two possible values. Then Lemma (22) can be used to decide which value holds (see Exercise 4.13).

Chapter 5 treats the codimension of a smooth function f. The codimension gives the minimal number of parameters needed for a versal unfolding of f. This is the context in which some linear algebra comes to bear. After showing that finite determinacy is equivalent to having a finite codimension (Corollary (8)), two lower bounds for the codimension are proved. The first is in terms of determinacy (Corollary (9)); the second is with respect to the corank (Theorem (19)) and is essential to classify the **7** elementary catastrophes.

The machinery developed up to this point, together with the classification of cubic forms on $\mathbf{R}^2$, now allows the proof of Thom's Classification of the Seven Elementary Catastrophes to be given in Chapter 6. This concludes the first part of the investigations.

Chapter 7 presents an introduction to the real goal of Catastrophe Theory— to classify smooth functions according to their behavior under perturbation or, in other words, to find their versal unfoldings. The fundamental theorem on universal unfoldings, which achieves this goal, is explained in Chapter 7, and the theorems needed to prove this are contained in the remaining three chapters.

Transversality, another notion from linear algebra, is the topic of Chapter 8; it is the major ingredient necessary to determine whether a smooth function possesses a versal unfolding or not and also to construct it (Lemma (2)). This notion is introduced in a pragmatic and direct way.

The Malgrange–Mather Preparation Theorem is proved in Chapter 9. The presentation here is derived basically from Golubitsky and Guillemin, *Stable Mappings and their Singularities* [GG]. This is the most difficult part of the proof of Thom's Fundamental Theorem on Universal Unfoldings. However, the result actually needed is easy to understand and is given in (1). Taking this for granted, the reader may omit Chapter 9.

In Chapter 10 it is shown that a smooth function has a versal unfolding exactly when it is finitely determined (Theorem (7)) and a criterion (6) for versality is proved using transversality. This gives an explicit method for constructing versal unfoldings. The culmination of the theory is the classification of versal unfoldings (8). Normal forms for the universal unfoldings of functions of codimension less than 7 are given in Chapter 7, Table 1, and in Exercise 7.6.

The authors have developed and tested the material in this book in three different courses on Catastrophe Theory. If the Malgrange–Mather preparation theorem is assumed, the remaining chapters can be covered in a 12-week course meeting four hours weekly; the exercises were treated in problem sessions held two hours a week. On the basis of this experience, we know that it is possible to offer advanced undergraduate students the opportunity to learn this mathematical theory. Actually, a course on Catastrophe Theory is a natural extension of a good calculus course. The subject inspires and motivates students, making teaching all the more pleasurable.

Aside from the original papers, in particular Thom's more philosophical than mathematical work *Stabilité Structurelle et Morphogénèse* [T1], the literature which we highly recommend to the reader are Trotman, *The Classification of Elementary Catastrophes of Codimension* ≤ 5 [Tro], Wasserman, *Stability of Unfoldings* [W], Bröcker and Lander, *Differentiable Germs and Catastrophes* [BL], Poston and Stewart, *Catastrophe Theory and Its Applications* [PS], Martinet, *Singularities of Smooth Functions and Maps* [M], and Demazure, *Bifurcations and Catastrophes* [D].

In concluding, we are happy to finally have the opportunity to acknowledge the students of our courses whose interest motivated us—Stephan Haas for the arduous and time-consuming task of producing the computer-generated graphics, Stephan Sautter for correcting the manuscript, and last but not least to thank Maria Jarisch for the excellently typed manuscript.

Domenico P. L. Castrigiano
Sandra A. Hayes

Preface to the Second Edition

Catastrophe Theory—a beautiful mathematical theory with many controversial applications—was introduced in the 1960s by René Thom and is an area of continual interest. Indeed, new applications of this theory to a wide range of fields including biology, economics, and chemical kinetics attest to its permanent usefulness for practical modelling. Catastrophe Theory is also fundamental to the rapidly expanding area of dynamical systems, treating in particular the phenomena of bifurcation and chaos.

For these reasons this new edition of the book *Catastrophe Theory* was indispensible, primarily because the presentation of the theory is easily comprehensible and complete. Thom himself considered the book unique, since its approach is pragmatic, the material self-contained, and the style is as elementary as possible, presuming only knowledge of calculus and linear algebra at an advanced undergraduate level in order to be accessible to a wide audience.

Perhaps more important, however, this edition includes two new chapters on the **genericity** and the **stability** of unfoldings.

In Chapter 11 it is shown that Thom's classification of the versal unfoldings of germs of codimension at most 4 pertains to a generic set, i.e., such unfoldings form an open and dense subset of the set of all unfoldings with at most 4 parameters (see (38)). Roughly speaking, almost all unfoldings with a maximum of 4 parameters are classified by Thom's theorem. It is only after proving this genericity that the relevance of Catastrophe Theory becomes evident. Surprisingly enough, however, a proof has never appeared before in any textbook.

Stability of an unfolding means a certain kind of persistence under the influence of small perturbations. The main result (12) in Chapter 12, which completes the theory of unfoldings, is that stability and versality imply each other. Thus, Catastrophe Theory reveals in a mathematically rigorous manner the true complementary nature of the seemingly unreconcilable notions of versality and stability, that is preserving identity in spite of development. This result has also never been available in textbook form and certainly enhances the depth of the theory.

Many sources were utilized for the conception of Chapter 11, but mainly Martinet, *Singularities of Smooth Functions and Maps* [M] and Trotman, *The*

Classification of Elementary Catastrophes of Codimension ≤ 5 [Tro]. The presentation of stability in Chapter 12 is derived basically from Wassermann, *Stability of Unfoldings* [W].

The original chapters have been revised and new material is included. Most important is the incorporation of the theorem on the uniqueness of the **residual singularity** (4.48), which is needed for the classification of critical points. Until now, no proof of the uniqueness was available in the literature, except for a rough sketch by Thom in *Stabilité Structurelle et Morphogénèse* [T1] using a result from an extensive theory developed by Tougeron [Tou]. The proof presented here due to Frank Hofmaier [H], a young colleague at the Technical University of Munich, does not require additional tools and has not been published before.

The material is still self-contained. Only Sard's Theorem, which enters the proof of the *Fundamental Lemma on Transversality* (11.18) and which is not a standard result of calculus, is not proved but an appropriate version of it is presented in Appendix A.6.

I am greatly indebted to *Frank Hofmaier*, who produced the computer-generated graphics and assisted me permanently in preparing this edition. Special thanks are due to Professor *David Chillingworth* for his critical reading of the new chapters.

Domenico P. L. Castrigiano

It is a particular pleasure for us to thank our publisher *Holly Hodder*. She is a dynamo, and we greatly appreciate her co-operation and courtesy. Thanks also go to *Jeffrey Robbins*, who encouraged us in the early stages.

Domenico P. L. Castrigiano   *Sandra A. Hayes*

Chapter 1

Nondegenerate Critical Points: The Morse Lemma

The most basic notion of Catastrophe Theory is introduced in this chapter, namely, that of a **critical point** of a smooth function. This notion is also referred to in the standard literature as a singular point or a singularity. An essential distinction is made by dividing these points into two classes—the nondegenerate and the **degenerate critical points**. The degenerate ones are more difficult to handle and will be discussed in later chapters, whereas the nondegenerate critical points can be completely treated here.

The main theorem about nondegenerate critical points is **Morse's Lemma**, which classifies all such points. Roughly speaking, the Morse Lemma states that locally around a nondegenerate critical point a smooth function of n real variables $x_1, \ldots, x_n$ can be transformed to a simple standard form by changing coordinates. There are exactly $n + 1$ such forms, and these are the **quadratic forms**

$$-x_1^2 - \cdots - x_s^2 + x_{s+1}^2 + \cdots + x_n^2,$$

where $s = 0, 1, \ldots, n$. To each function corresponds exactly one canonical quadratic form.

The proof of Morse's Lemma given here is based on the Taylor expansion of a smooth function, on the classification of quadratic forms on $\mathbf{R}^n$, and on a lemma locally diagonalizing a symmetric matrix of smooth functions. This lemma also plays a central role in proving the Reduction Lemma in Chapter 3. The classification of quadratic forms follows immediately from Sylvester's Law of Inertia, which will also be proved here for the sake of completeness.

The treatment of nondegenerate critical points in this chapter is an indispensable first step toward classifying degenerate critical points. Moreover, it exemplifies the content of what follows. The fundamental theme of Catastrophe Theory, the classification of critical points of smooth functions, is illustrated here by a simple case, and the central role the Taylor expansion plays in achieving this classification becomes apparent.

Definition and Convention. Let U be an open subset of $\mathbf{R}^n$. A function $f: U \to \mathbf{R}$ is **smooth** if it has derivatives of arbitrary order. To avoid boring repetitions, f will always denote a smooth function defined on an open subset U of $\mathbf{R}^n$.

If k is a nonnegative integer, then the kth derivative of f at a point p in U will be denoted by $D^k f(p)$, where $D^0 f(p) := f(p)$ and $D^1 f(p) := Df(p)$. For functions of one variable, the usual notation for derivatives will also be used, i.e., $f(p), f'(p), f''(p), \ldots, f^{(k)}(p)$. The fact that a function f is smooth is expressed in coordinates by saying that all partial derivatives of f of arbitrary order exist at every point $p \in U$. The notation $D_i^k f(p)$ as well as $(\partial^k f/\partial x_i^k)(p)$ or $\partial_i^k f(p)$ means the ith partial derivative of f at the point p of order k for $k \in \mathbf{N}$ and $i = 1, \ldots, n$. For $k = 1$ we simply write $D_i f(p)$ or $(\partial f/\partial x_i)(p) = \partial_i f(p)$; moreover, set $D_i^0 f(p) := f(p)$. Recall that $Df(p)$ is the linear map $\mathbf{R}^n \to \mathbf{R}$ given by $(y_1, \ldots, y_n) \mapsto D_1 f(p)y_1 + \cdots + D_n f(p)y_n$. The **gradient** of f at p is the row vector $(D_1 f(p), \ldots, D_n f(p))$ in $\mathbf{R}^n$. Frequently, $Df(p)$ will be identified with this vector.

A geometric picture of a smooth function emerges after its critical points are determined, since that is where the function can change its behavior.

Definition. A point $p \in U$ is called a **critical point** of f if the derivative $Df(p)$ of f at p vanishes.

If a smooth function has no critical point, then it looks locally like the projection onto the first coordinate, as will be proved later.

Example (1). $f: \mathbf{R}^n \to \mathbf{R}$, $f(x_1, \ldots, x_n) := x_1$, has no critical point.

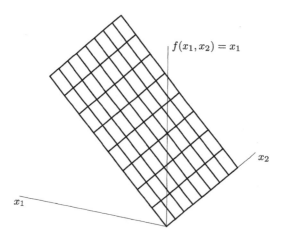

$f(x_1, x_2) = x_1$

The next two examples show smooth functions with exactly one critical point, which is at the origin.

Example (2). For $f: \mathbf{R} \rightarrow \mathbf{R}$, $f(x) := x^2$ ($f(x) := -x^2$), the critical point is a minimum (a maximum).

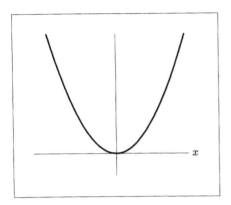 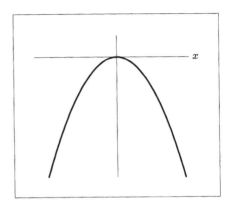

Example (3). For $f: \mathbf{R} \rightarrow \mathbf{R}$, $f(x) := x^3$, the critical point is a saddle point.

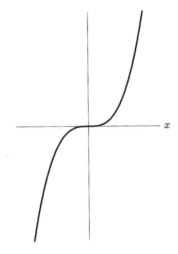

The critical points in the last examples are all isolated.

Definition. A critical point $p \in U$ of f is called **isolated** if there is a neighborhood V of p in U such that no other point in V is critical.

Critical points are not necessarily isolated:

Example (4). $f: \mathbf{R} \rightarrow \mathbf{R}$, $f(x) := \exp(-1/x)$ for $x > 0$ and $f(x) := 0$ for $x \leq 0$, is smooth with the nonpositive axis as the set of critical points.

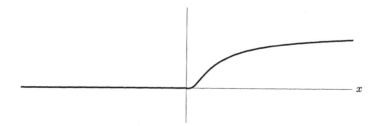

It will be shown that a smooth function of one variable looks locally like Example (2) around a point p if p is a critical point that is nondegenerate. Before we introduce this notion, recall that the **Hessian matrix** of f is the matrix

$$\begin{pmatrix} D_1^2 f(p) & D_2 D_1 f(p) \ldots & D_n D_1 f(p) \\ \vdots & & \\ D_1 D_n f(p) & D_2 D_n f(p) \ldots & D_n^2(p) \end{pmatrix}$$

of all second-order partial derivatives of f at p, where $D_i D_j f(p)$ is the mixed partial derivative

$$\frac{\partial^2 f}{\partial x_i \partial x_j}(p)$$

of f at the point p with respect to the jth and the ith coordinates. This $n \times n$ symmetric matrix can be canonically identified with the second derivative $D^2 f(p)$ of f at p. The Hessian matrix is used in elementary calculus to determine whether a critical point is a maximum, a minimum, or a saddle point.

> **Definition.**   A critical point $p \in U$ of f is **nondegenerate** if the Hessian matrix $D^2 f(p)$ of f at p is invertible. Otherwise the point p is called **degenerate**.

For one variable, a point p is a nondegenerate critical point if and only if $f'(p) = 0$ and $f''(p) \neq 0$ holds. Therefore, the only nondegenerate critical points in this case are isolated extrema, as in Example (2). On the other hand, there are isolated extrema that are degenerate:

> **Example (5).**   The following functions $f \colon \mathbf{R} \to \mathbf{R}$ have the origin as an isolated degenerate critical point.

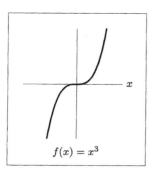
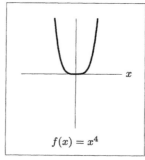
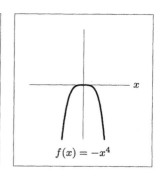

$$f(x) = x^3 \qquad\qquad f(x) = x^4 \qquad\qquad f(x) = -x^4$$

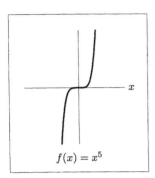

$f(x) = x^5$

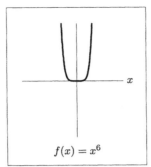

$f(x) = x^6$

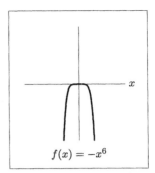

$f(x) = -x^6$

Notice that degenerate critical points need not be isolated, as Example (4) shows. In contrast to the one-dimensional case, nondegenerate critical points of functions of more than one variable are not necessarily extrema. This can be seen by the origin of the **hyperbolic paraboloid**, which is a saddle point:

Example (6). $f: \mathbf{R}^2 \to \mathbf{R}$, $f(x,y) := x^2 - y^2$, has only the origin as a critical point; this is nondegenerate but not an extremum.

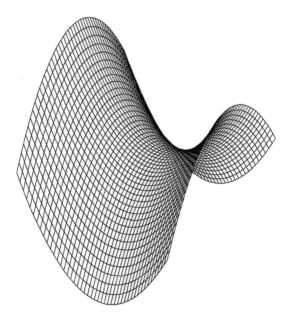

Nondegenerate critical points of functions of several variables are always isolated, however, just as in the one-dimensional case. This results from the Inverse Function Theorem, as will be shown in (30). As a matter of fact, it will follow from the Morse Lemma that in a neighborhood of a nondegenerate critical point, every smooth function of two variables looks either like Example (6) or like an **elliptic paraboloid**:

Example (7). For $f\colon \mathbf{R}^2 \to \mathbf{R}$, $f(x,y) := x^2 + y^2$ ($f(x,y) := -x^2 - y^2$), the critical point is nondegenerate and a minimum (a maximum).

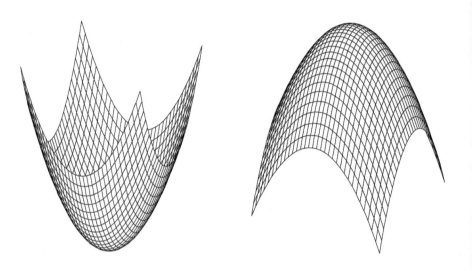

Examples of degenerate critical points of functions of two variables are given below.

Example (8). For $f\colon \mathbf{R}^2 \to \mathbf{R}$, $f(x,y) := x^2$, the set of all critical points is the y-axis, and every critical point is degenerate.

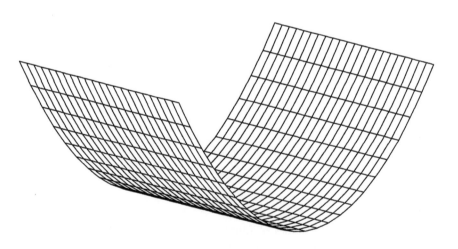

Example (9). $f\colon \mathbf{R}^2 \to \mathbf{R}$, $f(x,y) := x^3 + y^3$, has just the origin as a critical point, and this is degenerate.

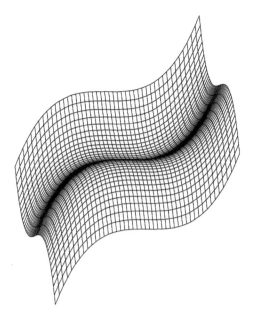

The graph of the next function is referred to as the **monkey saddle**.

Example (10). $f \colon \mathbf{R}^2 \to \mathbf{R}$, $f(x,y) := x^3 - xy^2$, has just the origin as a critical point, and this is degenerate. See left figure below.

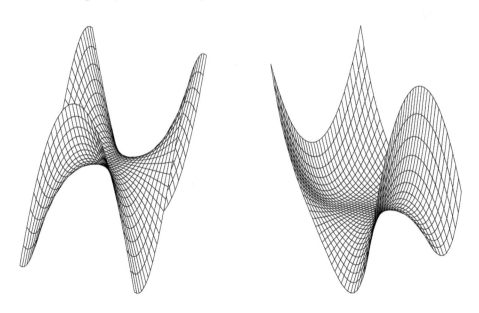

Example (11). $f \colon \mathbf{R}^2 \to \mathbf{R}$, $f(x,y) := x^2 y + y^4$, has just the origin as a critical point, and this is degenerate. See right figure above.

The seven examples,

$$(12) \qquad\qquad x^3, x^4, x^5, x^6, x^3 + y^3, x^3 - xy^2, x^2y + y^4,$$

(see (5),(9),(10),(11)) are basic to Catastrophe Theory; their critical points are called the **7** elementary catastrophes. At the heart of Catastrophe Theory are two theorems. The first theorem is René Thom's celebrated classification of degenerate critical points of those smooth functions f of n variables whose so-called codimension is at most 4. The classification of the universal unfoldings of critical points is the second major theorem of Catastrophe Theory; these notions are defined in Chapter 7. The main goal of this book is to give a complete proof of these two theorems. The first classification theorem states that after a smooth change of coordinates, f is the sum of two functions P and Q of different variables: P is one of the seven preceding polynominals (12) in one or two variables up to a change of sign, and Q is a nondegenerate quadratic form in the remaining $n-1$ respectively $n-2$ variables. The seven standard universal unfoldings in the second classification theorem are referred to as the **fold, cusp, swallowtail** or **dovetail, butterfly, elliptic umbilic, hyperbolic umbilic,** and **parabolic umbilic** (see Table 1, Chapter 7).

In order to classify quadratic forms on $\mathbf{R}^n$, Sylvester's Law of Inertia will be proved. For this purpose, a coordinate-free formulation is appropriate, so that an arbitrary n-dimensional real vector space V will be considered instead of $\mathbf{R}^n$. Recall that a **symmetric bilinear form** on V is a function $F: V \times V \to \mathbf{R}$ satisfying

$$F(x,y) = F(y,x) \quad \text{and} \quad F(x, y + \alpha z) = F(x,y) + \alpha F(x,z)$$

for all $x, y, z \in V$, and $\alpha \in \mathbf{R}$. It is **positive definite** on a subspace X of V if $F(x,x) > 0$ for all nonzero x in X and **negative definite** on a subspace Y of V if $F(y,y) < 0$ for all nonzero y in Y. Clearly, in this situation $X \cap Y = \{0\}$ holds, and for the subspace $N := \{x \in V : F(x,y) = 0 \text{ for all } y \in V\}$ of V, called the **degeneracy space** of F,

$$(13) \qquad\qquad (X \oplus Y) \cap N = \{0\}$$

is valid. To verify this, note that $F(x + y, x - y) = F(x,x) - F(y,y)$ vanishes for $x \in X$ and $y \in Y$ if and only if $x = y = 0$. Moreover, $F = 0$ if and only if $N = V$.

Example (14). The classical example of a symmetric bilinear form on $\mathbf{R}^n$ is the **(Euclidean) scalar product**

$$F(x,y) := \langle x, y \rangle := \sum_{i=1}^n x_i y_i$$

for $x = (x_1, \ldots, x_n)$, $y = (y_1, \ldots, y_n)$, which is positive definite on $\mathbf{R}^n$. More generally, let $C = (c_{ij})_{1 \le i,j \le n}$ be a real symmetric matrix and let Cy

be the vector with the ith component $\sum_{j=1}^{n} c_{ij} y_j$ for $1 \le i \le n$. Then

$$F(x, y) := \langle x, Cy \rangle = \sum_{i,j=1}^{n} c_{ij} x_i y_j$$

is also a symmetric bilinear form. Every symmetric bilinear form on $\mathbf{R}^n$ is of this type.

Sylvester's Law of Inertia (15). Let $F: V \times V \to \mathbf{R}$ be a symmetric bilinear form on an n-dimensional real vector space V. Then there are integers r and s with $0 \le s \le r \le n$ that are uniquely determined and a basis $b_1, \ldots, b_n$ of V such that for any vector $x = x_1 b_1 + \cdots + x_n b_n$ in V

$$(16) \qquad F(x, x) = -x_1^2 - \cdots - x_s^2 + x_{s+1}^2 + \cdots + x_r^2$$

holds, where $r = 0$ means $F = 0$. If $y = y_1 b_1 + \cdots + y_n b_n$ is another vector in V, then

$$(17) \qquad F(x, y) = -x_1 y_1 - \cdots - x_s y_s + x_{s+1} y_{s+1} + \cdots + x_r y_r$$

follows from (16).

Proof. Equation (17) follows immediately from Equation (16) due to the **polarization identity**,

$$(18) \qquad F(x, y) = \tfrac{1}{2}[F(x + y, x + y) - F(x, x) - F(y, y)].$$

To prove (16), consider the degeneracy space N of V. If $N = V$, then the theorem is trivial, because $r = s = 0$ follows. Assume therefore that $N \ne V$. Let W be a complement of N in V; i.e., $W \oplus N = V$. Set $r := \dim W$. Then $r > 0$ holds. By induction, $b_1, \ldots, b_r$ in W and $\varepsilon_1, \ldots, \varepsilon_r \in \{-1, 1\}$ will be defined so that

$$(19) \qquad F(b_i, b_j) = \varepsilon_i \delta_{ij}, \quad 1 \le i, j \le r,$$

is valid, where δ_{ij} is the Kronecker delta symbol. The $b_1, \ldots, b_r$ are then automatically linearly independent and constitute a basis of W.

Start by choosing $w \in W$ with $F(w, w) \ne 0$. Such a choice is possible, since otherwise F would vanish on $W \times W$ by (18). This would imply the contradiction $N = V$, since $F(x + w, x' + w') = F(x, x' + w') + F(w, x') + F(w, w') = F(w, w')$ for all $x, x' \in N$ and $w, w' \in W$. Define $b_1 := |F(w, w)|^{-1/2} w$. Then $\varepsilon_1 := F(b_1, b_1)$ is either -1 or 1.

Proceeding inductively, let $1 \le k < r$ and assume that $b_1, \ldots, b_k \in W$ and $\varepsilon_1, \ldots, \varepsilon_k \in \{-1, 1\}$ have already been defined, satisfying (19). Let $B := \langle b_1, \ldots, b_k \rangle$ be the subspace of W spanned by $b_1, \ldots, b_k$ and consider the subspace $B^0 := \{w \in W : F(w, b) = 0 \text{ for every } b \in B\}$ of W.

It will be shown that $y \in B^0$ exists with $F(y, y) \ne 0$. In this case, $b_{k+1} := |F(y, y)|^{-1/2} y$ and $b_1, \ldots, b_k$ together with $\varepsilon_1, \ldots, \varepsilon_k$ and $\varepsilon_{k+1} :=$

$F(b_{k+1}, b_{k+1})$ satisfy (19). To verify the existence of such a $y \in B^0$, note that for $w \in W$ the vector

$$w^0 := w - \sum_{j=1}^{k} \varepsilon_j F(b_j, w) b_j$$

lies in B^0. Furthermore, $w^0 \neq 0$ if $w \in W \backslash B$, so that $B^0 \neq \{0\}$ follows. Now if $F(y, y) = 0$ were valid for every $y \in B^0$, then F would vanish on $B^0 \times B^0$ by (18). Because $F(y, w + x) = F(y, w) = F(y, w^0)$ is true for $x \in N, y \in B^0$, and $w \in W$, the contradiction $B^0 \subset N$ would follow.

Consequently, a basis $b_1, \ldots, b_r$ of W exists as well as numbers $\varepsilon_1, \ldots, \varepsilon_r \in \{-1, 1\}$ satisfying (19). After a permutation, one obtains $\varepsilon_1, \ldots, \varepsilon_s = -1$ and $\varepsilon_{s+1}, \ldots, \varepsilon_r = 1$ with $0 \leq s \leq r$. Select any basis $b_{r+1}, \ldots, b_n$ of N. Then $b_1, \ldots, b_n$ is a basis of V and $F(b_i, b_j) = 0$ for $i = r + 1, \ldots, n$ and $j = 1, \ldots, n$, due to the definition of N. Together with (19), this implies the assertion (16).

It remains to show that s and r are independent of the choice of the basis $b_1, \ldots, b_n$ of V satisfying (16). For r this is obvious, since $r = n - \dim N$. The independence of s results from the fact that F is positive definite on $X := \langle b_{s+1}, \ldots, b_r \rangle$ and negative definite on $Y := \langle b_1, \ldots, b_s \rangle$. To see this, let $b'_1, \ldots, b'_n$ be another basis of V satisfying (16) with s replaced by s' and define X' and Y' analogously to X and Y. Then $(X' \oplus Y) \cap N = \{0\}$ and $(X \oplus Y') \cap N = \{0\}$ by (13), and one obtains $(r - s') + s + (n - r) \leq n$, and $(r - s) + s' + (n - r) \leq n$, from which $s' = s$ follows.

Restricting a symmetric bilinear form on $\mathbf{R}^n$ to the diagonal gives a quadratic form.

Definition. Let $C = (c_{ij})_{i,j=1,\ldots,n}$ be a real symmetric $n \times n$ matrix. Then the homogeneous polynomial of second degree

$$q(x) := \sum_{i,j=1}^{n} c_{ij} x_i x_j$$

for $x = (x_1, \ldots, x_n) \in \mathbf{R}^n$ is called a **quadratic form** on $\mathbf{R}^n$. It is **nondegenerate** if C is invertible and **degenerate** otherwise.

Using the scalar product notation, one has $q(x) = \langle x, Cx \rangle$ for $x \in \mathbf{R}^n$. A quadratic form q is clearly a smooth function. Notice that

$$Dq(x) = 2xC \quad \text{and} \quad D^2 q(x) = 2C.$$

This implies that C is uniquely determined by q, so that there is a unique correspondence between quadratic forms and symmetric bilinear forms. Moreover, it follows that the set of all critical points of q is the null space $\{x \in \mathbf{R}^n : Cx = 0\}$ of C. A critical point of q is nondegenerate if and only if C is invertible, which is equivalent to saying that q is nondegenerate. A nondegenerate quadratic form has just one critical point, namely, the origin.

Definition. A **linear coordinate transformation** of $\mathbf{R}^n$ is a bijective linear map from $\mathbf{R}^n$ onto $\mathbf{R}^n$.

According to elementary theorems in algebra, a linear map $\tau\colon \mathbf{R}^n \to \mathbf{R}^n$ can be uniquely represented by a real $n \times n$ matrix T, in the sense that $\tau(x) = Tx$ for $x \in \mathbf{R}^n$. Furthermore, τ is a linear coordinate transformation if and only if T is invertible. Note that under linear coordinate transformations, quadratic forms are transformed into quadratic forms.

Classification of Quadratic Forms on $\mathbf{R}^n$ (20). Let q be a quadratic form on $\mathbf{R}^n$. Then there are integers r and s with $0 \leq s \leq r \leq n$ that are uniquely determined and a linear coordinate transformation τ of $\mathbf{R}^n$ such that

$$q \circ \tau(x) = -x_1^2 - \cdots - x_s^2 + x_{s+1}^2 + \cdots + x_r^2$$

holds for $x = (x_1, \ldots, x_n) \in \mathbf{R}^n$, where $r = 0$ means $q = 0$. In particular, the quadratic form q is nondegenerate exactly when r is equal to n.

Proof. Let C be the real symmetric $n \times n$ matrix defining q, i.e., $q(x) = \langle x, Cx \rangle$. When one applies (15) to the symmetric bilinear form $F(x, y) := \langle x, Cy \rangle$, there are uniquely determined integers $0 \leq s \leq r \leq n$, and there is a basis $b_1, \ldots, b_n$ of $\mathbf{R}^n$ so that for $x = \xi_1 b_1 + \cdots + \xi_n b_n$ in $\mathbf{R}^n$,

$$q(x) = -\xi_1^2 - \cdots - \xi_s^2 + \xi_{s+1}^2 + \cdots + \xi_r^2$$

holds. Denote the **standard basis** of $\mathbf{R}^n$ by $e_1, \ldots, e_n$, i.e., the ith component $(e_k)_i$ of e_k is δ_{ki}. Let $\tau\colon \mathbf{R}^n \to \mathbf{R}^n$ be the linear coordinate transformation given by $\tau(e_k) := b_k, k = 1, \ldots, n$. Then τ satisfies the assertion.

It is worthwhile to look at the following interpretation of this classification theorem. A quadratic form on $\mathbf{R}^n$ is defined by a real symmetric $n \times n$ matrix, i.e., by $\frac{1}{2}n(n+1)$ real numbers. Identify the set Q_n of all quadratic forms on $\mathbf{R}^n$ with the $\frac{1}{2}n(n+1)$-dimensional vector space of all real symmetric $n \times n$ matrices. Call two quadratic forms q and $\tilde{q}$ on $\mathbf{R}^n$ **linearly equivalent** if a linear coordinate transformation τ of $\mathbf{R}^n$ exists satisfying $\tilde{q} = q \circ \tau$. Linear equivalence obviously is an equivalence relation on Q_n. Theorem (20) shows that there are exactly $\frac{1}{2}(n+1)(n+2)$ equivalence classes with representatives

(21) $$q_{sr}(x) := -x_1^2 - \cdots - x_s^2 + x_{s+1}^2 + \cdots + x_r^2$$

for $0 \leq s \leq r \leq n$, called the **normal quadratic forms**. When $q \circ \tau = q_{sr}$ holds, then r is referred to as the **rank** of q and s as its **index**. Furthermore, if one considers just nondegenerate quadratic forms on $\mathbf{R}^n$, there are $n + 1$ equivalence classes with representatives $q_{sn}, 0 \leq s \leq n$.

Note that if C is the matrix defining a quadratic form q on $\mathbf{R}^n$ and if τ is a linear transformation of $\mathbf{R}^n$ given by an invertible matrix T, then $\tilde{q} := q \circ \tau$ is defined by the matrix

(22) $$T^t C T,$$

where T^t denotes the transpose of T. Moreover, the normal quadratic forms q_{sr} are defined by the $n \times n$ matrices

(23)
$$E_{sr} := \begin{pmatrix} -E_s & & 0 \\ & E_{r-s} & \\ 0 & & 0 \end{pmatrix},$$

where E_k denotes the $k \times k$ identity matrix. The classification of quadratic forms (20) can be restated in the form that for every symmetric $n \times n$ matrix C there are uniquely defined integers s and r with $0 \le s \le r \le n$ and an invertible matrix T satisfying

(24)
$$T^t C T = E_{sr}.$$

The uniqueness means that if S is another invertible matrix with $S^t C S = E_{s'r'}$ for integers $0 \le s' \le r' \le n$, then $s' = s$ and $r' = r$ follows. Clearly, s (r) is the number of negative (nonzero) eigenvalues of C.

We now turn our attention to another essential ingredient for the proof of Morse's Lemma—the representation of a smooth function around a critical point as a homogeneous polynomial of second degree with coefficients that are smooth functions themselves. Actually, a more general result (28) will be proved that is of fundamental importance in the algebraic setting considered in later chapters (see also Exercise 4.18).

Recall that a subset M of $\mathbf{R}^n$ is **star-shaped with center** $p \in M$ if for any $x \in M$ the line segment

$$\{y \in \mathbf{R}^n : y = p + t(x - p), t \in [0, 1]\}$$

joining x and p lies in M. Obviously, every neighborhood of a point p in $\mathbf{R}^n$ contains a star-shaped open set with center p.

Theorem (25). Let U be star-shaped with center $p \in U$, and let $k \in \mathbf{N}$. Suppose that the derivatives of f at p satisfy $D^l f(p) = 0$ for $l = 0, \dots, k-1$. Then there are smooth functions

$$g_{i_1 \dots i_k} : U \to \mathbf{R}, \quad 1 \le i_1, \dots, i_k \le n,$$

with $g_{i_1 \dots i_k} = g_{j_1 \dots j_k}$ if $(j_1, \dots, j_k)$ is a permutation of $(i_1, \dots, i_k)$ such that

(26)
$$f(x) = \sum_{i_1, \dots, i_k = 1}^{n} g_{i_1 \dots i_k}(x)(x_{i_1} - p_{i_1}) \cdots (x_{i_k} - p_{i_k})$$

for $x \in U$. Consequently,

$$g_{i_1 \dots i_k}(p) = \frac{1}{k!} D_{i_k} \dots D_{i_1} f(p)$$

holds for $1 \le i_1, \dots, i_k \le n$.

Proof. It is no restriction to assume $p = 0$. Fix $x \in U$ and define $F: [0,1] \to \mathbf{R}$ by $F(t) := f(tx)$. Clearly, $F(0) = 0$ and $F(1) = f(x)$. The chain rule yields

$$F^{(l)}(t) = \sum_{i_1,\dots,i_l=1}^{n} D_{i_l} \dots D_{i_1} f(tx) x_{i_1} \dots x_{i_l}$$

for $l \in \mathbf{N}$. By hypothesis, $F^{(l)}(0) = 0$ for $l = 0,\dots,k-1$. Applying the fundamental theorem of calculus, it follows that

$$F^{(k-1)}(t_1) = \int_0^{t_1} F^{(k)}(t_0)dt_0, \quad F^{(k-2)}(t_2) = \int_0^{t_2} F^{(k-1)}(t_1)dt_1,$$

$$\dots, \quad F(t_k) = \int_0^{t_k} F'(t_{k-1})dt_{k-1}$$

for $0 \le t_0 \le t_1 \le \cdots \le t_k \le 1$. In particular, when $t_k = 1$, one obtains

$$f(x) = \int_0^1 \int_0^{t_{k-1}} \cdots \int_0^{t_1} F^{(k)}(t_0)dt_0 \cdots dt_{k-1}.$$

This gives the following representation of f:

$$f(x) = \sum_{i_1,\dots,i_k=1}^{n} g_{i_1\dots i_k}(x) x_{i_1} \cdots x_{i_k}$$

where

$$g_{i_1\dots i_k}(x) := \int_0^1 \int_0^{t_{k-1}} \cdots \int_0^{t_1} D_{i_k} \cdots D_{i_1} f(t_0 x)dt_0 \cdots dt_{k-1}.$$

Since f is smooth, its derivatives commute, and the first assertion of the theorem is proved.

Now consider a representation of f as in (26) with coefficients satisfying the permutation condition. Then

$$D_{j_k} \cdots D_{j_1} f(x) = \sum_{i_1,\dots,i_k=1}^{n} D_{j_k} \cdots D_{j_1}\big(g_{i_1\dots i_k}(x) x_{i_1} \dots x_{i_k}\big).$$

When the differentiation is carried out and evaluated at the point $x = 0$, every term of the sum vanishes except when $(j_1,\dots,j_k)$ is a permutation of $(i_1,\dots,i_k)$. This is because in any other case at least one factor x_{i_l} remains, due to the product rule. Since $k!$ is the number of permutations of the k elements $i_1,\dots,i_k$, there are $k!$ terms in the sum having the value $g_{i_1\dots i_k}(0)$.

For $k = 1$, Theorem (25) has the following form: If $f(p) = 0$, then f admits the representation

$$f(x) = \sum_{i=1}^{n} g_i(x)(x_i - p_i)$$

for $x = (x_1, \ldots, x_n) \in U$ as a first-degree polynomial with smooth functions $g_i \colon U \to \mathbf{R}$ as coefficients. It follows that

$$Df(p) = (g_1(p), \ldots, g_n(p)).$$

The statement of Theorem (25) for $k = 2$ is precisely what will be used to prove the Morse Lemma: If p is a critical point of f and $f(p) = 0$, then f admits a representation

$$f(x) = \sum_{i,j=1}^{n} g_{ij}(x)(x_i - p_i)(x_j - p_j)$$

for $x \in U$ as a quadratic polynomial where the coefficients are smooth functions $g_{ij} \colon U \to \mathbf{R}$ satisfying $g_{ij} = g_{ji}$. It follows that

$$\tfrac{1}{2}D^2 f(p) = (g_{ij}(p))_{i,j=1,\ldots,n}.$$

An important conclusion that can be deduced from Theorem (25) is Taylor's formula with integral remainder for a smooth function f. In showing this, the opportunity will be taken to introduce further notations.

If $\nu = (\nu_1, \ldots, \nu_n) \in \mathbf{N}_0^n$ is a **multi-index**, i.e., an n-tuple of nonnegative integers, set $|\nu| := \nu_1 + \cdots + \nu_n$ and $\nu! := \nu_1! \cdots \nu_n!$. When $\mu = (\mu_1, \ldots, \mu_n)$ is another such n-tuple, then $\mu \leq \nu$ will mean $\mu_1 \leq \nu_1, \ldots, \mu_n \leq \nu_n$. Furthermore, put

$$\mu + \nu := (\mu_1 + \nu_1, \ldots, \mu_n + \nu_n), \qquad \binom{\nu}{\mu} := \binom{\nu_1}{\mu_1} \cdots \binom{\nu_n}{\mu_n}.$$

For $x = (x_1, \ldots, x_n)$ in $\mathbf{R}^n$ let $x^\nu := x_1^{\nu_1} \ldots x_n^{\nu_n}$. Use $D^\nu f := D_1^{\nu_1} \cdots D_n^{\nu_n} f$ to designate higher-order mixed partial derivatives.

If k is a nonnegative integer, then the kth **Taylor polynomial of f at a point** $p \in U$ will be denoted by

$$T_{f,p}^k(x) := \sum_{0 \leq |\nu| \leq k} \frac{D^\nu f(p)}{\nu!} (x - p)^\nu.$$

For $p = 0$ set $T_f^k := T_{f,0}^k$.

Now let k be a positive integer and consider the **remainder function** $R \colon U \to \mathbf{R}$ defined by $R(x) := f(x) - T_{f,p}^{k-1}(x)$. Since the derivatives of R at the point $p \in U$ of orders less than k vanish, Theorem (25) can be applied to R, immediately giving Taylor's formula:

Taylor's Formula (27). Let U be star-shaped with center p. Then for any positive integer k,

$$f(x) = T_{f,p}^{k-1}(x) + R(x)$$

for $x \in U$ where

$$R(x) = \sum_{i_1,\dots,i_k=1}^{n} g_{i_1\dots i_k}(x)(x_{i_1} - p_{i_1})\cdots(x_{i_k} - p_{i_k})$$

and

$$g_{i_1\dots i_k}(x) := \int_0^1 \int_0^{t_{k-1}} \cdots \int_0^{t_1} D_{i_k}\cdots D_{i_1} f(p + t_0(x-p))dt_0 \cdots dt_{k-1}.$$

It follows that

$$|R(x)| \le \left(\sum_{i=1}^{n} |x_i - p_i| \right)^k \frac{1}{k!} \sup_{t,\nu} |D^\nu f(p + t(x-p))|,$$

where the supremum is taken over all $\nu \in \mathbf{N}_0^n$ with $|\nu| = k$ and all $t \in \mathbf{R}$ with $0 \le t \le 1$.

Proof. To verify the estimate for the remainder function R, use

$$\int_0^1 \int_0^{t_{k-1}} \cdots \int_0^{t_1} dt_0 dt_1 \dots dt_{k-1} = \frac{1}{k!}$$

and

$$\sum_{i_1,\dots,i_k=1}^{n} |x_{i_1} - p_{i_1}|\cdots|x_{i_k} - p_{i_k}| = \left(\sum_{i=1}^{n} |x_i - p_i| \right)^k.$$

The following straightforward generalization of Theorem (25) to partial maps will be needed for the proof of the Reduction Lemma (3.2), in Lemma (18) of Chapter 5, for the proof of the Main Lemma (10.1), and in (12.4), (12.10). Let U be as in (27), let V be an open subset of $\mathbf{R}^m$, and suppose that $\tilde{f}\colon U \times V \to \mathbf{R}$ is smooth such that for every y in V all derivatives of the map $x \mapsto \tilde{f}(x,y)$ of orders less than k vanish at the point $p \in U$. Then there are smooth functions

$$\tilde{g}_{i_1\dots i_k} : U \times V \to \mathbf{R}$$

with $\tilde{g}_{i_1\dots i_k} = \tilde{g}_{j_1\dots j_k}$ if $(j_1,\dots,j_k)$ is a permutation of $(i_1,\dots,i_k)$ such that

$$(28) \qquad \tilde{f}(x,y) = \sum_{i_1,\dots,i_k=1}^{n} \tilde{g}_{i_1\dots i_k}(x,y)(x_{i_1} - p_{i_1})\dots(x_{i_k} - p_{i_k})$$

for $(x,y) \in U \times V$ holds. Consequently,

$$\tilde{g}_{i_1\dots i_k}(p,y) = \frac{1}{k!} \frac{\partial^k \tilde{f}}{\partial x_{i_k} \dots \partial x_{i_1}}(p,y)$$

is true for $1 \leq i_1, \ldots, i_k \leq n$ and $y \in V$. The corresponding version of Taylor's formula (27) for $\tilde{f}$ is also valid. Notice that the coefficients of the Taylor polynomial in x of the partial function $x \mapsto \tilde{f}(x, y)$ are now smooth functions of y.

The objective is to classify critical points of smooth functions with respect to a class of coordinate transformations larger than just those that are linear, namely, the smooth ones.

> **Definition.** A map $F: U \to \mathbf{R}^m$ is **smooth** if each component $F_j :=$ $pr_j \circ F$ is smooth, $j = 1, \ldots, m$, where $pr_j: \mathbf{R}^m \to \mathbf{R}$ is the projection $(y_1, \ldots, y_n) \mapsto y_j$.

If $F: U \to \mathbf{R}^m$ is smooth, then the derivative $DF(p)$ of F at a point $p \in U$ is the linear map $\mathbf{R}^n \to \mathbf{R}^m$ with components $x \mapsto DF_j(p)x := \sum_{i=1}^{n} D_i F_j(p)x_i$. In other words, this map is given by the $m \times n$ Jacobi matrix $(D_i F_j(p))_{1 \leq j \leq m, 1 \leq i \leq n}$ of F at p, which will also be denoted by $DF(p)$. For $m = n$, its determinant is called the **Jacobian of F at p**.

> **Definition.** Let V be open in $\mathbf{R}^n$. A bijection $\varphi: U \to V$ is a **diffeomorphism** if both φ and φ^{-1} are smooth (considered as maps into $\mathbf{R}^n$).

Roughly speaking, a diffeomorphism distorts the coordinate lines in a smooth manner.

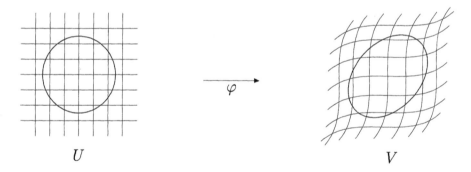

$$U \qquad\qquad\qquad \varphi \qquad\qquad\qquad V$$

Note that a surjective smooth map of an open set in $\mathbf{R}^n$ onto an open set in $\mathbf{R}^n$ need not be a diffeomorphism, even if it is injective. An example is given by $\mathbf{R} \to \mathbf{R}$, $x \mapsto x^3$. The example $\mathbf{R}_+ \times \mathbf{R} \to \mathbf{R}^2 \backslash \{0\}$, $(x, y) \mapsto (x \cos y, x \sin y)$ is also not a diffeomorphism, but it is at least locally a diffeomorphism at every point of $\mathbf{R}_+ \times \mathbf{R}$.

> **Definition.** A smooth map $\varphi: U \to \mathbf{R}^n$ is a **local diffeomorphism at a point** p in U if an open neighborhood of V of p in U exists such that $\varphi(V)$ is open in $\mathbf{R}^n$ and $V \to \varphi(V), x \mapsto \varphi(x)$, is a diffeomorphism. A **local inverse** of φ at $\varphi(p)$ is a smooth map $\psi: \mathbf{W} \to \mathbf{R}^n$ defined on an open neighborhood W of $\varphi(p)$ satisfying $\psi(\varphi(x)) = x$ for $x \in \varphi^{-1}(W) \cap V$. A

local diffeomorphism at a point $p \in U$ is also called a **smooth coordinate transformation at** p.

It is easy to see that the composition of two local diffeomorphisms $\varphi \colon U \to \mathbf{R}^n$ at $p \in U$ and $\psi \colon V \to \mathbf{R}^n$ at $q := \varphi(p)$ is again a local diffeomorphism at p defined on $\varphi^{-1}(V) \cap U$. The next theorem presents a practical condition to determine whether or not a smooth map $\varphi \colon U \to \mathbf{R}^n$ is a local diffeomorphism at a point $p \in U$.

Inverse Function Theorem (29). A smooth map $\varphi \colon U \to \mathbf{R}^n$ is a local diffeomorphism at a point $p \in U$ if and only if the Jacobi matrix $D\varphi(p)$ of φ at p is invertible.

Proof. Let φ be a local diffeomorphism at p. It is no restriction to assume that $V := \varphi(U)$ is open and that $\varphi \colon U \to V$ is a diffeomorphism. By the chain rule, $id_{\mathbf{R}^n} = D(\varphi^{-1} \circ \varphi)(p) = D\varphi^{-1}(\varphi(p))D\varphi(p)$ holds, implying that $D\varphi(p)$ is invertible.

To prove the converse, let $D\varphi(p)$ be invertible. The assertion will follow from the Implicit Function Theorem (see Appendix A.1). Consider the map $F \colon U \times \mathbf{R}^n \to \mathbf{R}^n$ defined by $F(x, y) := \varphi(x) - y$. Obviously, $F(p, \varphi(p)) = 0$. Let $D_1 F(x, y)$ denote the Jacobi matrix at x of the partial map $x \mapsto F(x, y)$. Then $D_1 F(p, \varphi(p)) = D\varphi(p)$ is invertible. According to the Implicit Function Theorem, there is an open ball V in $\mathbf{R}^n$ with center $q := \varphi(p)$ and a smooth map $\psi \colon V \to \mathbf{R}^n$ satisfying $\psi(V) \subset U$, $\psi(q) = p$, and $\varphi(\psi(y)) = y$ for $y \in V$. As a result, ψ is injective and by the chain rule $id_{\mathbf{R}^n} = D(\varphi \circ \psi)(q) = D\varphi(p)D\psi(q)$ holds. In particular, $D\psi(q)$ is invertible.

Now the foregoing argument can be applied to ψ instead of φ, proving the existence of an open ball W in $\mathbf{R}^n$ with center p and a smooth map $\tilde{\varphi} \colon W \to \mathbf{R}^n$ with $\tilde{\varphi}(W) \subset V$ such that $\tilde{\varphi}(p) = q$ and $\psi(\tilde{\varphi}(x)) = x$ for $x \in W$. Consequently, $W \subset U$ and $\tilde{\varphi}$ is injective. From the injectivity of ψ, one obtains $\tilde{\varphi}(W) = \psi^{-1}(W)$. Hence, $\tilde{\varphi}(W)$ is an open subset of V. If $x \in W$, then

$$\varphi(x) = \varphi(\psi(\tilde{\varphi}(x))) = \tilde{\varphi}(x)$$

because $\tilde{\varphi}(x) \in V$. It follows that $\varphi(W)$ is open and φ^{-1} equals ψ on $\varphi(W)$, which completes the proof.

The following global statement about smooth maps $\varphi \colon U \to \mathbf{R}^n$ is an immediate consequence of the inverse function theorem: The image $V := \varphi(U)$ of U under φ is open in $\mathbf{R}^n$, and the map $U \to V, x \mapsto \varphi(x)$, is a diffeomorphism if and only if φ is injective and $D\varphi(x)$ is invertible for all $x \in U$. Another application of the inverse function theorem, which is of particular interest in the context of classifying critical points, is the following corollary.

Corollary (30). Nondegenerate critical points are isolated critical points.

Proof. Let $p \in U$ be a nondegenerate critical point of f and set $\varphi := Df$. Then $D\varphi(p) = D^2 f(p)$ is invertible, and φ is a local diffeomorphism at p by (29). Therefore, an open subset V of U exists containing p such that φ is injective on V. This implies $Df(x) = \varphi(x) \neq \varphi(p) = Df(p) = 0$ for every x in V with $x \neq p$.

It is reasonable to classify critical points of smooth functions up to local diffeomorphisms at these points, since such transformations do not change their nature. More precisely, define the **index** and **rank of a smooth function** f **at a point** p to be the index and rank of the quadratic form given by the Hessian matrix $D^2 f(p)$ of f at p. For the sake of simplicity, assume that $p = 0$ and $f(0) = 0$ hold. Suppose now that ψ is a local diffeomorphism at the origin vanishing there. Set

(31) $$g(y) := f(\psi(y))$$

for small y. Then the origin is a critical point of f if and only if it is a critical point of g. Furthermore, when 0 is a critical point, the index and rank of f at 0 is the same as the index and rank of g at 0. In particular, the origin is a nondegenerate critical point of f if and only if it is a nondegenerate critical point of g. To verify this, recall that $D\psi(0)$ is invertible by (29). The chain rule gives

(32) $$Dg(0) = Df(0)D\psi(0),$$

and thus $Df(0) = 0$ holds exactly when $Dg(0) = 0$ is true. Moreover, the relationship between the Hessian matrices of f and g when $Df(0) = 0$ is valid is given by

(33) $$D^2 g(0) = (D\psi(0))^t D^2 f(0) D\psi(0).$$

The uniqueness of s and r in (24) proves the assertion.

A result exemplifying the subject matter of Catastrophe Theory is the following classification of those critical points of a smooth function of one variable that are not **flat**, i.e., the derivatives do not all vanish there.

Classification of Critical Points of One-Variable Functions (34). Let f be a function of one variable, and let f vanish at $0 \in U$. There is a local diffeomorphism ψ at $0 \in \mathbf{R}$ vanishing at the origin such that

$$f(\psi(y)) = \varepsilon^{k-1} y^k$$

for small y, where $k \in \mathbf{N}$ and $\varepsilon \in \{-1, 1\}$, if and only if k is the smallest positive integer satisfying $f^{(k)}(0) \neq 0$ and ε is the sign of $f^{(k)}(0)$.

Proof. Suppose first that $f(\psi(y)) = \varepsilon^{k-1} y^k$ holds. A local inverse of ψ at 0 has the form $x\chi(x)$ for smooth χ and small x with $\chi(0) \neq 0$ by Theorem (25). Thus,

$$f(x) = \varepsilon^{k-1} x^k (\chi(x))^k.$$

It follows that

$$f^{(k)}(0) = k! \varepsilon^{k-1}(\chi(0))^k \neq 0 \quad \text{and} \quad f^{(l)}(0) = 0$$

for $0 \leq l < k$. In particular, if k is even, ε is the sign of $f^{(k)}(0)$.

To prove the converse, let k be the smallest positive integer satisfying $f^{(k)}(0) \neq 0$, and denote the sign of $f^{(k)}(0)$ by ε. By Theorem (25), there is a smooth function g such that $f(x) = x^k g(x)$ for small x and

$$g(0) = \frac{1}{k!} f^{(k)}(0).$$

Since $\varepsilon g(0) > 0$ holds, the function $\varphi(x) := \varepsilon x(\varepsilon g(x))^{1/k}$ is defined and smooth for small x. Obviously, $\varphi(0) = 0$ and $f(x) = \varepsilon^{k-1}(\varphi(x))^k$. Due to $\varphi'(0) = \varepsilon(\varepsilon g(0))^{1/k} \neq 0$, φ is a local diffeomorphism at the origin according to (29). For a local inverse ψ of φ at the origin, one has $\psi(0) = 0$ and $f(\psi(y)) = \varepsilon^{k-1} y^k$ for small y.

Notice that if k is even, there is no smooth coordinate transformation ψ at the origin with $\psi(0) = 0$ transforming x^k into $-x^k$ by (34).

A standard nontrivial example of a flat point is the origin for the function $f : \mathbf{R} \to \mathbf{R}$ defined by $f(x) := \exp(-x^{-2})$ for nonzero $x \in \mathbf{R}$ and $f(0) := 0$.

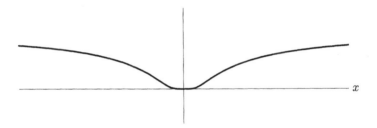

Points of smooth functions that are not critical are classified in the next theorem.

Classification of Noncritical Points (35). Let f vanish at $0 \in U$. Then $Df(0) \neq 0$ holds if and only if there is a local diffeomorphism ψ at 0 with $\psi(0) = 0$ such that

$$f(\psi(y)) = y_1$$

for all $y = (y_1, \dots, y_n)$ in a neighborhood of the origin.

Proof. If $f(\psi(y)) = y_1$ holds for small y, then the origin is not a critical point due to (32). On the other hand, if $Df(0) \neq 0$ is true, select an $i \in \{1, \dots, n\}$ with $D_i f(0) \neq 0$. Define $\varphi : U \to \mathbf{R}^n$ by $\varphi_1(x) := f(x), \varphi_i(x) := x_1$ if $i \neq 1$, and $\varphi_j(x) := x_j$ for all other $j \in \{1, \dots, n\}$. Then $|\det D\varphi(0)| = |D_i f(0)|$, and by Theorem (29), φ is a local diffeomorphism at 0. If ψ is a local inverse of φ at 0, then $f(x) = \varphi_1(x)$ is equivalent to $f(\psi(y)) = y_1$.

The main result of this chapter, i.e., Morse's Lemma, is a corollary of the following lemma, according to which an invertible symmetric matrix of smooth functions can be diagonalized locally. The proof of this lemma is based on the Implicit Function Theorem (for a more constructive proof see Exercise 1.12).

Diagonalization of Matrix-Valued Functions (36). Let U contain the origin, and let k be a positive integer. Suppose that $g_{ij} \colon U \to \mathbf{R}$ are smooth functions for $1 \le i, j \le k$ such that the matrix

$$G(x) := (g_{ij}(x))_{i,j=1,\ldots,k}$$

is symmetric for $x \in U$ and $G(0) = (\varepsilon_i \delta_{ij})$ holds with $\varepsilon_i \in \{-1, 1\}$, where δ_{ij} is the Kronecker delta symbol. Then there are smooth functions t_{ij} defined on an open set V in U containing the origin so that the matrix $T(x) := (t_{ij}(x))_{i,j=1, \ldots, k}$ satisfies $T(0) = E_k$ and

$$G(x) = T(x)^t G(0) T(x)$$

is valid for x in V.

Proof. Let $m := \frac{1}{2}k(k+1)$, and denote by $\boldsymbol{T}_k$ and $\boldsymbol{S}_k$ the m-dimensional vector space of all upper triangular and symmetric $k \times k$ matrices, respectively. These spaces are, of course, canonically isomorphic to $\mathbf{R}^m$. The map $F \colon U \times \boldsymbol{T}_k \to \boldsymbol{S}_k$,

$$F(x, T) := G(x) - T^t G(0) T$$

is smooth, and $F(0, E_k) = 0$ holds. Consider the partial map $T \mapsto F(0, T)$. Its derivative at E_k is the linear map $\boldsymbol{T}_k \to \boldsymbol{S}_k$ given by

$$T \to -T^t G(0) - G(0)T,$$

which can be verified by calculating the partial derivatives at E_k. Due to T being upper triangular, this linear map is obviously injective and hence bijective. The assertion now follows from the Implicit Function Theorem (see Appendix A.1).

Combining this lemma with the classification of quadratic forms on $\mathbf{R}^n$ (20), one easily obtains a classification of nondegenerate quadratic forms depending on parameters. In matrix form analogous to (24), this means for the matrix $G(x)$, where $G(0)$ is now just invertible, that a matrix $S(x)$ of smooth functions exists satisfying

$$S(x)^t G(x) S(x) = E_{sn}$$

where s is the index of $G(0)$.

Let 0 be a critical point of f, and let τ be a linear coordinate transformation of $\mathbf{R}^n$ given by a matrix T. If $g(y) := f(\tau(y))$ for small y, one obtains

$$D^2 g(0) = T^t D^2 f(0) T$$

by (33). According to (24), τ can be chosen to satisfy

$$\tfrac{1}{2}D^2 g(0) = E_{sr}.$$

By definition, s and r are the index and rank of f.

Morse's Lemma (37).   Let f vanish at $0 \in U$. The origin is a nondegenerate critical point of f if and only if a local diffeomorphism ψ at 0 exists with $\psi(0) = 0$ such that

(38) $$f(\psi(y)) = -y_1^2 - \cdots - y_s^2 + y_{s+1}^2 + \cdots + y_n^2$$

holds around the origin, where s denotes the index of f at 0.

Proof. If f satisfies (38), then the origin is a nondegenerate critical point due to (33). The converse will be verified now. By the foregoing considerations, the classification of quadratic forms (24) enables one to assume that $D^2 f(0) = 2E_{sn}$ holds, where E_{sn} is defined as in (23). Now Theorem (25) implies

$$f(x) = \langle x, G(x)x \rangle \quad \text{and} \quad D^2 f(0) = 2G(0),$$

where $G(x) := (g_{ij}(x))_{i,\, j=1,\ldots,n}$ is a symmetric matrix of smooth functions. Hence, $G(0) = E_{sn}$. According to Lemma (36)

$$f(x) = \langle T(x)x, E_{sn}T(x)x \rangle \quad \text{and} \quad T(0) = E_n,$$

where $T(x) = (t_{ij}(x))_{i,j=1,\ldots,n}$ is a matrix of smooth functions. Let $\varphi(x) := T(x)x$. Then $\varphi(0) = 0, D\varphi(0) = T(0) = E_n$, and $f(x) = q_{sn}(\varphi(x))$ follow. It results from (29) that φ is a local diffeomorphism at 0. Then a local inverse ψ of φ satisfies (38).

Another way of formulating Morse's Lemma is to say that

$$f(x) = \langle \varphi(x), \tfrac{1}{2}D^2 f(0)\varphi(x) \rangle,$$

holds for some local diffeomorphism φ at 0 with $\varphi(0) = 0$. This means that after a suitable choice of coordinates f is locally equal to its own second Taylor polynomial at the origin. Obviously, this result describes the *qualitative* features of the function f rather than the *quantitative* ones. For this reason, it is interesting to compare this statement to the well-known result implied by Taylor's formula (27) that f is approximated by its second Taylor polynomial with an error that vanishes to the second order:

$$\left| f(x) - \langle x, \tfrac{1}{2}D^2 f(0)x \rangle \right| / \langle x, x \rangle \to 0 \quad \text{for } x \to 0.$$

The significance of Morse's Lemma is in reducing the family of all smooth functions vanishing at the origin in $\mathbf{R}^n$, with the origin as a nondegenerate critical point, to just $n+1$ simple stereotypes, namely,

$$-x_1^2 - \cdots - x_s^2 + x_{s+1}^2 + \cdots + x_n^2, \quad 0 \le s \le n.$$

A question that naturally arises is whether functions with just nondegenerate critical points really do occur frequently. This chapter concludes by briefly treating this question.

A smooth function without degenerate critical points is called a **Morse function**. The set of all Morse functions on $\mathbf{R}^n$ is an open and dense subset of the set $C^\infty(\mathbf{R}^n)$ of all smooth functions on $\mathbf{R}^n$, if $C^\infty(\mathbf{R}^n)$ is supplied with the **Whitney C^∞-topology**. This topology is finer than the topology of uniform convergence on compact sets in $\mathbf{R}^n$ of all derivatives. It has a lot of open sets, implying that a dense subset of $C^\infty(\mathbf{R}^n)$ must be very large. $C^\infty(\mathbf{R}^n)$ is not a topological vector space in this topology, because the scalar multiplication is not a continuous operation. Furthermore, $C^\infty(\mathbf{R}^n)$ does not satisfy the first axiom of countability in the Whitney C^∞-topology. For further details see [GG].

The definition of the Whitney C^∞-topology on $C^\infty(\mathbf{R}^n)$ is given in Chapter 11. There the theorem on Topological Genericity (11.38) is proven, a corollary of which is the above-mentioned denseness of the set of Morse functions in $C^\infty(\mathbf{R}^n)$. This is shown in Exercise 11.6.

Exercises

1.1.

    (a)   Show that every symmetric bilinear form F on $\mathbf{R}^n$ is of the form

$$F(x,y) = \langle x, Cy \rangle$$

        where C is a symmetric matrix uniquely determined by F.

    (b)   Let A be any real $n \times n$ matrix. Show that there is a quadratic form q on $\mathbf{R}^n$ such that
$$q(x) = \langle x, Ax \rangle.$$

1.2.   Let C be a real symmetric $n \times n$ matrix and q be the quadratic form on $\mathbf{R}^n$ given by
$$q(x) = \langle x, Cx \rangle.$$
Recall that there is an orthogonal matrix U with

$$U^t C U = \begin{pmatrix} \lambda_1 & & 0 \\ & \ddots & \\ 0 & & \lambda_n \end{pmatrix},$$

where λ_i, $1 \leq i \leq n$, are the eigenvalues of C. Deduce from this that the index of q is equal to the number of negative eigenvalues of C and that the rank of q is equal to the number of nonzero eigenvalues of C.

1.3. Consider the quadratic homogeneous polynomial p on $\mathbf{R}^3$ given by $p(x) := \langle x, Ax \rangle$,

$$A := \begin{pmatrix} -8 & 3 & 20 \\ 1 & 13 & 18 \\ 20 & 26 & 4 \end{pmatrix}.$$

Show that there is a linear coordinate transformation τ of $\mathbf{R}^3$ such that

$$p(\tau(x)) = -x_1^2 + x_2^2$$

for $x \in \mathbf{R}^3$. Compute such a τ explicitly.

1.4.

(a) The quadratic forms on $\mathbf{R}$ are given by $q(x) := ax^2$ for $a \in \mathbf{R}$. Show that their normal forms are 0, x^2, and $-x^2$ for $a = 0, a > 0$, and $a < 0$, respectively.

(b) The quadratic forms on $\mathbf{R}^2$ are given by $q(x, y) := ax^2 + 2bxy + cy^2$ for $a, b, c \in \mathbf{R}$. Show that their normal forms are

$$
\begin{array}{lll}
0 & \text{for} & a = b = c = 0; \\
x^2 & \text{for} & b^2 = ac, \quad a + c > 0; \\
-x^2 & \text{for} & b^2 = ac, \quad a + c < 0; \\
x^2 + y^2 & \text{for} & b^2 < ac, \quad a + c > 0; \\
-x^2 + y^2 & \text{for} & b^2 > ac; \\
-x^2 - y^2 & \text{for} & b^2 < ac, \quad a + c < 0.
\end{array}
$$

1.5. A quadratic form q on $\mathbf{R}^n$ is called **positive definite** if $q(x) > 0$ for all nonzero x in $\mathbf{R}^n$, and **positive semidefinite** if $q(x) \geq 0$ for all $x \in \mathbf{R}^n$. Let C be a symmetric $n \times n$ matrix such that

$$q(x) = \langle x, Cx \rangle$$

for $x \in \mathbf{R}^n$. Show that the following statements are equivalent:

(i) q is positive semidefinite.

(ii) The index s of q is zero.

(iii) The eigenvalues of C are nonnegative.

Furthermore, show the equivalence of the following statements:

(iv) q is positive definite.

(v) q is positive semidefinite and the rank r of q is n.

(vi) The eigenvalues of C are positive.

(vii)   For all $k \in \{1, \ldots, n\}$,

$$\det \begin{pmatrix} C_{11} & \cdots & C_{1k} \\ \vdots & & \vdots \\ C_{k1} & \cdots & C_{kk} \end{pmatrix} > 0$$

holds (criterion of Jacobi and Hurwitz).

Hint: Use Exercise 1.2 and, for (vii), induction on n.

1.6.   Let $p \in U$ be a critical point of f, and denote by q the quadratic form given by

$$q(x) := \langle x, \tfrac{1}{2} D^2 f(p) x \rangle.$$

Show that p is an isolated minimum if q is positive definite and that q is positive semidefinite if p is a minimum.

1.7.   Let f be a smooth function on an open set $U \subset \mathbf{R}^n$. Let k be a nonnegative integer. The kth **total derivative** $D^k f(p)$ of f at $p \in U$ is the **multilinear map** on $(\mathbf{R}^n)^k$ given by

$$D^k f(p)(a^1, \cdots, a^k) := \sum_{i_1, \ldots, i_k = 1}^{n} D_{i_1} \cdots D_{i_k} f(p) a_{i_1}^1 \cdots a_{i_k}^k.$$

Show that

$$T_{f,p}^m(x) = \sum_{k=0}^{m} \frac{1}{k!} D^k f(p)(x - p, x - p, \cdots, x - p)$$

holds for the mth Taylor polynomial of f at p. Clearly, the assertion amounts to proving the equality

$$\sum_{|\nu|=k} \frac{1}{\nu!} D^\nu f(p)(x - p)^\nu = \frac{1}{k!} \sum_{i_1, \ldots, i_k = 1}^{n} D_{i_1} \cdots D_{i_k} f(p)(x_{i_1} - p_{i_1}) \cdots (x_{i_k} - p_{i_k}).$$

1.8.   Let U be the n-dimensional rectangle $\{x : |x_i - p_i| < r_i, 1 \leq i \leq n\}$ with center $p = (p_1, \ldots, p_n)$ and radii $r_i > 0$. Let f be an $\mathbf{R}$-analytic function given by a power series $\sum_{|\nu|=0}^{\infty} a_\nu (x - p)^\nu$, which is convergent in U; in particular, f is smooth. Referring to Taylor's formula (27), show that

(i) $T_{f,p}^{k-1}(x) = \sum_{|\nu|=0}^{k-1} a_\nu (x - p)^\nu$

and

(ii) $g_{i_1 \ldots i_k}(x) = \sum_{|\mu|=0}^{\infty} \frac{|\mu|!(\mu + \gamma)!}{\mu!(|\mu| + k)!} a_{\mu + \gamma}(x - p)^\mu,$

with $\gamma_j := \operatorname{card}\{l\colon 1 \le l \le k, i_l = j\}$ for $1 \le j \le n$.

1.9. Show that $f\colon \mathbf{R} \to \mathbf{R}$, $f(x) := \exp(-x^{-2})\sin^2(x^{-1})$ for $x \ne 0$ and $f(0) := 0$, is smooth and that the origin is a flat, nonisolated critical point of f.

1.10. To illustrate Theorem (35), transform

$$f(x,y) := \sqrt{1 - x^2 - y^2}$$

for (x,y) in $U := \{(x,y) \in \mathbf{R}^2\colon x^2 + y^2 < 1\}$ to a linear function around every noncritical point of f by a suitable local diffeomorphism.

1.11. Let $U \subset \mathbf{R}^2$ be open, and let $F\colon U \to \mathbf{R}^3$ be smooth. $F(U)$ is called a **smooth surface** in $\mathbf{R}^3$ if the vector product

$$(D_1 F)(x) \times (D_2 F)(x)$$

does not vanish for $x \in U$. The **metric tensor corresponding to** F is the 2×2 matrix G of smooth functions g_{ij} on U defined by

$$g_{ij}(x) := \langle D_i F(x), D_j F(x) \rangle, \qquad 1 \le i,\ j \le 2.$$

Let $0 \in U$ and let ψ be a local diffeomorphism of $\mathbf{R}^2$ at 0 with $\psi(0) = 0$. Then $\tilde{F}(x) := F(\psi(x))$ for small x defines a smooth surface in $\mathbf{R}^3$.

   (a)   Compute the metric tensor $\tilde{G}$ corresponding to $\tilde{F}$.

   (b)   Show that a ψ exists such that $\tilde{G}(0)$ is equal to E_2.

It should be noted that a similar statement is true for arbitrary smooth surfaces in $\mathbf{R}^n$.

1.12. Give a constructive proof of lemma (36) by induction on k. *Hint:* To proceed from $k-1$ to k, consider the $k \times k$ matrix

$$G(x) = \begin{pmatrix} \tilde{G}(x) & H(x) \\ H(x)^t & g(x) \end{pmatrix},$$

where $\tilde{G}(x)$ is a $(k-1) \times (k-1)$ matrix. Assuming

$$\tilde{G}(x) = \tilde{T}(x)\tilde{G}(0)\tilde{T}(x)$$

and $\tilde{T}(0) = E_{k-1}$, define

$$T(x) := \begin{pmatrix} \tilde{T}(x) & F(x) \\ 0 & t(x) \end{pmatrix}$$

for

$$F(x) := \tilde{G}(0)(\tilde{T}(x)^{-1})^t H(x) \quad \text{and}$$
$$t(x) := (g(0)g(x) - g(0)F(x)^t \tilde{G}(0)F(x))^{1/2}.$$

Using this exercise, a constructive proof of Morse's Lemma (37) can be given.

1.13. Consider the smooth function

$$f: \mathbf{R}^2 \to \mathbf{R}, \quad f(x,y) := x^3 + xy \cos xy.$$

Verify that f vanishes at the origin and that the origin is a nondegenerate critical point of f. According to Morse's Lemma (37), there is a local diffeomorphism φ at the origin that vanishes there and a unique normal quadratic form q on $\mathbf{R}^2$ such that

$$f(x,y) = q(\varphi(x,y))$$

for (x, y) in an open set containing the origin. Find q and an explicit expression for φ.

Hint: Follow the constructive proof of (36) outlined in Exercise 1.12 and the proof of Morse's Lemma (37).

Solutions

1.1.

(a)   Let $e_1, \ldots, e_n$ be the standard basis of $\mathbf{R}^n$, and set

$$c_{ij} := F(e_i, e_j)$$

for $1 \le i, j \le n$. The symmetry of F implies that of $C := (c_{ij})$. The bilinearity of F yields

$$F(x,y) = F\left(\sum_{i=1}^n x_i e_i, \sum_{j=1}^n y_j e_j\right) = \sum_{i,j=1}^n x_i y_j c_{ij} = \langle x, Cy \rangle.$$

(b)   Let $C := \frac{1}{2}(A + A^t)$ and consider the quadratic form q on $\mathbf{R}^n$ given by $q(x) := \langle x, Cx \rangle$. Then $q(x) = \frac{1}{2}\langle x, Ax \rangle + \frac{1}{2}\langle x, A^t x \rangle = \langle x, Ax \rangle$, since $\langle x, A^t x \rangle = \langle Ax, x \rangle = \langle x, Ax \rangle$.

1.2.   It is no restriction to assume that the eigenvalues $\lambda_1, \ldots, \lambda_n$ of C are already ordered in the sense that the first s ones are negative and the last $n - r$ ones are zero. Then the normal form of q is achieved by the linear coordinate transformation $\tau(x) := UDx$, where D is the diagonal matrix with entries $|\lambda_1|^{-1/2}, \ldots, |\lambda_r|^{-1/2}, 1, \ldots, 1$. This means that

$$q \circ \tau(x) = -x_1^2 - \cdots - x_s^2 + x_{s+1}^2 + \cdots + x_r^2,$$

which proves the assertion.

1.3. The symmetric part $C := \frac{1}{2}(A + A^t)$ of A is

$$\begin{pmatrix} -8 & 2 & 20 \\ 2 & 13 & 22 \\ 20 & 22 & 4 \end{pmatrix}.$$

Then $p(x) = \langle x, Cx \rangle$ holds. First C will be diagonalized. To achieve this, an orthonormal basis of eigenvectors will be computed. The eigenvalues of C, i.e., the zeros of the characteristic polynomial $\det(C - \lambda E_3)$, are the zeros of $-\lambda^3 + 9\lambda^2 + 4 \cdot 3 \cdot 81\lambda$ by the rule of Sarrus. Clearly, one zero is $\lambda_3 := 0$. The other two zeros are $\lambda_1 := -27$ and $\lambda_2 := 36$.

According to Exercise 1.2, the index and rank of the quadratic form $\langle x, Cx \rangle$ are 1 and 2, respectively. Hence, the existence of a linear coordinate transformation τ as claimed follows from the classification theorem of quadratic forms (20). Solving the linear system $(C - \lambda E_3)x = 0$ for $x \in \mathbf{R}^3$ by the Gaussian elimination procedure, one obtains for the eigenvalues $\lambda_1, \lambda_2, \lambda_3$, the orthonormal basis of eigenvectors

$$x_1 = \frac{1}{3}\begin{pmatrix} 2 \\ 1 \\ -2 \end{pmatrix}, \quad x_2 = \frac{1}{3}\begin{pmatrix} 1 \\ 2 \\ 2 \end{pmatrix}, \quad x_3 = \frac{1}{3}\begin{pmatrix} 2 \\ -2 \\ 1 \end{pmatrix}.$$

Then $U := (x_1, x_2, x_3)$ satisfies $U^t C U = (\lambda_1 e_1, \lambda_2 e_2, \lambda_3 e_3)$, where e_1, e_2, e_3 denotes the standard basis of $\mathbf{R}^3$. Thus, the linear coordinate transformation τ defined by

$$T = U\left(|\lambda_1|^{-1/2} e_1, \ |\lambda_2|^{-1/2} e_2, \ e_3\right)$$

satisfies the assertion.

1.4.

(a)   The solution is trivial.

(b)   Since $q(x, y) = \left\langle (x, y), C\begin{pmatrix} x \\ y \end{pmatrix} \right\rangle$ with $C = \begin{pmatrix} a & b \\ b & c \end{pmatrix}$, one has to determine the number of negative and zero eigenvalues of C according to Exercise 1.2. Let λ and μ be the eigenvalues of C. Evaluating the determinant and the trace of C, one has

$$\lambda\mu = ac - b^2, \quad \lambda + \mu = a + c.$$

The case $a = b = c = 0$ obviously implies $\lambda = \mu = 0$. If $b^2 = ac$ and $a + c > 0$ hold, then $\lambda\mu = 0$ and $\lambda + \mu > 0$ follow; i.e., one eigenvalue is zero and the other is positive. The case $b^2 = ac$ and $a + c < 0$ is analogous. From $b^2 < ac$ and $a > 0$, one obtains $c > 0$, and hence both eigenvalues are positive. The case $b^2 > ac$ means that both eigenvalues are nonzero with different signs. Finally, from $b^2 < ac$, it follows that both eigenvalues are nonzero with the same sign; hence, using $a < 0$, they must be negative.

1.5. Let τ be a linear coordinate transformation. Then a quadratic form q is positive definite or semidefinite exactly when $q \circ \tau$ is. Since the equivalence of (i) and (ii) obviously holds for the normal forms

$$-x_1^2 - \cdots - x_s^2 + x_{s+1}^2 + \cdots + x_r^2,$$

it holds generally. The equivalence of (ii) and (iii) follows from Exercise 1.2. The proof of the equivalence of (iv), (v), and (vi) is analogous to proving the equivalence of (i), (ii), and (iii).

It will now be shown that (iv) implies (vii). For $k \in \{1, \ldots, n\}$, $y \in \mathbf{R}^k$, and $0 \in \mathbf{R}^{n-k}$, put $p(y) := q(y, 0)$. Then p is a positive definite quadratic form on $\mathbf{R}^k$ and

$$p(y) = \langle y, By \rangle \quad \text{for} \quad B := \begin{pmatrix} C_{11} & \cdots & C_{1k} \\ \vdots & & \vdots \\ C_{k1} & \cdots & C_{kk} \end{pmatrix}$$

By (vi), the eigenvalues of B are positive and therefore $\det B > 0$. This proves (vii).

It remains to verify that (vii) implies (iv). Using induction on n, the case $n = 1$ is clear. To proceed from $n - 1$ to n, note that the above matrix B (with k replaced by $n-1$) satisfies the induction hypothesis. Thus, p is positive definite. Therefore, there is an invertible $(n-1) \times (n-1)$ matrix S such that $S^t B S = E_{n-1}$. Write

$$C = \begin{pmatrix} B & c \\ c^t & a \end{pmatrix},$$

and consider an orthogonal $(n-1) \times (n-1)$ matrix U satisfying $c^t S U = (\gamma, 0, \ldots, 0)$, where γ is the Euclidean length of $c^t S$. Then the $n \times n$ matrix

$$T := \begin{pmatrix} S & 0 \\ 0 & 1 \end{pmatrix} \begin{pmatrix} U & 0 \\ 0 & 1 \end{pmatrix} \begin{pmatrix} E_{n-1} & \nu \\ 0 & 1 \end{pmatrix}$$

with $\nu := (-\gamma, 0, \ldots, 0)$ is invertible. It is easily verified that

$$T^t C T = \begin{pmatrix} E_{n-1} & 0 \\ 0 & a - \gamma^2 \end{pmatrix}.$$

Taking the determinant, one obtains $\det C (\det T)^2 = a - \gamma^2$. By hypothesis, $\det C > 0$ and $a - \gamma^2 > 0$ follows. Therefore, q is positive definite, since $x \mapsto q(Tx)$ is.

1.6. It is no restriction to assume $p = 0$, $f(0) = 0$, and that q is a normal form. According to (27),

$$f(x) = q(x) + R(x),$$

where R is a smooth function on an open set containing the origin and vanishing to the second order; i.e.,

$$R(x)/|x|^2 \to 0 \quad \text{for} \quad x \to 0$$

with $|x|^2 = x_1^2 + \cdots + x_n^2$. Now suppose q is positive definite. Then $q(x) = |x|^2$ and therefore $f(x)/|x|^2 = 1 + R(x)/|x|^2$ so that $f(x)$ has to be positive on a neighborhood of the origin. Assume now that q is not positive semidefinite. Then $q(\alpha e_1) = -\alpha^2$ for $e_1 := (1, 0, 0, \ldots, 0)$, so that

$$f(\alpha e_1)/\alpha^2 = -1 + R(\alpha e_1)/\alpha^2.$$

Hence $f(\alpha e_1)$ is negative for α small enough.

1.7. The asserted equality concerns two homogeneous polynomials of degree k. It suffices to show that all their partial derivatives of order k coincide at the point p. Without restriction let $p = 0$. Then for $|\mu| = k$ clearly

$$D^\mu \sum_{|\nu|=k} \frac{1}{\nu!} D^\nu f(0) x^\nu \big|_{x=0} = D^\mu f(0)(0)$$

holds. On the other hand, using the convention that whenever indices appear twice in a product the sum is to be taken over those indices, it follows by symmetry

$$D_1(D_{i_1} \cdots D_{i_k} f(0))\, x_{i_1} \cdots x_{i_k} =$$
$$D_{i_1} \cdots D_{i_k} f(0)\, (\delta_{1 i_1} x_{i_2} \cdots x_{i_k} + \cdots + \delta_{1 i_k} x_{i_1} \cdots x_{i_{k-1}}) =$$
$$k\, D_1 D_{i_2} \cdots D_{i_k} f(0)\, x_{i_2} \cdots x_{i_k}.$$

More generally one has

$$D_1^{\mu_1}(D_{i_1} \cdots D_{i_k} f(0))\, x_{i_1} \cdots x_{i_k} =$$
$$k(k-1) \cdots (k - \mu_1 + 1)\, D_1^{\mu_1} D_{i_{\mu_1+1}} \cdots D_{i_k} f(0)\, x_{i_{\mu_1+1}} \cdots x_{i_k},$$

whence

$$D^\mu(D_{i_1} \cdots D_{i_k} f(0))\, x_{i_1} \cdots x_{i_k} = k!\, D^\mu f(0).$$

1.8. It is no restriction to assume $p = 0$. The formula

$$a_\nu = \frac{1}{\nu!} D^\nu f(0)$$

for the Taylor coefficients of a function given by a power series proves (i). To prove (ii), fix $1 \le i_1, \ldots, i_k \le n$ and define γ as in (ii). Then

$$D_{i_k} \ldots D_{i_1} f(x) = D^\gamma f(x) = \sum_{|\kappa|=0}^{\infty} a_\kappa D^\gamma x^\kappa$$
$$= \sum_{|\mu|=0}^{\infty} a_{\mu+\gamma} D^\gamma x^{\mu+\gamma} = \sum_{|\mu|=0}^{\infty} a_{\mu+\gamma} \frac{(\mu+\gamma)!}{\mu!} x^\mu.$$

Moreover
$$\int_0^1 \int_0^{t_{k-1}} \cdots \int_0^{t_1} t_0^m \, dt_0 \ldots dt_{k-1} = \frac{m!}{(m+k)!}$$

holds for every nonnegative integer m. When these two results are combined, the assertion (ii) follows. It should be noted that the interchange of summation with differentiation or integration is allowed since the power series converges uniformly on every compact subset of U.

1.9. Using induction on $n = 1, 2, \ldots$, it will first be shown that $f^{(n)}(0) = 0$ and that
$$f^{(n)}(x) = e^{-1/x^2} P_n\left(\frac{1}{x}, \sin\frac{2}{x}, \cos\frac{2}{x}\right)$$

for $x \neq 0$, where P_n is some polynomial in three variables. Indeed, the assertion holds for $n = 0$, since
$$\sin^2(1/x) = \tfrac{1}{2}[1 - \cos(2/x)].$$

To proceed from n to $n+1$, note that the derivative of a monomial in $1/x$, $\sin(2/x)$, and $\cos(2/x)$ is a polynomial of these functions.

Then the assertion about $f^{(n+1)}(x)$ at $x \neq 0$ follows. Moreover,
$$\frac{1}{x}(f^{(n)}(x) - f^{(n)}(0)) = \exp(-x^{-2})x^{-1} P_n(x^{-1}, \; \sin(2/x), \; \cos(2/x))$$

vanishes as x tends to zero, so that $f^{(n+1)}(0)$ exists and is equal to zero. Hence, f is smooth, and the origin is flat.

Now the nonzero critical points of f will be studied. Since
$$f'(x) = 2e^{-1/x^2}\left(x^{-3}\sin\frac{1}{x} - x^{-2}\cos\frac{1}{x}\right)\sin\frac{1}{x},$$

such a point x is given by
$$\sin(1/x) = 0 \quad \text{or} \quad \tan(1/x) = x.$$

In particular, $(1/n\pi)_n$ is a sequence of critical points that tends to zero. This proves that zero is a nonisolated critical point.

1.10. The only critical point of f is the origin, since
$$Df(x, y) = -(1 - x^2 - y^2)^{-1/2}(x, y).$$

Consider a point (x_0, y_0) in U outside the origin.

First case: $x_0 \neq 0$. Then $D_1 f(x_0, y_0) \neq 0$, and $\varphi(x, y) := (f(x, y), y) =: (u, v)$ for $(x, y) \in U$ is a local diffeomorphism at (x_0, y_0), because
$$D\varphi(x_0, y_0) = \begin{pmatrix} D_1 f(x_0, y_0) & D_2 f(x_0, y_0) \\ 0 & 1 \end{pmatrix}$$

is invertible. Solving $\varphi(x, y) = (u, v)$ for (x, y) yields a local inverse ψ of φ at $\varphi(x_0, y_0)$. By elementary computations,

$$\psi(u, v) = (\varepsilon_0(1 - u^2 - v^2)^{1/2}, v)$$

follows for $(u, v) \in U$, where ε_0 denotes the sign of x_0. As shown in the proof of Theorem (35),

$$f(\psi(u, v)) = u$$

holds, which can also be easily verified explicitly. Of course, the above relation is true only for those $(u, v) \in U$ with $u > 0$, as is evident from the definition of φ.

Second case: $y_0 \neq 0$. Then $D_2 f(x_0, y_0) \neq 0$, and $\varphi(x, y) := (f(x, y), x)$ $=: (u, v)$ for $(x, y) \in U$ defines a local diffeomorphism at (x_0, y_0). A local inverse ψ of φ at $\varphi(x_0, y_0)$ is given by

$$\psi(u, v) = (v, \varepsilon_0(1 - u^2 - v^2)^{1/2}),$$

where ε_0 denotes the sign of y_0. Finally,

$$f(\psi(u, v)) = u$$

follows for $(u, v) \in U$ with $u > 0$.

1.11.

(a) $D_i \tilde{F}(x) = \Sigma_{k=1}^2 D_k F(\psi(x)) D_i \psi_k(x)$, and hence

$$\tilde{g}_{ij}(x) = \langle D_i \tilde{F}(x), D_j \tilde{F}(x) \rangle$$

$$= \sum_{k,l=1}^2 \langle D_k F(\psi(x)) D_i \psi_k(x), D_l F(\psi(x)) D_j \psi_l(x) \rangle$$

$$= \sum_{k,l=1}^2 D_i \psi_k(x) g_{kl}(\psi(x)) D_j \psi_l(x).$$

Therefore, $\tilde{G}(x) = (D\psi(x))^t G(\psi(x))(D\psi)(x)$, which is the transformation law of the metric tensor.

(b) Let $a_i := D_i F(0)$. Since $a_1 \times a_2 \neq 0$, the vectors a_1 and a_2 are linearly independent. Hence the **Gram matrix** of scalar products $g_{ij}(0) = \langle a_i, a_j \rangle$ is positive definite; i.e., their eigenvalues are positive. In the present two-dimensional case, this follows immediately from Exercise 1.5(vii), since $\langle a_1, a_1 \rangle > 0$ and $\langle a_1, a_1 \rangle \langle a_2, a_2 \rangle - (\langle a_1, a_2 \rangle)^2 > 0$ by the Cauchy–Schwarz inequality. Thus, an invertible 2×2 matrix T exists so that $T^t G(0) T = E_2$, as stated in (24).
Define $\psi(x) := Tx$, $x \in \mathbf{R}^2$. Then, by (a),

$$\tilde{G}(0) = T^t G(0) T = E_2.$$

1.12. For $k = 1$, take $t_{11}(x) := (\varepsilon_1 g_{11}(x))^{1/2}$. Due to $g_{11}(0) = \varepsilon_1$, t_{11} is smooth around the origin. Now assume the assertion to be valid for $k - 1$ with $k \geq 2$. Since $\tilde{G}(x)$ satisfies the hypothesis of (36), a smooth map $\tilde{T}$ exists with

$$\tilde{G}(x) = \tilde{T}(x)^t G(0)\tilde{T}(x), \qquad \tilde{T}(0) = E_{k-1}.$$

Because the map $x \mapsto \det \tilde{T}(x)$ is smooth and has the value 1 at the origin, it does not vanish in a neighborhood V of the origin. Hence, $\tilde{T}(x)$ is invertible for $x \in V$. Cramer's rule implies that $x \mapsto \tilde{T}(x)^{-1}$ is smooth, and consequently F is smooth on V.

Since $F(0) = 0$, the function given by

$$g(0)g(x) - g(0)F(x)^t \tilde{G}(0)F(x)$$

is smooth and positive around the origin. Thus t is smooth. It is easy to show that T satisfies the assertion.

1.13. Since f and Df vanish at the origin, according to (25) there exists a symmetric 2×2 matrix G of smooth functions defined on a neighborhood of the origin with

$$f(x, y) = (x, y)G(x, y)\begin{pmatrix} x \\ y \end{pmatrix}.$$

Hence,

$$G(x, y) = \begin{pmatrix} x & \frac{1}{2}\cos xy \\ \frac{1}{2}\cos xy & 0 \end{pmatrix}.$$

Due to $D^2 f(0,0) = 2G(0,0) = \begin{pmatrix} 0 & 1 \\ 1 & 0 \end{pmatrix}$, one has $\det D^2 f(0,0) = -1$. It follows that the origin is a nondegenerate critical point of f, and that

$$q(x, y) := -x^2 + y^2$$

is the normal form of the quadratic form determined by half of the Hessian matrix of f at the origin. According to the proof of Morse's Lemma (37), one needs a 2×2 matrix $T(x, y)$ of smooth functions defined on a neighborhood of the origin with

$$G(x, y) = T(x, y)^t \begin{pmatrix} -1 & 0 \\ 0 & 1 \end{pmatrix} T(x, y).$$

Then $\varphi(x, y) := T(x, y)\begin{pmatrix} x \\ y \end{pmatrix}$ defines a local diffeomorphism at the origin, as desired.

In order to apply Exercise 1.12, $G(0, 0) = \frac{1}{2}\begin{pmatrix} 0 & 1 \\ 1 & 0 \end{pmatrix}$ has to be transformed to $\begin{pmatrix} -1 & 0 \\ 0 & 1 \end{pmatrix}$, which is accomplished by

$$\begin{pmatrix} 1 & -1 \\ 1 & 1 \end{pmatrix} G(0,0) \begin{pmatrix} 1 & 1 \\ -1 & 1 \end{pmatrix} = \begin{pmatrix} -1 & 0 \\ 0 & 1 \end{pmatrix}.$$

Therefore, consider

$$\begin{pmatrix} 1 & -1 \\ 1 & 1 \end{pmatrix} G(x,y) \begin{pmatrix} 1 & 1 \\ -1 & 1 \end{pmatrix} = \begin{pmatrix} x - \cos\ xy & x \\ x & x + \cos\ xy \end{pmatrix},$$

and $R(x,\ y)$ can be constructed, satisfying

$$\begin{pmatrix} x - \cos\ xy & x \\ x & x + \cos\ xy \end{pmatrix} = R(x,y)^t \begin{pmatrix} -1 & 0 \\ 0 & 1 \end{pmatrix} R(x,y)$$

by Exercise 1.12: For $|x| < \frac{1}{2}$, $|y| < \frac{1}{2}$ set $w(x,y) := (-x + \cos\ xy)^{1/2}$. A short computation yields

$$R(x,y) = \begin{pmatrix} w & -x/w \\ 0 & (\cos\ xy)/w \end{pmatrix}.$$

From this, $T(x,y) = R(x,y)\frac{1}{2} \begin{pmatrix} 1 & -1 \\ 1 & 1 \end{pmatrix}$, and

$$\varphi_1(x,y) = \tfrac{1}{2}(-x + \cos\ xy)^{-1/2}(-2x^2 + (x - y)\cos\ xy)$$
$$\varphi_2(x,y) = \tfrac{1}{2}(-x + \cos\ xy)^{-1/2}(x + y)\cos\ xy$$

for $|x| < \frac{1}{2}, |y| < \frac{1}{2}$, follows.

Chapter 2

The Fold and the Cusp

The development of the theory needed to classify degenerate critical points of smooth functions will be continued in Chapters 3–6. After that, the remaining six chapters are devoted to the local stability of smooth functions under small perturbations, a central aspect of Catastrophe Theory that is particularly important for applications. As a matter of fact, the name "catastrophe" was chosen for this theory because of its applications. In this chapter, stability will be treated heuristically by considering examples, in order to give an insight into the complete theory now and not to have to wait until the last six chapters.

As René Thom pointed out in his book *Stabilité Structurelle et Morphogénèse* [T1], the essential characteristics of a smooth function can be recognized by studying its embedding in a smooth family of functions. This fact is of extreme importance for applications, since natural phenomena are always subjected to perturbations. When a natural system is described by a function f of the **state variables** $x = (x_1, \ldots, x_n)$, then the perturbations are represented by parameters $u = (u_1, \ldots, u_r)$, called **external** or **control parameters**, on which the function depends. The potential energy and the entropy of a system are examples of such a function f. This is how a smooth family of functions arises in the study of natural phenomena. An **unfolding** F of a function f is such a family; it is a smooth function $F(x, u)$ of the state variables and the parameters satisfying $F(x, 0) = f(x)$. The goal of Catastrophe Theory is to detect properties of a function by studying its unfoldings.

As an illustration, consider the function x^2. It can be shown that any unfolding of x^2, as for example,

$$x^2 + u_1 x + u_2 x^2 + u_3 x^3 + \text{higher order terms},$$

has just one critical point near the origin that is a minimum (see Exercise 7.10). More precisely, there is a neighborhood U of the origin so that the unfolding has only one critical point in U for all sufficiently small u_i, and this point is a minimum.

In contrast to this, the one-parameter perturbation

$$x^4 - 2u^2 x^2$$

of x^4 exhibits three critical points around the origin as long as u does not vanish, namely, a maximum at 0 and two minima at u and $-u$, whereas x^4 has just one critical point around the origin, namely, a minimum exactly at 0. The origin is a stable critical point of x^2 but an unstable one for x^4; it is also an unstable critical point for x^3 (see below). More generally, every degenerate critical point is unstable.

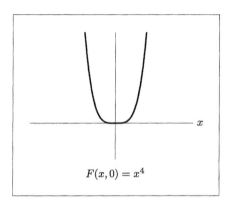

$$F(x,0) = x^4$$

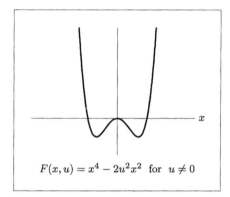

$$F(x,u) = x^4 - 2u^2x^2 \quad \text{for} \quad u \neq 0$$

Obviously, there is a multitude of different unfoldings of a given function f. A natural question is whether an unfolding of f exists containing the essential information about all unfoldings of f. Such an unfolding, which describes qualitatively the complete spectrum of all possible behaviors of the system represented by f under perturbations, is called **versal**. It will be referred to as **universal** when the number of parameters is minimal; in Chapter 7 the precise definitions will be given. According to the above remarks, it is clear that x^2 is its own universal unfolding. The unfoldings of x^3 and x^4 yield the simplest nontrivial unfoldings. The objective of this chapter is to introduce the fold and the cusp, which are universal unfoldings of x^3 and x^4, and to discuss some of their applications.

Generally speaking, Catastrophe Theory can be applied whenever **quasistatic forces**, i.e., forces not generating momenta, cause sudden changes. A wide range of applications is given by considering the potential energy $F(x, u)$ of a **mechanical system** depending on n state variables $x = (x_1, \ldots, x_n)$ and r external parameters $u = (u_1, \ldots, u_r)$. When the system is at rest in a position of **equilibrium**, then the value of x determines the state of the system. The parameters u describe the dependence of the system on external forces. These forces cause a change in the equilibria of the system. As a consequence, even though just quasistatic forces operate, and the external parameters vary continuously, the system can jump from a stable equilibrium position, which is a minimum of the potential energy, to another. A sudden transition resulting from a continuous parameter change is referred to as a **catastrophe**. Catastrophe Theory can also be applied to nonphysical systems, e.g., **biological**, **ecological**, **sociological**, **economic**, etc., whenever such discontinuities are observed. Most of these phenomena have, of course, been well understood in the respective disciplines for a long time. The innovative contribution of Catastrophe Theory is in establishing a unifying view of these changes. It provides an abstract and systematic treatment without reference to the underlying laws or principles governing the specific system under consideration. With respect to the applications, it is truly interdisciplinary. The often cited controversial aspect of some applications, particularly to the **humanities**, involves the more fundamental question of when mathematical models can be appropriately applied [Gu, Sma, SZ].

The usefulness of Catastrophe Theory in analyzing mathematical models is undisputed (whereas its appropriateness for producing think tools is controversial). The theory culminates in a rather surprising classification of catastrophes—there are only seven basic catastrophes if the number of essential parameters is limited to four.

The Fold

The simplest of all catastrophes is the **fold**, which is the function

(1) $$F(x, u) := x^3 + ux.$$

In Chapter 10 it will be shown that F is a universal unfolding of x^3. The set

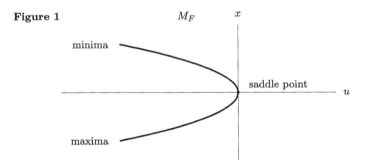

Figure 1

of critical points of all partial functions $F_u(x) := F(u, x)$ of an arbitrary F, called the **catastrophe surface**, is here simply the parabola $M_F = \{(x, u)\colon 3x^2 + u = 0\}$ in the left half plane, see Fig. 1. There are no critical points of F_u if u is positive. When u vanishes, there is exactly one critical point at the origin, which is a saddle point. The function F_u has exactly two critical points for negative u, namely, a maximum at $x = -|u/3|^{1/2}$ and a minimum at $x = |u/3|^{1/2}$.

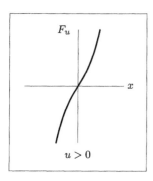
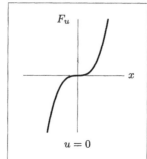
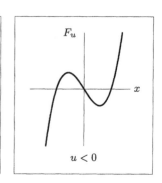

If F is interpreted to be the potential energy of a system, then its minima give the **stable equilibria**, whereas all other critical points represent **unstable equilibria**. For the fold, the only stable equilibrium is at $|u/3|^{1/2}$ for negative u. However, as u approaches 0, the minimum becomes flatter until it disappears for $u = 0$. This represents a sudden transition from a stable equilibrium to nonequilibrium—where the system does not rest at all—resulting from a continuous parameter change, i.e., a catastrophe.

The existence of an unstable equilibrium position for negative u is typical for the fold. That position is unstable in the sense that the smallest perturbation causes the system either to jump to the stable equilibrium or to move and not rest at all. The separation of the stable equilibria from the unstable occurs where the second derivative of F_u also vanishes, i.e., at the degenerate critical points. Generally, the subset C_F of the catastrophe surface M_F consisting of the degenerate critical points of all partial functions F_u is called the **catastrophe set**, and its projection onto the parameter space is called the **bifurcation set** B_F. For the fold, C_F is simply the origin in $\mathbf{R}^2$ and B_F is the origin in $\mathbf{R}$.

Figure 2

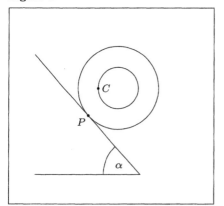

Figure 3

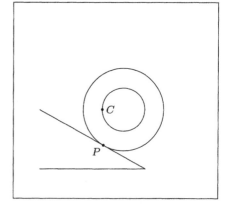

The simplest device demonstrating the fold is given by a wheel, whose center of gravity C is on the inner circumference and which is placed on a slope. Call α the angle of the slope, and denote the point of contact of the wheel by

P. The equilibrium positions of the wheel will be studied as α varies. These positions occur whenever C lies directly over P. Depending on α, there are three possibilities, which are sketched in Figs. 2–4. Either C never lies over P, as in Fig. 2, or C lies over P in just one position of the wheel (Fig. 3), or there are exactly two positions where C is over P (Fig. 4). In the last case, one of these two positions is a stable equilibrium—namely, when C is below the geometric center of the wheel—and the other is unstable. In the unstable position, a slight touch of the wheel can cause it to roll away. When this is compared to the catastrophe surface of the fold in Fig. 1, the following similarity is apparent. When a certain critical value of the angle α is exceeded, there is no equilibrium at all (Fig. 2), and the wheel simply rolls down the slope. Below this critical value, however, there are two positions at which the wheel can come to a standstill, a stable and an unstable one.

Figure 4

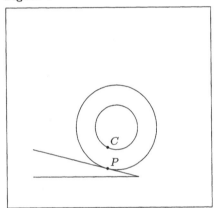 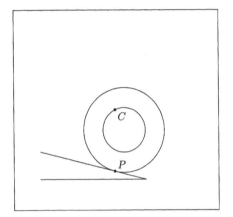

For an explicit calculation, let the wheel have unit radius, and denote the distance between the center of gravity C and the center of the wheel M by r. The state variable will be taken to be the angle Θ between the horizontal line and the line connecting C to the center of the wheel, which uniquely determines the position of the wheel. The role of the external parameter is assumed by the angle α. Up to an irrelevant proportionality factor, the potential energy V is the height of C above a preassigned zero level. If the wheel is moved uphill from $\Theta = 0$ to $\Theta = \Theta_0$, the point P moves a distance Θ_0 on the slope, and the position of the center of the wheel increases by $\Theta_0 \sin \alpha$. On the other hand, the center of gravity moves down on the inner circle and is decreased by $r \sin \Theta_0$. This yields

$$V(\Theta, \alpha) = \Theta \sin \alpha - r \sin \Theta + c(\alpha),$$

where $c(\alpha)$ is an irrelevant constant not depending on Θ. Because of

$$\frac{\partial V_\alpha}{\partial \Theta} = \sin \alpha - r \cos \Theta,$$

the catastrophe surface is

$$M_V = \left\{ (\Theta, \alpha) : \cos \Theta = \frac{\sin \alpha}{r} \right\}.$$

Figure 5

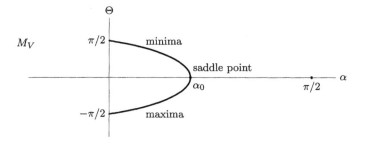

See Fig. 5, where $\alpha_0 := \arcsin r$. The condition $r \cos \Theta = \sin \alpha$ is equivalent to C lying directly over P. The positions sketched in Figs. 2, 3, and 4 occur when $\sin \alpha > r$, $\sin \alpha = r$, and $\sin \alpha < r$, respectively.

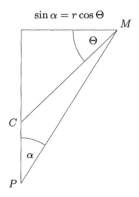

It is at this point that Catastrophe Theory enters the scene. Develop the potential energy $V(\Theta, \alpha)$ in its Taylor series around the critical point ($\Theta = 0, \alpha = \alpha_0$), giving

$$V(\Theta, \alpha) = \frac{r}{6}\Theta^3 + (\alpha - \alpha_0)\Theta \cos(\alpha_0) + c(\alpha)$$
$$+ \text{ higher order terms in } \Theta \text{ and } (\alpha - \alpha_0).$$

According to the theory, V is an example of a fold; see Exercise 10.1(a). More precisely, this means that there is a smooth function ω of Θ and α as well as smooth functions ψ and γ of α with the following properties holding for small Θ and $\alpha - \alpha_0$:

(i) $\omega(\cdot, \alpha_0)$ is a local diffeomorphism at the origin with $\omega(0, \alpha_0) = 0$,

(ii) ψ is a local diffeomorphism at α_0 with $\psi(\alpha_0) = 0$, and

(iii) $V(\Theta, \alpha) = F(\omega(\Theta, \alpha), \psi(\alpha)) + \gamma(\alpha)$, where F denotes the fold (1).

In short, V is locally a fold with respect to the new state variable $x := \omega(\Theta, \alpha)$ and the new control parameter $u := \psi(\alpha)$.

The Cusp

The catastrophe most frequently occurring in applications is the **cusp**, which is the function

(2) $$F(x; u, v) := x^4 - ux^2 + vx.$$

In spite of its simplicity, it exhibits all basic properties of higher catastrophes, such as **modality** and **hysteresis**, which are not present in the fold. This section includes a discussion of these properties. The cusp is a universal unfolding of x^4, as will be shown in Chapter 10.

The **dual cusp** is $-F$, and it is the universal unfolding of $-x^4$. Recall that x^4 and $-x^4$ are not equivalent; see (1.34). Of course, the catastrophe surface of the dual cusp equals that of the cusp; however, the positions of the maxima and minima are interchanged.

In contrast to the fold, the cusp has two external parameters u and v; again x denotes the state variable. The catastrophe surface

$$M_F = \{(x, u, v) : 4x^3 - 2ux + v = 0\}$$

is the following surface in $\mathbf{R}^3$ (Fig. 6):

Figure 6

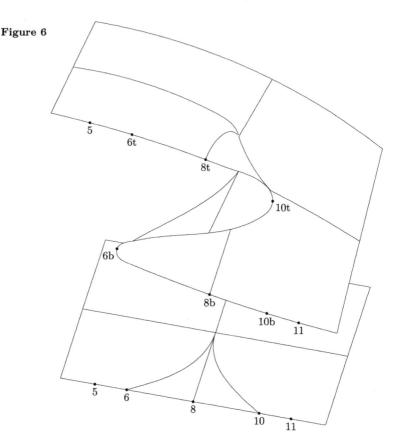

A simple calculation shows that the catastrophe set

$$C_F = \{(x, 6x^2, 8x^3) : x \in \mathbf{R}\}$$

is the image of a smooth curve in M_F with a vertical tangent at the origin.

The name cusp comes from the bifurcation set

$$B_F = \{(6x^2, 8x^3) : x \in \mathbf{R}\}$$

which is a **Neil parabola** in $\mathbf{R}^2$; the origin is a nondifferentiable point. To illustrate the cusp catastrophe geometrically, the projection $M_F \to \mathbf{R}^2$, $(x, u, v) \mapsto (u, v)$ will be considered as a covering map. Let G denote the domain in the half plane of $\mathbf{R}^2$ bounded by B_F and containing the positive u-axis. That part of M_F lying over the complement of $B_F \cup G$ corresponds to the minima of the partial functions $F_{u,v}$ and lies schlicht over this complement, i.e., every point in $\mathbf{R}^2 \backslash (B_F \cup G)$ has just one preimage. Every point (x, u, v) of C_F corresponds to a saddle point of $F_{u,v}$ with the exception of the origin, which is a fourth-order zero. The most interesting part of the surface M_F lies over G and is composed of three sheets. The top and bottom sheets comprise the minima of $F_{u,v}$, whereas the maxima are all on the middle sheet.

Figure 7

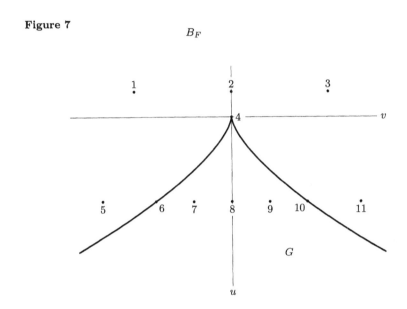

When the cusp F is interpreted to be the potential energy of a physical system, the minima of F are the positions of stable equilibria. By varying the parameters u, v continuously and staying outside of $B_F \cup G$, e.g., going from position 1 to 2 and then to 3 in Fig. 7, the system remains in a stable equilibrium. However, if u and v are changed so that B_F is transversed, something unusual happens. To see this, start in position 5 of Fig. 7, where the system is in a stable equilibrium. This state corresponds to a position on the top sheet of the surface M_F in Fig. 6. Moving parallel to the v-axis toward position 6 and then on to positions 7, 8, and 9, nothing changes, qualitatively speaking. When position 10 is reached, the system becomes unstable; its state corresponds to a point on C_F that is on the rim of M_F. The state variable falls parallel to the x-axis onto the bottom sheet of M_F. There the system is stable again and remains so while

Figure 8

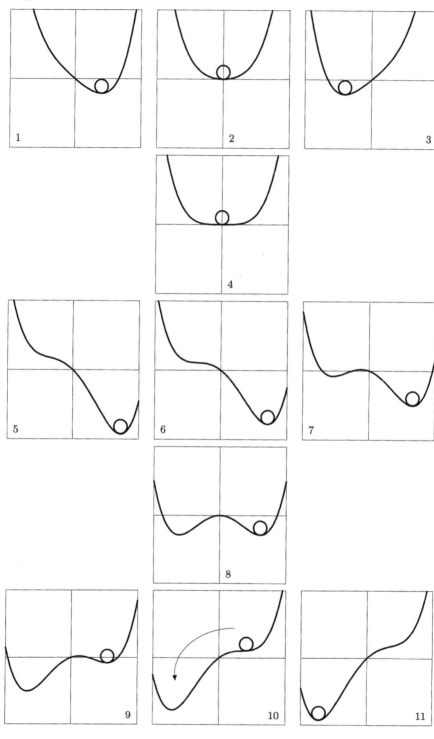

moving onward to position 11. It is important to note that the discontinuity in
the behavior of the system occurs exactly when leaving G. Fig. 8 gives a qualitative picture of what is happening. For the parameters (u, v) of positions 1–11,
the graphs of $F_{u,v}$ are shown, and the ball represents the state of the system.

By starting in position 11 of the parameter space and reversing the above
route, a catastrophic jump in the system also occurs when leaving G at position
6. This time, however, the jump is upward from the bottom sheet onto the
top sheet.

The middle sheet of M_F in Fig. 6 represents a different type of equilibrium.
This equilibrium corresponds to a maximal potential energy and is unstable.
The slightest perturbation of the system will cause it to jump into one of the
neighboring minima on the top or bottom sheet of the surface M_F.

The cusp catastrophe exhibits two phenomena, called modality and hysteresis, which the fold does not have. For the same parameter values $(u, v) \in G$ the
cusp has two states of stable equilibrium, namely, x_t and x_b such that (u, v, x_t)
and (u, v, x_b) lie on the top and bottom sheet of M_F, respectively. This is
called **bimodality**. Which state the system is in depends on the history of how
that state was reached, meaning what route was taken in the parameter plane.
Hysteresis is the name used to describe this dependence on the past.

Two applications of the cusp catastrophe, which come from mechanics and
biology, are discussed now. The first, due to E. C. Zeeman [Z1, pp. 409–415],
is a simple mechanical device illustrating the cusp, called **Zeeman's catastrophe machine**.

Let the center C of a cardboard disk be pinned onto a piece of wood so that
it can turn freely. Attach two elastics to a point S on the circumference of the
disk. Now tack the end of one of the elastics to the wood at point A so that the
elastic is taut (see Fig. 9). The end P of the other elastic will be moved slowly
in the plane of the disk. As long as P is outside the area G of Fig. 9, the point
S responds continuously to the movements of P. However, once P is inside G,
the disk will jump as soon as P leaves G. This **jump** is called the catastrophe.
For example, starting with a negative value of v, move the point P with coordinates (u, v) along a line parallel to the v-axis. The point S on the disk lies
to the left of the AC-axis, even when P enters G. When P leaves G, the disk
suddenly jumps clockwise, and S is now to the right of the AC-axis at the same
level as it was before entering G. For parameter values in G, the disk has two
positions of stable equilibrium that are reflections of each other with respect to
the AC-axis. Which of these positions it assumes depends on how it got there.

To obtain an explicit calculation of the phenomenon just mentioned, let the
diameter of the disk have unit length, and let the distance AC be twice the unit
length. Denote the angle between AC and CS by Θ. This serves as the state
variable and uniquely determines the position of the disk. The coordinates $(u,
v)$ of the point P are the control parameters. The potential energy V of the
system is in the elastic bands and is given by **Hooke's law** as

$$V = (SA - 1)^2 + (SP - 1)^2$$

up to an irrelevant proportionality constant.

Figure 9

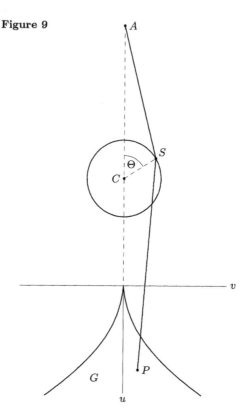

Simple geometrical reasoning yields

$$V(\Theta, u, v) = \left(\left[\frac{17}{4} - 2\cos\Theta \right]^{1/2} - 1 \right)^2$$

$$+ \left(\left[(u+a)^2 + v^2 + \frac{1}{4} + (u+a)\cos\Theta - v\sin\Theta \right]^{1/2} - 1 \right)^2,$$

where the origin 0 of the parameter plane is placed at the tip of the cusp and a denotes the distance from 0 to C. A straightforward calculation shows that

$$V(\Theta, 0, 0) = b + \left(\frac{1}{3} - \frac{a}{2} \cdot \frac{a - (1/2)}{a + (1/2)} \right) \Theta^2 + c\Theta^4$$

$$+ \text{ terms of higher order in } \Theta$$

for constants b and c depending on a. For $a := (7 + \sqrt{97})/12$, the coefficient of Θ^2 vanishes, and c turns out to be positive.

It follows now from Catastrophe Theory that V is locally a cusp. The precise meaning of this statement has been explained above for the fold—see (i)–(iii)—and the computations will be carried out for the cusp in Exercise 10.1(b).

Another interesting illustration of the cusp catastrophe comes from ecology and is given by studying the population growth of a species. This example also elucidates the link between Catastrophe Theory and **dynamical systems**. Let $N(t)$ be the size of a population at time t. The simplest mathematical model, called the **Malthusian law of population growth**, is to assume that the population changes at a rate proportional to the given population,

$$(3) \qquad \frac{dN}{dt} = rN,$$

where r is the **fertility coefficient** representing birth rate minus death rate. The uniquely determined solution of the autonomous differential equation (3) is

$$N(t) = N_0 e^{rt},$$

where $N_0 := N(0)$ is the initial population of the species. In this simple model, $r < 0$ implies extinction, for $r = 0$ the population remains unchanged, and $r > 0$ results in a population explosion. These results do not always comply with reality, particularly for long-term behavior.

To obtain a more sophisticated model, limiting factors such as food supply and the effects of parasites and predators must be taken into consideration. To illustrate the dependence on self-limiting factors like food supply, let k be a positive constant representing the **carrying capacity of the environment**, and consider the so-called **logistical differential equation**,

$$(4) \qquad \frac{dN}{dt} = rN\left(1 - \frac{N}{k}\right),$$

also referred to as **Pearl–Verhulst's equation**. The solution,

$$N(t) = k\left(1 + \frac{k - N_0}{N_0}e^{-rt}\right)^{-1},$$

has been used to explain the actual human population growth in various countries; see [Bra, 1.5]. Notice that despite a positive fertility coefficient r, the population now tends to a limiting value as t approaches infinity, which is a more realistic situation than before.

The effect of other limiting factors, such as predation, will be taken into account by subtracting a term $p(N)$ from the right-hand side of (4). Since a predation effect saturates at high prey densities, it is reasonable to assume that $p(N)$ approaches an upper limit as N approaches infinity, e.g., $p(N) = aN^2(b^2 + N^2)^{-1}$ for positive constants a and b; see [LJH p. 317]. The exact form of the new term $p(N)$ is not important for a qualitative analysis, as long as it meets the biologically necessary criteria. The reason for taking the above expression for $p(N)$ is explained in [LJH]. The resulting differential equation is

$$(5) \qquad \frac{dN}{dt} = rN\left(1 - \frac{N}{k}\right) - a\frac{N^2}{b^2 + N^2}.$$

By setting

$$x(t) := \frac{1}{b}N\left(\frac{b}{a}t\right),$$

the number of parameters is reduced, yielding

$$(6) \qquad \frac{dx}{dt} = G(x) := ux\left(1 - \frac{x}{v}\right) - \frac{x^2}{1 + x^2},$$

where $u := rb/a$ and $v := k/b$ are now dimensionless.

Note that for every **equilibrium point** x_0 of (6), i.e., for every zero x_0 of G, the constant function $x(t) = x_0$ is trivially a solution of (6). If the derivative $(dG/dx)(x_0)$ is negative, all solutions of (6) sufficiently close to x_0 strive exponentially toward x_0, and hence x_0 is called a **sink**. On the other hand, if $(dG/dx)(x_0)$ is positive, x_0 is a **source**; see [HS]. Generally, the system modeled by (6) approaches the sinks asymptotically and eventually comes to a standstill due to friction and other dissipative processes not taken into account by (6). When the main focus is on the final states of a system, it is sufficient to consider just the x values for which G vanishes.

The connection to Catastrophe Theory is established by observing that the negative of the primitive of G,

$$F(x; u, v) := -\frac{1}{2}ux^2 + \frac{1}{3}\frac{u}{v}x^3 + x - \arctan x,$$

can be thought of as an unfolding if the parameters are allowed to vary. The catastrophe surface M_F of F corresponds to the zero set of G and is of particular interest in understanding the solutions to (6) qualitatively, as has just been mentioned.

To analyze F with the machinery of Catastrophe Theory, find first all critical points that locally are zeros of $\partial F/\partial x$ with maximal order, the so-called **organization centers** of F. For such a critical point x_0 of the partial function $x \mapsto F(x; u_0, v_0)$, F is an unfolding of this function. Second, determine what kind of an unfolding F is for each x_0. There are essentially two cases,

$$(x_0, u_0, v_0) = (0, 0, v_0) \text{ with } v_0 \neq 0 \text{ and}$$

$$(x_0, u_0, v_0) = \pm\left(\sqrt{3}, \frac{3}{8}\sqrt{3}, 3\sqrt{3}\right),$$

which will be treated separately.

The Taylor expansion of F around $(0, 0, v_0)$ gives

$$F(x; u, v) = \frac{1}{3}x^3 - \frac{1}{2}ux^2 + \text{higher order terms.}$$

Notice that

$$\frac{1}{3}\frac{u}{v}x^3 = \frac{1}{3}\frac{u}{v_0}x^3 - \frac{1}{3}\frac{u(v - v_0)}{v_0^2}x^3 + \cdots$$

Figure 10

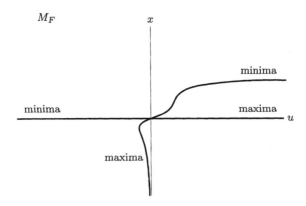

is higher order term. Clearly, F is a two-parameter unfolding of x^3 with the irrelevant parameter v. For fixed v_0, the one-parameter unfolding $(x, u) \mapsto F(x; u, v_0)$ does not represent a fold, as is readily seen by looking at its catastrophe surface. This surface is given by the u-axis and the relation

$$u = \frac{x}{(1 - x/v_0)(1 + x^2)}$$

shown in Fig. 10 (see also Exercise 10.1(c)). Concerning applications, just $v_0 > 0$ is of interest. For positive u_0 there are two possible population sizes. Initially, $x = 0$ is an unstable equilibrium, corresponding to a maximum of F_{u_0, v_0}. Then the population begins to grow and approaches the size bx_0 given by solving

$$u_0 = \frac{x_0}{(1 - x_0/v_0)(1 + x_0^2)}$$

for x_0. The carrying capacity $k_0 = bv_0$ is an upper bound for this size. Note that in general the size bx of the population depends monotonically on $u = r\frac{b}{a}$, which is the quotient of the growth rate r and the "aggression" a/b of the limiting factors.

To calculate the other organization centers of F, set

$$Q(x; u, v) := \frac{u}{v}x^3 - ux^2 + \left(1 + \frac{u}{v}\right)x - u.$$

Then

$$\frac{\partial}{\partial x}F(x; u, v) = \frac{x}{1 + x^2}Q(x; u, v).$$

Since $x = 0$ was already treated, assume $x \neq 0$. The critical points of $F_{u,v}$ of maximal order are then given by the common roots of the three equations $Q = 0, \partial Q/\partial x = 0$, and $\partial^2 Q/\partial x^2 = 0$, because Q is a cubic polynomial. The first equation yields

$$u = \frac{u}{v}x - \frac{x}{x^2 + 1},$$

which, when substituted in the second equation, gives

$$\frac{u}{v} = \frac{x^2 - 1}{(x^2 + 1)^2} \quad \text{and hence} \quad u = \frac{2x^3}{(x^2 + 1)^2}.$$

Using the third equation, it follows that the only possibility for a positive population is

$$(x_0, u_0, v_0) = \sqrt{3}(1, 3/8, 3).$$

Thus, F is an unfolding of the function

$$F(x; u_0, v_0) = \frac{-3\sqrt{3}}{16}x^2 + \frac{1}{24}x^3 + x - \arctan x.$$

Since the derivative of this function is

$$\frac{x}{8(1+x^2)}(x - \sqrt{3})^3,$$

the function $F(x; u_0, v_0) - F(x_0; u_0, v_0)$ has a fourth-order zero at x_0, and F represents a cusp (see also Exercise 10.1(d)). The catastrophe surface and set are

$$M_F = \{(x; u, v) : Q(x, u, v) = 0\}, C_F = \left\{\left(x, \frac{2x^3}{(x^2+1)^2}, \frac{2x^3}{x^2-1}\right) : x > 1\right\}.$$

The bifurcation set

$$B_F = \left\{\left(\frac{2x^3}{(x^2+1)^2}, \frac{2x^3}{x^2-1}\right) : x > 1\right\}$$

is sketched in Fig. 11. The tip of the cusp is at (u_0, v_0) corresponding to the state x_0.

Figure 11

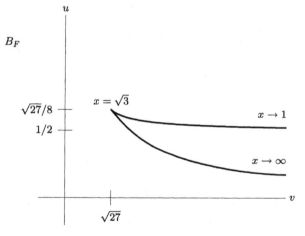

As is generally true for the cusp, continuous parameter changes can result in sudden jumps of the associated states. Here this gives an explanation of the outbreaks and collapses of some insect populations. An illustrative example is the study of the spruce budworm on the basis of Catastrophe Theory as presented in [C1] and [LJH].

The range of applications of Catastrophe Theory includes not only most of the natural sciences and engineering fields but also the humanities. In physics, the major fields where applications have been treated are classical mechanics, structural mechanics (elasticity), fluid dynamics, optics, thermodynamics, and meteorology [Be, BU, Gil, J, PS, Z1]. In addition to biology [Z1], which gave René Thom the original impulses for developing this theory [T1], Catastrophe Theory has been used to analyze problems in ecology and medicine [C1, LJH, PS, Sei, Z1]. Applications in the humanities include sociology [C2, Z1, Z2], economics [U, Z1], and linguistics [Sin]. Some of these applications are not convincing and are extremely controversial [Gu, Sma, SZ], but it should be emphasized again that, of course, the mathematical theory itself is indisputable.

There are two predominant reasons for the large variety of fields to which Catastrophe Theory has been applied. First of all, an investigation of the inner mechanism leading to the phenomena one wants to understand in applying Catastrophe Theory is frequently not regarded as necessary. The fact that phenomena exhibiting certain aspects of a catastrophe are observed is often sufficient to warrant an attempt to utilize the theory—in particular, whenever **bimodality**, or higher **modality, hysteresis,** and **sudden jumps** in a system appear. The second reason why it is tempting to apply Catastrophe Theory is that this theory reduces the number of possibilities drastically. If the number of relevant parameters for a system is not more than four, then there are just seven prototypes of catastrophic phenomena, the **7** elementary catastrophes. This certainly explains part of the fascination of Catastrophe Theory, above all in the early 1970s.

Exercises

2.1. Let a horizontal elastic beam be subjected to a vertical load α and a horizontal compression β. When the forces α and β are varied quasistatically, the beam stays in equilibrium until a certain value of α and β is attained. Then

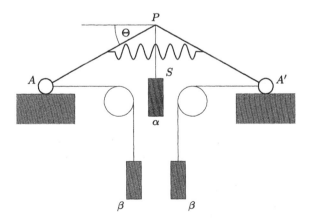

it suddenly snaps to a position buckling upward or downward. The details of this collapse show the characteristics of the cusp catastrophe. A simple model of such an elastic structure is the **Euler arch**. This consists of two massless rigid rods of unit length connected by a pivot P. A spring S of modulus k keeps the unstressed system in a straight position. Now two forces act on the system: a vertical force α and horizontal compression β. In the figure they are represented by weights. In equilibrium, the state of the system is described by the angle Θ between the rods and the horizontal. The forces α and β play the role of parameters.

Analyze the Euler arch just as Zeeman's catastrophe machine was analyzed in the section on the cusp. Recall that the potential energy of the spring is equal to $2k\Theta^2$ according to **Hooke's law**.

2.2. Roughly speaking, **caustics** are sharp, bright regions in space, where the light intensity is considerably higher than in the environment. Their appearance is a result of the refractive and reflecting properties of matter. In geometrical optics, caustics have a simple explanation—they occur as the **envelopes** of light paths. In a medium of constant refractive index, these paths are straight lines. Light paths obey **Fermat's principle of least time**. According to this principle, they are extremal (more precisely: stationary) in time among all possible paths having the same endpoints.

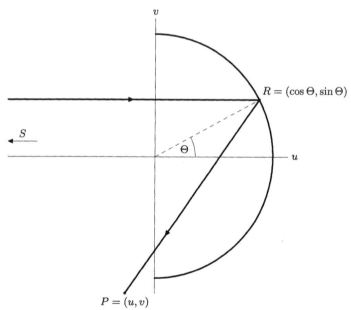

In this exercise, a special caustic will be studied that appears in the plane when light from a far source S is reflected off a circularly shaped mirror. It can be observed on the surface of coffee in a cup where light is reflected off the sides of the cup. The length of the ray from the source S to the endpoint P hitting the half circle at $R = (\cos\varphi, \sin\varphi)$ is a function F of the angle Θ and of the coordinates (u, v) of P.

Show that F defines a dual cusp catastrophe, where Θ is the state variable and u, v are the control parameters. Deduce from Fermat's principle that the caustic is just the bifurcation set. Consider only the limiting case of an infinitely remote light source S.

Similar in principle is the most familiar caustic, the **rainbow**. Each raindrop reflects the sun's rays at its internal sphere to form a cone-like caustic surface, whose axis points toward the sun and which corresponds to a fold catastrophe, see [PS],[Z1]. However, for each wave length, a colored cone emerges from the drop. This additional phenomenon is due to the refracting nature of water and does not occur in the previous caustic.

2.3.   The catastrophes following the cusp are the **swallowtail**,

$$F(x; u, v, w) = x^5 + ux^3 + vx^2 + wx$$

and the **butterfly**,

$$F(x; t, u, v, w) = x^6 + tx^4 + ux^3 + vx^2 + wx$$

(see Table 1, Chapter 7). Determine their modality, i.e., the maximal number of stable states corresponding to the same parameter values.

Solutions

2.1.   The potential energy V of the arch is the sum of the potential energies of the spring and of the weights. Since the potential energy of a weight is proportional to its height, one has

$$V(\Theta; \alpha, \beta) = 2k\Theta^2 + \alpha \sin \Theta - 2\beta(1 - \cos \Theta)$$

up to an irrelevant constant. The Taylor expansion of V with respect to Θ at $\Theta = 0$ yields

$$\alpha\Theta + (2k - \beta)\Theta^2 - \frac{\alpha}{6}\Theta^3 + \frac{\beta}{12}\Theta^4 + \text{higher orders in } \Theta.$$

Hence for $\alpha = 0$ and $\beta = 2k$ the first nonvanishing term of the Taylor expansion is of fourth order with the positive coefficient $k/6$. Then $V(\Theta; \alpha, \beta)$ is an unfolding of $V(\Theta; 0, 2k) = (k/6)\Theta^4 + \cdots$. As will be shown in Exercise 10.2, V represents a cusp catastrophe.

2.2.   Let the light source S be on the negative u-axis at $(-s, 0)$. The length F_S of the straight line segments from S to $R = (\cos \Theta, \sin \Theta)$ and from R to

$P = (u, v)$ is

$$F_S(\Theta; u, v) = |SR| + |RP| =$$
$$[(s + \cos \Theta)^2 + \sin^2 \Theta]^{1/2} + [(u - \cos \Theta)^2 + (v - \sin \Theta)^2]^{1/2}.$$

Since the medium between source and mirror is presumed to be homogeneous, F_S is proportional to the time required by the light to get from S to P via R; therefore, by Fermat's principle, the path actually traversed by the light is determined by the equation

$$\frac{\partial}{\partial \Theta} F_S(\Theta, u, v) = \frac{-s \sin \Theta}{|SR|} + \frac{u \sin \Theta - v \cos \Theta}{|RP|} = 0.$$

The limit of $\partial F_S / \partial \Theta$ exists as s approaches ∞, since $-s \sin \Theta / |SR| \to -\sin \Theta$. Therefore, in the case of an infinitely remote light source, F_S is replaced by

$$(1) \qquad F(\Theta; u, v) = \cos \Theta + [(u - \cos \Theta)^2 + (v - \sin \Theta)^2]^{1/2}.$$

Denote by $f_\Theta(u, v)$ the partial derivative of F with respect to Θ at $(\Theta; u, v)$, i.e.,

$$(2) \qquad f_\Theta(u, v) := -\sin \Theta + \frac{u \sin \Theta - v \cos \Theta}{\sqrt{(u - \cos \Theta)^2 + (v - \sin \Theta)^2}}.$$

Then the equation

$$(3) \qquad\qquad\qquad f_\Theta(u, v) = 0$$

determines the actual light paths, parametrized by Θ. Of course, these paths are the rays starting at R and going through P, because $f_\Theta(u, v) = f_\Theta((\cos \Theta, \sin \Theta) + t((u, v) - (\cos \Theta, \sin \Theta)))$ for all $t > 0$. Notice that according to (3),

$$\langle (1, 0), (\sin \varphi, -\cos \varphi) \rangle = \left\langle \frac{(u - \cos \varphi, v - \sin \varphi)}{|(u - \cos \varphi, v - \sin \varphi)|}, (\sin \varphi, -\cos \varphi) \right\rangle$$

holds, showing that the tangent vector $(\sin \varphi, -\cos \varphi)$ to the circle at R forms the same angle with the incident ray as with the reflected ray; this is the well-known law of reflection.

Given a one-parameter family of curves as in (3), the envelope of that family satisfies (3) and

$$(4) \qquad\qquad\qquad \frac{\partial}{\partial \Theta} f_\Theta(u, v) = 0.$$

In the present case, the envelope of (3) is the caustic formed by the light rays $\vec{RP}$. Hence the caustic is the bifurcation set B_F of the unfolding F.

In order to get an explicit expression for the caustic it is convenient to use

$$(5) \qquad\qquad\qquad \tan(2\Theta) = \frac{v - \sin \Theta}{u - \cos \Theta}$$

instead of (3). Equation (5) easily follows from (3) and is geometrically an immediate consequence of the law of reflection. Rewriting (5) as $(v - \sin\Theta)\cos(2\Theta) = (u - \cos\Theta)\sin(2\Theta)$, differentiating with respect to Θ, and then solving for u and v yields the caustic

(6) $$u = \cos\Theta - \tfrac{1}{2}\cos\Theta\cos 2\Theta, \quad v = \sin\Theta - \tfrac{1}{2}\cos\Theta\sin 2\Theta.$$

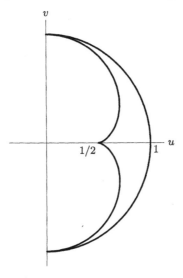

This curve is referred to as the **nephroid**. The tip of its cusp is at $\Theta = 0$ with coordinates $(u, v) = (\tfrac{1}{2}, 0)$. Expanding $F(\Theta; u, v)$ at the cusp point for small angles Θ yields

(7)
$$F\left(\Theta; \frac{1}{2}, 0\right) = \cos\Theta + \left[\frac{5}{4} - \cos\Theta\right]^{1/2}$$
$$= \frac{3}{2} - \frac{1}{4}\Theta^4 \text{ plus higher order terms,}$$

revealing its nature as an unfolding of a dual cusp.

Finally, it should be remarked that the foregoing considerations are valid for incident rays where $|\Theta| \le \pi/3$. When $\pi/3 < \Theta < \pi/2$, the rays are reflected more than once. Geometrically, this is obvious, and analytically it follows from the condition $u \le 1$, which the solution u of $f_\Theta(u, 0) = 0$ has to satisfy.

2.3.   The modality of the swallowtail catastrophe F is the maximal number of minima of the partial map $x \mapsto F(x; u, v, w)$ for fixed u, v, w. The modality of the butterfly catastrophe is defined analogously.

A polynomial p of degree $n \ge 2$,

$$x^n + a_{n-1}x^{n-1} + \cdots + a_1 x + a_0,$$

has at most $n - 1$ extrema, and for certain values of the coefficients it actually possesses $n - 1$ extrema. Then, if n is even, $n/2$ extrema are minima, and if n is odd, there are $(n-1)/2$ minima. A translation $x \mapsto x - b$ along the x-axis with a real b has no effect on the extrema. The same is true when adding a constant. After two such transformations, the polynomial becomes $p(x - a_{n-1}/n) - p(-a_{n-1}/n)$, which is of the desired form

$$x^n + a'_{n-2}x^{n-2} + \cdots + a'_1 x,$$

i.e., the constant term and the $(n-1)$st coefficient are zero. It remains to show that the maximal number of minima occurs for arbitrarily small coefficients $a'_{n-2}, \ldots, a'_1$. To see this, note that there are two more transformations that do not alter the number of minima, namely, the scaling $x \mapsto cx$ of the variable x by a positive factor c and the multiplication of the polynomial by a positive constant d. For $d := c^{-n}$ one obtains the polynomial

$$x^n + a'_{n-2}c^{-2}x^{n-2} + \cdots + a'_1 c^{1-n} x,$$

whose coefficients tend to zero as c approaches infinity. For $n = 5$ and 6, these considerations show that the swallowtail is bimodal like the cusp and that the butterfly is trimodal.

Chapter 3

Degenerate Critical Points: The Reduction Lemma

The investigation of degenerate critical points of smooth functions is initiated in this chapter by proving the **Reduction Lemma**. This lemma, which is also referred to in the literature as the splitting lemma, states that around a degenerate critical point p, a smooth function f of n variables can be expressed in suitable coordinates as a sum of two functions q and g of different variables. The function q is a nondegenerate normal quadratic form in r variables, whereas g is a smooth function of $n - r$ variables that is **totally degenerate** at the corresponding critical point, i.e., g has rank zero there, so that g is a representative of the **residual singularity** of f at p. The number r is the rank of f at p, and the index of f at p is that of q. Due to the Reduction Lemma, the study of degenerate critical points is reduced to analyzing the behavior of the totally degenerate ones. Now it is possible to be more precise about the content of Thom's first major theorem of Catastrophe Theory. This theorem classifies the totally degenerate critical points of functions of codimension at most 4. The notation codimension will be explained in Chapter 5.

As already mentioned, the proof of the Reduction Lemma uses the diagonalization of symmetric matrices of smooth functions shown in (1.36). Before commencing with the proof of the Reduction Lemma, recall that $f: U \to \mathbf{R}$ always denotes a smooth function defined on an open set U of $\mathbf{R}^n$. If r is the rank of f at a point p in U, then $n - r$ will be called the **corank** of f at p. In case the point p is critical, the corank is nonzero exactly when p is degenerate. The corank is a measure of the degree of degeneracy of critical points, since it is the dimension of the null space of the Hessian matrix at such points. Like the rank and index, the corank of a critical point is invariant under local diffeomorphisms. This follows from (1.33).

Again, for the sake of simplicity, let the origin be the point under consideration.

Definition. Let r be an integer with $0 \le r < n$, and let V be open in

$\mathbf{R}^{n-r}$ with $0 \in V$. Then f is **reducible** at the origin to a smooth function
$g: V \to \mathbf{R}$ if a local diffeomorphism ψ at $0 \in \mathbf{R}^n$ with $\psi(0) = 0$ exists as
well as a normal quadratic form q_{sr} on $\mathbf{R}^r$ such that $\psi(y) \in U$ and

(1) $$f(\psi(y)) = q_{sr}(y_1, \ldots, y_r) + g(y_{r+1}, \ldots, y_n)$$

hold for all $y \in W \times V$, where W is open in $\mathbf{R}^r$ with $0 \in W$.

Suppose now that the origin is a critical point of f. From (1) it follows that
the rank of f at the origin is the sum of r and the rank of g at the origin.
Likewise, the index of f at the origin is s plus the index of g at the origin, and
the corank of f at the origin is that of g there. In particular, if g has rank zero
at 0, then r and s are the rank and index, respectively, of f at 0, and the corank
of f at 0 is the number of variables of g. Moreover, in this case, using (1.25),
smooth functions g_{ijk} on a neighborhood V' of the origin contained in V exist
for $r + 1 \leq i, j, k \leq n$ with $g_{ijk} = g_{jik} = \cdots$ for all permutations of the indices
such that

$$g(y_{r+1}, \ldots, y_n) = f(0) + \sum_{i,j,k=r+1}^{n} g_{ijk}(y_{r+1}, \ldots, y_n) y_i y_j y_k$$

holds for $(y_{r+1}, \ldots, y_n) \in V'$. In particular,

$$g_{ijk}(0, \ldots, 0) = \frac{1}{6} D_i D_j D_k g(0, \ldots, 0)$$

holds for $r + 1 \leq i, j, k \leq n$.

Reduction Lemma (2). If the origin is a degenerate critical point of
f, then f is reducible there to a smooth function g with rank zero at the
origin, for which the origin is also a critical point. Furthermore, f and g
have the same value at the origin.

Proof. By the remark preceding (1.37), it is no restriction to assume that
$\frac{1}{2} D^2 f(0) = \begin{pmatrix} A & 0 \\ 0 & 0 \end{pmatrix}$, where

$$A := \begin{pmatrix} -E_s & 0 \\ 0 & E_{r-s} \end{pmatrix}.$$

Let $W \subset \mathbf{R}^r$ and $V \subset \mathbf{R}^{n-r}$ be open with $(0,0) \in W \times V \subset U$. The proof
will be divided into four steps.

(i) Consider first the smooth map $F: W \times V \to \mathbf{R}^r$ given by

$$F(x, y) := D_1 f(x, y) = \left(\frac{\partial f}{\partial x_1}(x, y), \ldots, \frac{\partial f}{\partial x_r}(x, y) \right)$$

which vanishes at the origin and satisfies $D_1 F(0) = 2A$. Using the
Implicit Function Theorem (see Appendix A.1), one obtains a smooth
function h defined on an open subset V_1 of V containing the origin
with $h(V_1) \subset W$ and $h(0) = 0$ such that $F(h(y), y) = 0$ holds for
$y \in V_1$.

(ii) A local diffeomorphism $\chi\colon W \times V_1 \to \mathbf{R}^n$ vanishing at the origin is given by $\chi(x, y) := (x + h(y), y)$, since

$$D\chi(0,0) = \begin{pmatrix} E_r & Dh(0) \\ 0 & E_{n-r} \end{pmatrix}$$

is invertible. Hence, there are open balls $W_1 \subset \mathbf{R}^r$ and $V_2 \subset \mathbf{R}^{n-r}$ with $(0,0) \in W_1 \times V_2 \subset W \times V_1$ so that $\chi(W_1 \times V_2) \subset U$.

(iii) Define $k(x, y) := f(\chi(x, y)) - f(h(y), y)$ for $(x, y) \in W_1 \times V_2$. Then $k(0, y) = 0$ and $D_1 k(0, y) = F(h(y), y) = 0$ for $y \in V_2$. Furthermore, $D_1^2 k(0, 0) = D_1^2 f(0, 0) = 2A$, and therefore (1.28) implies $k(x, y) = \langle x, K(x,y)x \rangle$ for a smooth map K with the value A at the origin. Applying (1.36) to K, it follows that

$$k(x, y) = \langle \varphi_1(x, y), A\varphi_1(x, y) \rangle$$

for $\varphi_1(x, y) := T(x, y)x$, where T is smooth on an open subset $W_2 \times V_3$ of $W_1 \times V_2 \subset \mathbf{R}^n$ containing the origin and $T(0, 0) = E_r$ holds. The map φ_1 will now be used to define the desired coordinate transformation.

(iv) Set $\varphi(x, y) := (\varphi_1(x, y), y)$ for $(x, y) \in W_2 \times V_3$. Clearly $\varphi(0, 0) = 0$. Due to $D\varphi(0, 0) = E_n$, φ is a local diffeomorphism at $0 \in \mathbf{R}^n$. A local inverse ψ of φ at the origin is defined on an open set Z in $\mathbf{R}^n$ with $0 \in Z$ satisfying $\psi(Z) \subset W_2 \times V_3$ and has the form $\psi(u, y) = (\psi_1(u, y), y)$. Replacing the old coordinates (x, y) of k by the new ones (u, y) and after setting $g(y) := f(h(y), y)$,

$$f(\chi(\psi(u, y))) = \langle u, Au \rangle + g(y)$$

follows for $(u, y) \in Z$. The rest of the assertion is obvious.

The Reduction Lemma was proved by D. Gromoll and W. Meyer in 1969 in a much more general setting, namely, for smooth maps between Banach spaces [GM].

The function g in (2) is a representative of the so-called **residual singularity of f at 0**. What is not a consequence of the Reduction Lemma and what is also needed for the classification of critical points is that the residual singularity is unique up to a smooth coordinate transformation around the critical point. This nontrivial result will be proved at the end of the next chapter.

An example that illustrates the Reduction Lemma and the constructive character of its proof concludes this chapter. For a constructive proof of Lemma (1.36), which was used to prove the Reduction Lemma, see Exercise 1.12. The Implicit Function Theorem is the main tool to derive the reduced function g (see (i) in the above proof), and thus it is not always possible to find an explicit expression for g. However, by differentiating h implicitly in $D_1 f(h(y), y) = 0$, one obtains at least the Taylor expansion of $g(y) = f(h(y), y)$, see Exercise 3.3. This is a crucial point for the applicability of Catastrophe Theory, because only the Taylor expansions are needed to classify critical points. The following three chapters are devoted to such a classification.

Example (3). The origin is a degenerate critical point of the polynomial $f: \mathbf{R}^2 \to \mathbf{R}$, $f(x,y) := x^2 + 2xy^2 + x^2y^2$, which is reducible there to the function $g(y) = -y^4(1+y^2)^{-1}$, $y \in \mathbf{R}$. Clearly, g has zero rank at the origin.

Proof. To see this, observe first that $\frac{1}{2}D^2f(0) = \begin{pmatrix} 1 & 0 \\ 0 & 0 \end{pmatrix}$, i.e., the rank r of f at $0 \in \mathbf{R}^2$ is 1. Since $(\partial f/\partial x)(x,y) = 2x + 2y^2 + 2xy^2$ is true, the function $h(y) := -y^2(1+y^2)^{-1}$ solves $(\partial f/\partial x)(h(y), y) = 0$ for all $y \in \mathbf{R}$. The proof of the Reduction Lemma shows that $g(y) := f(h(y), y) = -y^4(1+y^2)^{-1}$ gives a splitting of f, i.e., f becomes

$$u^2 - \frac{y^4}{1+y^2}$$

in the new coordinates (u, y).

In order to obtain an explicit expression for the coordinate transformation $(x,y) \mapsto (u,y)$, note that $k(x,y) := f(x+h(y),y) - g(y) = x^2(1+y^2)$ and thus $\varphi_1(x,y) = x\sqrt{1+y^2}$. With the new coordinate $u := x\sqrt{1+y^2}$, one has

$$f\left(\frac{u}{\sqrt{1+y^2}} - \frac{y^2}{1+y^2}, y\right) = u^2 - \frac{y^4}{1+y^2}.$$

A further simplification is possible, since by (1.34) changing the coordinate y to another coordinate $v := y(1+y^2)^{-1/4}$ transforms g into the function $-v^4$. Solving the last equation for y, it follows that

$$f\left(\left(\sqrt{1+v^4/4} - v^2/2\right)(u - v^2),\; v\sqrt{\sqrt{1+v^4/4} + v^2/2}\right)$$
$$= u^2 - v^4$$

for all $(u, v) \in \mathbf{R}^2$.

Exercises

3.1. Find a local diffeomorphism at 0 by which $f: \mathbf{R}^2 \to \mathbf{R}$, $f(x,y) := x - x\cos y - x^2 \cos y$, is reduced to y^4.

3.2. Show that the function $f: \mathbf{R}^3 \to \mathbf{R}$, $f(x,y,z) := x^2 + y^2 + y^3 - (x+y)z^2$, is reducible at the origin to $-z^4$. Find a local diffeomorphism at 0 by which the reduction is achieved.

3.3. Let the origin be a degenerate critical point of f and assume that $\frac{1}{2}D^2 f(0)$ $= \begin{pmatrix} A & 0 \\ 0 & 0 \end{pmatrix}$ with A as in the proof of the Reduction Lemma (2). Recall that h is defined implicitly by $D_1 f(h(y), y) = 0$ for small y and $h(0) = 0$. Show that the fifth Taylor polynomial of $g(y) := f(h(y), y)$ is given by

$$T_g^5(y) = f(0) + \frac{1}{3!}F_{\alpha\beta\gamma}y_\alpha y_\beta y_\gamma + \left(\frac{1}{4!}F_{\alpha\beta\gamma\delta} - \frac{1}{16}A_{kl}F_{k\alpha\beta}F_{l\gamma\delta}\right)y_\alpha y_\beta y_\gamma y_\delta$$
$$+ \left(\frac{1}{5!}F_{\alpha\beta\gamma\delta\varepsilon} - \frac{1}{24}A_{ij}F_{i\alpha\beta\gamma}F_{j\delta\varepsilon} + \frac{1}{32}A_{ik}A_{jl}F_{i\alpha\beta}F_{j\gamma\delta}F_{kl\varepsilon}\right)y_\alpha y_\beta y_\gamma y_\delta y_\varepsilon$$

for $i, j, k, l \in \{1, \ldots, r\}$ and $\alpha, \beta, \gamma, \delta \in \{1, \ldots, n-r\}$. The **Einstein convention** is used here meaning that whenever indices appear twice in a product the sum is to be taken over those indices. Furthermore,

$$F_{\alpha\beta\gamma} := \frac{\partial^3}{\partial y_\alpha \partial y_\beta \partial y_\gamma}f(0) \quad \text{and} \quad F_{i\alpha\beta} := \frac{\partial^3}{\partial x_i \partial y_\alpha \partial y_\beta}f(0) \text{ etc.,}$$

i.e., the Latin indices designate partial derivatives with respect to the x coordinates and the Greek indices with respect to the y coordinates.

It may be of interest to note that the Taylor expansion of g can be obtained by **implicit automatic differentiation** and that the Reduction Lemma is a useful tool in **optimization theory**, see [Sch].

Solutions

3.1. Follow the proof of the Reduction Lemma (2), and use the same notations. Then $A = -1$ and $F(x, y) = 1 - \cos y - 2x \cos y$, so that

$$h(y) := \frac{1 - \cos y}{2 \cos y}$$

satisfies $h(0) = 0$ and $F(h(y), y) = 0$. By (2) this yields that f is reducible at the origin to a function g given by

$$g(y) := f(h(y), y) = \frac{(1 - \cos y)^2}{4 \cos y}.$$

The first nonvanishing term of the Taylor expansion of g is $y^4/16$. Due to (1.34), f can be reduced to y^4.

The proof of (2) shows how to find a local diffeomorphism giving the desired reduction. First introduce the local diffeomorphism $\chi(x, y) := (x + h(y), y)$. Next consider $k(x, y) := f(\chi(x, y)) - g(y) = -x^2 \cos y$. This gives rise to the local diffeomorphism $\varphi(x, y) = (x\sqrt{\cos y}, y)$. A local inverse of φ is given by $\psi(u, y) = (u/\sqrt{\cos y}, y)$. According to (2), one obtains $f(\chi(\psi(u, y))) = -u^2 + g(y)$.

Now g will be transformed into the function v^4 by a local diffeomorphism at the origin. The equation $g(y) = v^4$ is easily solved for y, yielding

$$y = \arccos(1 + 2v^4 - 2v^2\sqrt{1 + v^4}).$$

Replacing y by this expression in

$$\chi(\psi(u, y)) = \left(\frac{u}{\sqrt{\cos y}} + \frac{1 - \cos y}{2\cos y}, y\right),$$

one arrives at the desired local diffeomorphism Φ of the new variables u and v satisfying $f(\Phi(u, v)) = -u^2 + v^4$.

3.2. Again, follow the proof of the Reduction Lemma (2) and use its notations. Then $A = \begin{pmatrix} 1 & 0 \\ 0 & 1 \end{pmatrix}$ and $F(x, y, z) = (2x - z^2, 2y + 3y^2 - z^2)$, so that

$$h(z) := \left(\frac{1}{2}z^2, -\frac{1}{3} + \frac{1}{3}\sqrt{1 + 3z^2}\right)$$

satisfies $h(0) = 0$ and $F(h(z), z) = 0$. It follows from (2) that f is reducible at the origin to the function g given by

$$g(z) := f(h(z), z) = \frac{2}{27} + \frac{1}{3}z^2 - \frac{1}{4}z^4 - \frac{2}{27}(1 + 3z^2)^{3/2}.$$

The first nonvanishing term of the Taylor expansion of g is $-\frac{1}{2}z^4$. Hence the assertion results from (1.34).

According to (2), one has $f(\chi(\psi(u, v, z))) = u^2 + v^2 + g(z)$, where the local diffeomorphisms χ and ψ at the origin are determined as follows. First, χ is given by

$$\chi(x, y, z) := (x + h_1(z), y + h_2(z), z)$$
$$= \left(x + \frac{1}{2}z^2, y - \frac{1}{3} + \frac{1}{3}\sqrt{1 + 3z^2}, z\right).$$

Next consider $k(x, y, z) := f(\chi(x, y, z)) - g(z) = x^2 + y^2\sqrt{1 + 3z^2} + y^3$, which obviously is equal to

$$(x, y) \begin{pmatrix} 1 & 0 \\ 0 & y + \sqrt{1 + 3z^2} \end{pmatrix} \begin{pmatrix} x \\ y \end{pmatrix}.$$

This gives rise to the local diffeomorphism

$$\varphi(x, y, z) := \left(x, y\sqrt{y + \sqrt{1 + 3z^2}}, z\right)$$

at the origin, which is a local inverse of ψ. Hence $\psi(u, v, z) = (u, y, z)$, where y is the unique solution of the cubic equation $y^3 + y^2\sqrt{1 + 3z^2} - v^2 = 0$, which has the same sign as v for small z and v. Explicitly, y is the real part of

$$(1 - i\sqrt{3})\sqrt[3]{(1/27)b^3 - (1/2)v^2 + iv\sqrt{(1/27)b^3 - (1/4)v^2}} - (1/3)b$$

with $b := \sqrt{1 + 3z^2}$, where that cubic root is chosen that becomes real as v approaches zero.

Finally, according to (1.34), the function that is 0 at the origin and equal to $-z(-z^{-4}g(z))^{1/4}$ for small $z \neq 0$ is a local diffeomorphism. A local inverse γ of it satisfies $g(\gamma(w)) = -w^4$. Then $\Phi(u, v, w) := \chi(\psi(u, v, \gamma(w)))$ is a local diffeomorphism at the origin vanishing there and satisfying $f(\Phi(u, v, w)) = u^2 + v^2 - w^4$.

3.3. Since $h(0) = 0$, differentiating $D_1 f(h(y), y) = 0$ with respect to y yields $D_1^2 f(0)Dh(0) + D_2 D_1 f(0) = 0$. By the assumption on $D^2 f(0)$, the second term is zero and $D_1^2 f(0)$ is invertible. Hence, $Dh(0) = 0$ follows. Therefore, the only terms of the Taylor expansion of f contributing to $T_g^5(y)$ are

(1)
$$f(0) + A_{ij}x_i x_j + \frac{3}{3!}F_{ij\alpha}x_i x_j y_\alpha$$

$$+ \frac{3}{3!}F_{i\alpha\beta}x_i y_\alpha y_\beta + \frac{1}{3!}F_{\alpha\beta\gamma}y_\alpha y_\beta y_\gamma$$

$$+ \frac{4}{4!}F_{i\alpha\beta\gamma}x_i y_\alpha y_\beta y_\gamma + \frac{1}{4!}F_{\alpha\beta\gamma\delta}y_\alpha y_\beta y_\gamma y_\delta$$

$$+ \frac{1}{5!}F_{\alpha\beta\gamma\delta\varepsilon}y_\alpha y_\beta y_\gamma y_\delta y_\varepsilon.$$

From $D_1 f(h(y), y) = 0$ it follows for all l that

(2)
$$0 \equiv 2A_{li}h_i + F_{li\alpha}h_i y_\alpha + \frac{1}{2}F_{l\alpha\beta}y_\alpha y_\beta + \frac{1}{6}F_{l\alpha\beta\gamma}y_\alpha y_\beta y_\gamma + \cdots .$$

Differentiating (2) with respect to y_ζ and then y_η and y_ϑ gives

(3) $\quad 0 \equiv 2A_{li}h_{i,\zeta} + F_{li\alpha}h_{i,\zeta}y_\alpha + F_{li\zeta}h_i + F_{l\zeta\alpha}y_\alpha + \frac{1}{2}F_{l\zeta\alpha\beta}y_\alpha y_\beta + \cdots ,$

(4) $\quad 0 \equiv 2A_{li}h_{i,\zeta\eta} + F_{li\alpha}h_{i,\zeta\eta}y_\alpha + F_{li\eta}h_{i,\zeta} + F_{li\zeta}h_{i,\eta} + F_{l\zeta\eta} + F_{l\zeta\eta\alpha}y_\alpha + \cdots ,$

and

(5)
$$0 \equiv 2A_{li}h_{i,\zeta\eta\vartheta} + F_{li\alpha}h_{i,\zeta\eta\vartheta}y_\alpha + F_{li\vartheta}h_{i,\zeta\eta} + F_{li\eta}h_{i,\zeta\vartheta}$$
$$+ F_{li\zeta}h_{i,\eta\vartheta} + F_{l\zeta\eta\vartheta} + \cdots .$$

Here $h_{i,\zeta} := \partial_\zeta h_i$, $h_{i,\zeta\eta} := \partial_\zeta\partial_\eta h_i$, and $h_{i,\zeta\eta\vartheta} := \partial_\zeta\partial_\eta\partial_\vartheta h_i$. Due to $h(0) = 0$ and $Dh(0) = 0$, the evaluation of (4) and (5) at $y = 0$ yields

(6) $\qquad\qquad 0 = 2A_{li}h_{i,\zeta\eta}(0) + F_{l\zeta\eta},$

(7) $\qquad\qquad 0 = 2A_{li}h_{i,\zeta\eta\vartheta}(0) + F_{li\vartheta}h_{i,\zeta\eta}(0) + F_{li\eta}h_{i,\zeta\vartheta}(0)$
$$+ F_{li\zeta}h_{i,\eta\vartheta}(0) + F_{l\zeta\eta\vartheta}.$$

From this one obtains

(8) $$h_{i,\varsigma\eta}(0) = -\frac{1}{2}A_{ij}F_{j\varsigma\eta}$$

and

(9) $$h_{i,\varsigma\eta\vartheta}(0) = \frac{1}{4}A_{ij}A_{kl}(F_{jk\vartheta}F_{l\varsigma\eta} + F_{jk\eta}F_{l\vartheta\varsigma} + F_{jk\varsigma}F_{l\eta\vartheta}) - \frac{1}{2}A_{ij}F_{j\varsigma\eta\vartheta}.$$

The third Taylor polynomial of h_i is then

(10) $$T^3_{h_i}(y) = -\frac{1}{4}A_{ij}F_{j\alpha\beta}y_\alpha y_\beta$$
$$+ \left(\frac{1}{8}A_{ij}A_{kl}F_{jk\alpha}F_{l\beta\gamma} - \frac{1}{12}A_{ij}F_{j\alpha\beta\gamma}\right)y_\alpha y_\beta y_\gamma.$$

From this the assertion follows by substituting (10) in (1) for x_i, and the corresponding expression for $T^3_{h_j}(y)$ from (10) for x_j.

Chapter 4

Determinacy

In this chapter the following question, which is fundamental to Catastrophe Theory, is investigated: When is a smooth function f **determined** in a neighborhood of a point p by one of its Taylor polynomials at p in the sense that every other function having the same Taylor polynomial coincides with f around p up to a smooth change of coordinates?

In general a function does not have this property at an arbitrary point; for instance, a function cannot be determined by its Taylor polynomial in the above sense at a flat point p, unless it vanishes on a neighborhood of p. In 1968 John Mather [M III] gave two conditions, one sufficient and the other necessary, for a function to be determined by its kth Taylor polynomial. As a consequence, Mather was able to characterize those functions that are determined by at least one of their Taylor polynomials. The goal of this chapter is to prove these results. The proof given here of the necessary criterion is based on the Linearization Lemma (37), which addresses a general property of Lie transformation groups needed in Mather's original proof as well as in all other known proofs and which has not been treated.

The determinacy question has been already answered in Chapter 1 for special cases. For simplicity, consider smooth functions f around the origin vanishing there. If f is a function of only one variable, then f is determined by the first nonvanishing term of its Taylor expansion around the origin by (1.34). Functions f of several variables that are not critical at the origin are determined by the linear part of their Taylor expansion. This follows immediately from (1.35). The Morse lemma (1.37) says that when the origin is a nondegenerate critical point, f is determined by its second Taylor polynomial at the origin. What remains to be considered are the nonflat degenerate critical points of smooth functions of more than one variable. At such points these functions are not necessarily determined by their Taylor polynomials, as the following Example (1) shows.

The Reduction Lemma of Chapter 3 left the question of the uniqueness of the residual singularity open. Using the Sufficient Criterion for Equivalence of Germs (22), it is now possible to prove the uniqueness at the end of this chapter.

Example (1). Let k and l be nonnegative integers. The functions $f \colon \mathbf{R}^2 \to \mathbf{R}$, $f(x,y) := x^2$, and $g \colon \mathbf{R}^2 \to \mathbf{R}$, $g(x,y) := x^2 - y^{2l}$, have the same kth Taylor polynomial at $0 \in \mathbf{R}^2$ when $l > k/2$ holds, but if $\varphi := (\varphi_1, \varphi_2)$ is any local diffeomorphism at $0 \in \mathbf{R}^2$, then

$$f(\varphi(0,y)) = (\varphi_1(0,y))^2 \neq -y^{2l} = g\,(0,y)$$

is true for nonzero $y \in \mathbf{R}$. Thus, f is not determined by any of its Taylor polynomials.

Studying the behavior of a function locally, i.e., in an arbitrarily small neighborhood of a point p, means looking at what is called the germ of the function at p. More generally, the notion of a germ of a smooth map into $\mathbf{R}^m$ is introduced.

Definition.   Let $m, n \in \mathbf{N}$ and $p \in \mathbf{R}^n$ be fixed. On the set

$$\{F \colon U \to \mathbf{R}^m : U \text{ open in } \mathbf{R}^n \text{ with } p \in U \text{ and } F \text{ smooth}\}$$

consider the following equivalence relation: If $F_1 \colon U_1 \to \mathbf{R}^m$ and $F_2 \colon U_2 \to \mathbf{R}^m$ are in this set, then $F_1 \sim F_2$ will mean that F_1 coincides with F_2 on an open subset of $U_1 \cap U_2$ containing p. The equivalence class $[F]$ of an element F is called the **germ of** F **at** p. Let $\mathcal{E}_{n,m}$ be the set of all such germs at the origin of $\mathbf{R}^n$. If $m = 1$, just $\mathcal{E}_n$, or more simply $\mathcal{E}$, will be used instead of $\mathcal{E}_{n,1}$.

In a natural way, operations on $\mathcal{E}$ are defined making $\mathcal{E}$ a commutative and associative algebra over $\mathbf{R}$: For $\alpha \in \mathbf{R}$ and $[f_1], [f_2]$ in $\mathcal{E}$ with representatives $f_1 \colon U_1 \to \mathbf{R}$ and $f_2 \colon U_2 \to \mathbf{R}$, set

$$[f_1] + \alpha[f_2] := [f_1|V + \alpha f_2|V]$$

and

$$[f_1][f_2] := [(f_1|V)(f_2|V)]$$

with $V := U_1 \cap U_2$. The algebra $\mathcal{E}$ has an **identity**, namely, the germ of the constant function $\mathbf{R}^n \to \mathbf{R}$, $x \mapsto 1$, which will be denoted by $[1]$. Notice that a germ $[f] \in \mathcal{E}$ is invertible if and only if f does not vanish at the origin. This condition is trivially necessary, and when it is satisfied, then a neighborhood V of the origin exists on which f is nonzero. The germ of the function $g := 1/f$, which is well defined and smooth on V, is obviously an inverse of $[f]$, i.e., $[f][g] = [1]$ holds.

As a real vector space, $\mathcal{E}$ has infinite dimension, since the monomials $x^\nu = x_1^{\nu_1}, \ldots, x_n^{\nu_n}$, $\nu \in \mathbf{N}_0^n$, are smooth functions on $\mathbf{R}^n$ defining linearly independent germs at the origin. This follows immediately from Taylor's formula (1.27). As a matter of fact, it is interesting to notice that any germ $[f] \in \mathcal{E}$ is the germ of a globally defined smooth function, i.e., the equivalence map

$$C^\infty(\mathbf{R}^n) \to \mathcal{E}, \qquad g \mapsto [g],$$

is surjective. To verify this, let $f\colon U \to \mathbf{R}$ be a smooth function defined on an open neighborhood U of the origin in $\mathbf{R}^n$. Then U contains a ball around 0 with a positive radius R. By means of the function h defined below, a function $g \in C^\infty(\mathbf{R}^n)$ will be given whose germ is $[f]$, namely, $g(x) := h(|x|)f(x)$ for x in U and $g(x) := 0$ otherwise, with $|x|$ denoting the Euclidean length $(x_1^2 + \cdots + x_n^2)^{1/2}$.

Example (2). Suppose that r and R are positive with $r < R$. Then the following function h is smooth on the positive real axis. For $r < t < R$, set

$$h(t) := \left[1 + \exp\left(\frac{1}{r-t} + \frac{1}{R-t} \right) \right]^{-1}$$

and let $h(t) := 1$ for $0 \le t \le r$ and $h(t) := 0$ for $t \ge R$.

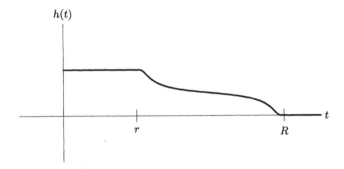

The Lemma of E. Borel, which will be proved next, illustrates the fact that the vector space $\mathcal{E}$ is indeed large. Let (a_ν) be any family of real numbers, when ν varies over all n-tuples $\mathbf{N}_0^n$ of nonnegative integers. Then a smooth function f will be shown to exist on some neighborhood of the origin in $\mathbf{R}^n$ whose Taylor expansion there is given by just this family, i.e., $a_\nu = (1/\nu!)D^\nu f(0)$ for all ν. Denote by $\mathbf{R}[[x_1, \ldots, x_n]]$ the set of **formal power series in n variables**,

$$\sum_\nu a_\nu x^\nu$$

in $\mathbf{R}$, meaning that $x = (x_1, \ldots, x_n) \in \mathbf{R}^n$ and $a_\nu \in \mathbf{R}$ for all ν in $\mathbf{N}_0^n$. Endowed with coefficientwise addition and scalar multiplication,

$$\sum_\nu a_\nu x^\nu + \alpha \sum_\nu b_\nu x^\nu := \sum_\nu (a_\nu + \alpha b_\nu) x^\nu,$$

as well as with the **Cauchy product**,

$$\left(\sum_\lambda a_\lambda x^\lambda \right) \left(\sum_\mu b_\mu x^\mu \right) := \sum_\nu \left(\sum_{\lambda+\mu=\nu} a_\lambda b_\mu \right) x^\nu,$$

this set becomes a commutative and associative algebra over $\mathbf{R}$.

Lemma of E. Borel (3).   The map $T\colon \mathcal{E} \to \mathbf{R}[[x_1, \ldots, x_n]]$ defined by

$$T[f] := \sum_{\nu \in \mathbf{N}_0^n} \frac{1}{\nu!} D^\nu f(0) x^\nu$$

is a surjective algebra homomorphism.

Proof. T is well defined, because if $g \in [f]$, then f and g coincide on a neighborhood of the origin and thus $D^\nu f(0) = D^\nu g(0)$ for all ν. The linearity of T is obvious. The Leibniz rule of derivation says that

(4) $$D^\nu(uv)(p) = \sum_{0 \le \lambda \le \nu} \binom{c\nu}{\lambda} D^\lambda u(p) D^{\nu-\lambda} v(p)$$

is valid for the product uv of smooth functions u and v on a neighborhood of a point p. Applying this, one obtains for $[h] := [f][g]$ that

$$T[h] = \sum_\nu \frac{1}{\nu!} D^\nu h(0) x^\nu = \sum_\nu \left[\sum_{\kappa + \lambda = \nu} \frac{1}{\kappa! \lambda!} D^\lambda f(0) D^\kappa g(0) \right] x^\nu$$

holds, which is the Cauchy product $T[f]T[g]$. Thus T is an algebra homomorphism.

The nontrivial assertion in (3) is that T is surjective. To show this, start with an arbitrary family (a_ν) in $\mathbf{R}$. Consider any smooth function $u\colon \mathbf{R}^n \to \mathbf{R}$ such that $0 \le u(x) \le 1$, $u(x) = 1$ for $|x| \le \frac{1}{2}$ and $u(x) = 0$ for $|x| \ge 1$. For instance, take $u(x) := h(|x|)$ with h as in Example (2) for $R := 1$ and $r := \frac{1}{2}$. Moreover, put $t_\nu := \frac{1}{2}$ if $|a_\nu| \le 1$ and $t_\nu := \frac{1}{2}|a_\nu|^{-1}$ if $|a_\nu| > 1$. For x in $\mathbf{R}^n$, set

$$f(x) := \sum_\nu \frac{1}{\nu!} a_\nu u(x/t_\nu) x^\nu,$$

where ν varies over all n-tuples of nonnegative integers. It will be shown that f is a smooth function on $\mathbf{R}^n$ with $D^\mu f(0) = a_\mu$ for all μ, from which the surjectivity of T follows. Put

$$g_\nu(x) := \frac{1}{\nu!} a_\nu u(x/t_\nu) x^\nu, \quad \text{for } x \in \mathbf{R}^n.$$

In order to show that the sum $\sum_\nu g_\nu$ defines a smooth function f whose partial derivatives $D^\mu f$, $\mu \ge 0$, are given by the series $\sum_\nu D^\mu g_\nu$, it suffices to show that these series converge uniformly on $\mathbf{R}^n$. Indeed, in this case, the theorem about the Derivative of a Limiting Function (see Appendix A.2) applies to the partial sums $f_m := \sum_{|\nu| \le m} g_\nu, m \in \mathbf{N}_0$, and their partial derivatives, proving the assertion.

By (4) and the fact that $u(x/t_\nu)$ vanishes for $|x| \ge t_\nu$, it follows for $\mu < \nu$ that

$$|D^\mu g_\nu(x)| \le \mu! |a_\nu| K_\mu \sum_{0 \le \kappa \le \mu} t_\nu^{-|\mu - \kappa|} t_\nu^{|\nu - \kappa|}$$

holds, since

$$K_\mu := \max\{|D^\lambda u(x)| : 0 \le \lambda \le \mu, x \in \mathbf{R}^n\}$$

is an upper bound for $|D^{\mu-\kappa}u(x/t_\nu)|$ and $t_\nu^{|\nu-\kappa|}$ dominates $|x^{\nu-\kappa}|$ if $|x| < t_\nu$, because of $|x_i| < t_\nu$. Then $|a_\nu|t_\nu^{-|\mu-\kappa|}t_\nu^{|\nu-\kappa|} = |a_\nu|t_\nu^{-|\mu|+|\nu|} \le 2^{|\mu|-|\nu|}$ implies

$$|D^\mu g_\nu(x)| \le C_\mu 2^{-|\nu|} \quad \text{for} \quad \nu > \mu,$$

where

$$C_\mu := \mu! K_\mu \sum_{0 \le \kappa \le \mu} 2^{|\mu|}$$

does not depend on ν or x. Since the series $\sum_\nu 2^{-|\nu|}$ converges (namely, to 2^n), the uniform convergence of $\sum_\nu D^\mu g_\nu$ for every μ is verified.

Due to $D^\kappa u(0) = 0$ for $\kappa \ne 0$, one obtains $D^\mu f(0) = \sum_\nu D^\mu g_\nu(0) = D^\mu g_\mu(0) = a_\mu$ by (4). Hence, T is surjective, which completes the proof of Borel's Lemma.

The following straightforward generalization of Borel's Lemma (3) will be needed to prove the Malgrange–Mather Preparation Theorem in Chapter 9.

Lemma (5). For $\nu \in \mathbf{N}_0^n$ let (a_ν) be a family of smooth functions defined on an open neighborhood V of the origin in $\mathbf{R}^m$. Let V_1 be a bounded open neighborhood of the origin in $\mathbf{R}^m$ whose closure is contained in V. Then there exists a smooth function f on $\mathbf{R}^n \times V_1$ such that

$$D_1^\nu D_2^\mu f(0, y) = D^\mu a_\nu(y)$$

for $y \in V_1$ and $(\nu, \mu) \in \mathbf{N}_0^n \times \mathbf{N}_0^m$, where D_1^ν and D_2^μ denote the partial derivatives with respect to the first n variables $x_1, \dots, x_n$ and the last m variables $y_1, \dots, y_m$, respectively.

Proof. Following the proof of (3), set

$$f(x, y) := \sum_\nu \frac{1}{\nu!} a_\nu(y) u(x/t_\nu) x^\nu,$$

but now let t_ν be defined by replacing the modulus of a_ν with

$$\sup\{|D^\kappa a_\nu(y)| : y \in V_1, |\kappa| \le |\nu|\}$$

for every $\nu \in \mathbf{N}_0^n$. Then the uniform convergence on $\mathbf{R}^n \times V_1$ of the series

$$\sum_\nu D_1^\kappa D_2^\lambda g_\nu, \quad \text{where} \quad g_\nu(x, y) := \frac{1}{\nu!} a_\nu(y) u(x/t_\nu) x^\nu,$$

follows from

$$|D_1^\kappa D_2^\lambda g_\nu(x, y)| \le C_\kappa 2^{-|\nu|} \quad \text{for} \quad \kappa < \nu, |\lambda| \le |\nu|.$$

This estimate can be verified by considerations analogous to those in the proof of (3), and with the same constants C_κ.

Notice that the set of germs $[f] \in \mathcal{E}$ whose representatives f have a flat point at the origin is exactly the kernel of the homomorphism T in (3). This will be denoted by m^∞ and is an **ideal** of $\mathcal{E}$. Recall that a subset $\mathcal{A}$ of $\mathcal{E}$ is an ideal if $\mathcal{A}$ is a linear subspace of $\mathcal{E}$ and if the product of an element in $\mathcal{A}$ and an element in $\mathcal{E}$ yields an element in $\mathcal{A}$. According to Borel's Lemma (3), T induces an isomorphism between $\mathbf{R}[[x_1, \ldots, x_n]]$ and the quotient algebra $\mathcal{E}/m^\infty$. An example of an element in m^∞ is given by the germ of $f(x) := \exp(-|x|^{-2})$ for nonzero x in $\mathbf{R}^n$ and $f(0) := 0$.

Whether or not a smooth function is determined by one of its Taylor polynomials has no relevance to the convergence of its Taylor expansion. The Taylor series of any function f with a flat point at the origin converges everywhere but does not give a qualitative description of f. Even if f is a polynomial, it need not determine itself, as Example (1) shows. On the other hand, although the power series

$$\sum_{k=1}^{\infty} 2^{(k^2)} x^k$$

of one variable converges only at the origin, a function having this Taylor expansion is uniquely determined around the origin up to a smooth change of coordinates. Even the first term $2x$ of this series, i.e., the first Taylor polynomial of such a function, is sufficient to determine the function by (1.35). Of course, for several variables the situation is more involved and will be investigated in the following.

The next definitions formalize the concept of a function f being determined by its kth Taylor polynomial T_f^k at the origin. Identifying all functions with the same kth Taylor polynomial as a given function means looking at what is called the k-jet of that function.

Definition. Let k be a nonnegative integer. Two germs $[f]$ and $[g]$ in $\mathcal{E}$ are said to be **k-equivalent** if $D^\nu f(0) = D^\nu g(0)$ for all $\nu \in \mathbf{N}_0^n$ with $|\nu| \leq k$. The class of all germs k-equivalent to $[f]$ is called the **k-jet** of $[f]$ and is denoted by $j^k[f]$. The set of all k-jets of germs in $\mathcal{E}$ is denoted by J_n^k or simply by J^k.

Two germs $[f]$ and $[g]$ are k-equivalent exactly when their kth Taylor polynomials T_f^k and T_g^k coincide. In particular, the germ $[f]$ of a smooth function f around the origin is k-equivalent to the germ $[T_f^k]$ of its kth Taylor polynomial at 0, i.e., $j^k[f] = j^k[T_f^k]$. In fact, T_f^k is the unique polynomial of degree at most k whose germ is k-equivalent to $[f]$. In order to state this more formally, let $\mathbf{R}[x_1, \ldots, x_n]$ be the **algebra over $\mathbf{R}$ of polynomials in n variables**, and denote by $\deg p$ the degree of such a polynomial $p(x) := \sum_\nu a_\nu x^\nu$, $x = (x_1, \ldots, x_n) \in \mathbf{R}^n$. If

$$P^k := \{p \in \mathbf{R}[x_1, \ldots, x_n] : \deg p \leq k\},$$

then the map

$$J^k \to P^k, \quad j^k[f] \mapsto T_f^k$$

is bijective. The polynomials in P^k form a real vector space with

(6) $$N := \dim P^k = \binom{n+k}{k};$$

see Exercise 4.1. P^k is canonically isomorphic to $\mathbf{R}^N$ via the map

$$\sum_{0 \leq |\nu| \leq k} a_\nu x^\nu \mapsto (a_\nu)_{0 \leq |\nu| \leq k}.$$

In this way, the linear and topological structures of $\mathbf{R}^N$ are transferred to the **jet space** J^k. Then $\dim J^k = N$, and the above map $J^k \to P^k$ as well as the **jet map**

$$j^k : \mathcal{E} \to J^k, \quad [f] \mapsto j^k[f]$$

are obviously linear.

A further equivalence relation on $\mathcal{E}$ will be needed in order to make the concept of determinacy precise. Toward this end, let $\mathcal{G}_n$ — or, for short, $\mathcal{G}$ when n is understood—be the **group of germs $[\varphi]$ of local diffeomorphisms** φ at the origin in $\mathbf{R}^n$ **leaving the origin invariant**. The natural group operation is given by the composition $[\varphi][\psi]$ of germs $[\varphi], [\psi] \in \mathcal{G}$, where $[\varphi][\psi]$ is the germ of the composition $\varphi \circ \psi$ defined on a sufficiently small neighborhood of the origin. The identity in $\mathcal{G}$ is the germ of the map $\mathbf{R}^n \to \mathbf{R}^n$, $x \mapsto x$, which is denoted by id_n or simply id. The inverse of a germ $[\varphi] \in \mathcal{G}$ is the germ $[\psi]$ of a local inverse ψ at the origin of one of its representatives φ.

The group $\mathcal{G}$ acts on the set $\mathcal{E}$ to the right in the following way. Let $[f] \in \mathcal{E}$ and $[\varphi] \in \mathcal{G}$. Then the composition $f \circ \varphi$ is defined on an appropriate neighborhood of the origin. Its germ will be denoted by $[f][\varphi]$ and does not depend on the representatives of the respective germs. In this way an **action** $\mathcal{E} \times \mathcal{G} \to \mathcal{E}$, $([f], [\varphi]) \mapsto [f][\varphi]$, **of $\mathcal{G}$ on $\mathcal{E}$** is given, signifying that

$$([f][\varphi])([\psi]) = [f]([\varphi][\psi]) \quad \text{and} \quad [f][id] = [f]$$

hold for all $[f] \in \mathcal{E}$ and $[\varphi], [\psi] \in \mathcal{G}$. This action naturally defines an equivalence relation on $\mathcal{E}$. (Be aware of the ambiguity of the notation $[\alpha][\beta]$, which for $[\alpha] \in \mathcal{E}_1$ and $[\beta] \in \mathcal{G}_1$ could also mean the product of the germs $[\alpha], [\beta] \in \mathcal{E}_1$.)

Definition. Two germs $[f]$ and $[g]$ in $\mathcal{E}$ are **equivalent** if there is a $[\varphi] \in \mathcal{G}$ such that $[g] = [f][\varphi]$ is valid; in this case one writes $[f] \sim [g]$. The **orbit through $[f]$ under $\mathcal{G}$** is the set

$$[f]\mathcal{G} := \{[f][\varphi] : [\varphi] \in \mathcal{G}\}$$

of all germs in $\mathcal{E}$ equivalent to $[f]$.

Using the last definition, the results of Chapter 1 for smooth functions f vanishing at the origin can be reformulated. If f depends on one variable and the origin is not flat, its germ is equivalent to the germ of a monomial $\pm x^k$ for some $k \in \mathbf{N}$ after (1.34); furthermore, if k is even, then x^k is not equivalent to $-x^k$, whereas the opposite is true for uneven k. When the origin is not a critical point of f, its germ is equivalent to the germ of the projection x_1 onto the first coordinate; see (1.35). Finally, by the Morse Lemma (1.37), the germ of f is equivalent to the germ of the normal quadratic form $-x_1^2 - \cdots - x_s^2 + x_{s+1}^2 + \cdots + x_n^2$ if the origin is a nondegenerate critical point and s denotes the index of f at the origin.

There is no obvious relationship between the notions k-*equivalence* and *equivalence* for germs in $\mathcal{E}$. In particular, one notion does not imply the other. This can be seen by considering any nonzero germ $[f]$ where f has a flat critical point at the origin. In this case, $[f]$ is trivially k-equivalent but not equivalent to the zero function for any $k \geq 0$. Another example is furnished by the germs $[f]$ and $[g]$ with representatives given in Example (1); these germs are k-equivalent but not equivalent. On the other hand, the germ of $f \colon \mathbf{R} \to \mathbf{R}$, $f(x) := 2(e^x - 1)$, is equivalent to $[id_1]$ by (1.35) but not k-equivalent for $k \geq 1$.

> **Definition.** Let k be a positive integer. A germ $[f] \in \mathcal{E}$ is called **k-determined** if every germ that is k-equivalent to $[f]$ is equivalent to $[f]$. A germ is **finitely determined** if it is k-determined for some $k \in \mathbf{N}$. If a germ $[f] \in \mathcal{E}$ is finitely determined, then the smallest k for which $[f]$ is k-determined is referred to as the **determinacy** of $[f]$ and denoted by $\det[f]$. The notation $\det[f] = \infty$ means that $[f]$ is not finitely determined.

The central question treated in this chapter can now be phrased as follows: When is the germ $[f] \in \mathcal{E}$ of a smooth function finitely determined? Moreover, for a given positive integer k, when is $[f]$ k-determined?

Obviously, every germ k-equivalent to a k-determined germ is itself k-determined. A germ $[f] \in \mathcal{E}$ is k-determined exactly when every function with the same kth Taylor polynomial at 0 coincides with f around the origin after a smooth coordinate transformation. In particular, f is then locally equal to its own kth Taylor polynomial up to a smooth change of coordinates, i.e., $[f] \sim [T_f^k]$. Being k-determined, however, means more than this, as Example (1) shows. Notice that if $[f]$ is k-determined, then $T_f^k \neq 0$, because otherwise the contradiction $[x_1^{k+1}] = [0]$ would follow. Therefore, if the origin is a critical point of a finitely determined germ, it is never flat. If a germ $[f] \in \mathcal{E}$ is k-determined, then trivially it is also l-determined for every $l \geq k$. Furthermore, $\det[f] = \det[-f]$ obviously holds.

Clearly, the determinacy of the germ of $f \colon \mathbf{R} \to \mathbf{R}$, $f(x) := bx^k$, with a nonzero real b, is k, and $\det[f] = 1$ for $f \colon \mathbf{R}^n \to \mathbf{R}$, $f(x_1, \ldots, x_n) := x_1$. Moreover, the germ of a nondegenerate normal quadratic form q_{sn} has determinacy 2. Before considering further examples, it is convenient to have the following notation. If k is a positive integer, define

$$m_n^k := \{[f] \in \mathcal{E}_n \colon D^\nu f(0) = 0 \quad \text{for all} \quad |\nu| < k\}.$$

For $k = 1$ simply write

$$m_n = \{[f] \in \mathcal{E}_n : f(0) = 0\},$$

and for $k = 0$ put $m_n^0 := \mathcal{E}_n$. Moreover, when n is understood, just m^k will be written. Because of (4), m^k is an ideal of $\mathcal{E}$. The ideal m is the only **maximal ideal** of $\mathcal{E}$, since it coincides with the set of all noninvertible elements of $\mathcal{E}$. Obviously, $m^{k+1} \subset m^k$ holds. If k is a nonnegative integer and $f(x_1, \ldots, x_n) := x_1^k$, then $[f]$ is in m^k but not in m^{k+1}. Thus

(7)
$$\mathcal{E} \supsetneq m \supsetneq m^2 \supsetneq m^3 \supsetneq \cdots$$

is true. Moreover,

$$m^\infty = \bigcap \{m^k : k \geq 0\}$$

holds. Note that

$$m^k = \{[f] \in \mathcal{E} : j^{k-1}[f] = 0\}.$$

For any $[f] \in \mathcal{E}$, the germ of the remainder function $R := f - T_f^{k-1}$ lies in m^k. Let $c \in \mathbf{R}$ be a constant and $[f] \in \mathcal{E}$ be a germ. Then $[f]$ is k-determined if and only if $[f + c]$ is k-determined. Due to $[f - f(0)] \in m$, it suffices to consider just germs in m when investigating determinacy.

The algebraic structure of germs will be emphasized here. Evidently, the map

(8)
$$\mathcal{E}/m^{k+1} \to J^k, \quad [f] + m^{k+1} \mapsto j^k[f]$$

is a linear bijection, which implies that the dimension of the vector space $\mathcal{E}/m^{k+1}$ is $N = \binom{n+k}{k}$ by (6). The map (8) enables the operations on the quotient algebra $\mathcal{E}/m^{k+1}$ to be transferred to J^k. Thus, J^k becomes a commutative and associative algebra over $\mathbf{R}$, and the linear structure coincides with the previous one introduced after (6). The algebra multiplication is as follows:

$$j^k[f]j^k[g] := j^k([f][g]).$$

The jet map $j^k : \mathcal{E} \to J^k$, $[f] \mapsto j^k[f]$, is now an **algebra homomorphism**, i.e., it is linear and compatible with the algebra multiplication.

On the set P^k of polynomials of degree not greater than k, the algebra multiplication is given by the Cauchy product, where, however, the monomials of degree greater than k are omitted. Hence, the product of two polynomials p and q is given by T_{pq}^k. This algebraic structure is the natural one if P^k is identified with the quotient algebra

$$\mathbf{R}[x_1, \ldots, x_n]/\{p \in \mathbf{R}[x_1, \ldots, x_n] : T_p^k = 0\}.$$

Note again that $\mathcal{E}$ is a real vector space. If A is an arbitrary subset of $\mathcal{E}$, then the **subspace of $\mathcal{E}$ spanned by** A will be denoted by $\langle A \rangle$. This is the set of all finite linear combinations of elements in A. When B is another subset of $\mathcal{E}$, then AB is the set of all products ab with $a \in A$ and $b \in B$. Note that when

A and B are ideals of $\mathcal{E}$, then $\langle AB \rangle$ yields the usual ideal product of A and B. The **ideal of $\mathcal{E}$ generated by** A will be referred to by $\langle A \rangle_{\mathcal{E}}$. It is the subspace $\langle \mathcal{E}A \rangle$ spanned by the set $\mathcal{E}A$. When $\mathcal{E}A = A$ holds, then $\langle A \rangle_{\mathcal{E}} = \langle A \rangle$ follows. If A is a finite set $\{[f_1], \ldots, [f_k]\}$, then for the sake of brevity one writes

$$\langle f_1, \ldots, f_k \rangle := \langle \{[f_1], \ldots, [f_k]\} \rangle \quad \text{and}$$
$$\langle f_1, \ldots, f_k \rangle_{\mathcal{E}} := \langle \{[f_1], \ldots, [f_k]\} \rangle_{\mathcal{E}}.$$

An immediate consequence of (1.25) is

(9) $$m^k = \langle x^{\nu} : |\nu| = k \rangle_{\mathcal{E}}.$$

In particular, $m = \langle x_1, \ldots, x_n \rangle_{\mathcal{E}}$. If l is another nonnegative integer, then (9) implies

(10) $$m^{k+l} = \langle m^k m^l \rangle.$$

A consequence of the following lemma is that equivalent germs have the same determinacy.

Lemma (11). Let k be a nonnegative integer. If $[f]$ and $[g]$ are in $\mathcal{E}$ and $[\varphi]$ is in $\mathcal{G}$, then

$$j^k([f][\varphi]) = j^k[T_f^k \circ T_{\varphi}^k]$$

follows, where $T_{\varphi}^k := (T_{\varphi_1}^k, \ldots, T_{\varphi_n}^k)$ and φ_j is the jth component of φ for $1 \leq j \leq n$. Moreover, $j^k[f] = j^k[g]$ holds if and only if $j^k([f][\varphi]) = j^k([g][\varphi])$ is true.

Proof. Let $[f] = [T_f^k] + [\tilde{f}]$ with $[\tilde{f}]$ in m^{k+1}. Because $[\varphi_i]$ is in m, it follows that $[\varphi]^{\nu} = [\varphi_1]^{\nu_1} \cdots [\varphi_n]^{\nu_n}$ is in m^{k+1} for $|\nu| = k+1$, using (9) and (10), and hence that $[\tilde{f}][\varphi] \in m^{k+1}$ by (9). Therefore, $j^k([f][\varphi]) = j^k([T_f^k][\varphi])$ holds.

Now let $[\varphi_i] = [T_{\varphi_i}^k] + [\tilde{\varphi}_i]$ with $[\tilde{\varphi}_i]$ in m^{k+1}. Then $[\varphi_i]^l \in [T_{\varphi_i}^k]^l + m^{k+1}$ for $l \geq 1$. Therefore, $j^k([T_f^k][\varphi]) = j^k([T_f^k \circ T_{\varphi}^k])$ holds. This proves $j^k([f][\varphi]) = j^k[T_f^k \circ T_{\varphi}^k]$.

For the proof of the second assertion, it suffices to consider the case $[g] = 0$ because of the linearity of the jet map j^k. This case, however, follows immediately from the first assertion, since $j^k[f] = 0$ holds if and only if $T_f^k = 0$ is true.

Lemma (11) shows that the k-jet of the composition $[f][\varphi]$ depends only on the k-jets of $[f]$ and $[\varphi]$. Another consequence of Lemma (11) is that the action of $\mathcal{G}$ on $\mathcal{E}$ induces an action of $\mathcal{G}$ on m^k for every $k \in \mathbf{N}$.

Corollary (12). Suppose that $[f] \in \mathcal{E}$ is k-determined and $[g] \in \mathcal{E}$ is equivalent to $[f]$. Then $[g]$ is k-determined.

Proof. By assumption, a $[\varphi] \in \mathcal{G}$ exists with $[g] = [f][\varphi]$. Let $[h] \in \mathcal{E}$ be k-equivalent to $[g]$. Then $j^k[f] = j^k([g][\varphi]^{-1}) = j^k([h][\varphi]^{-1})$ by (11). Since $[f]$ is k-determined, there is a $[\psi] \in \mathcal{G}$ with $[f] = [h][\varphi]^{-1}[\psi]$, implying $[g] = [h][\varphi]^{-1}[\psi][\varphi]$.

A consequence of (12) is that equivalent germs are either both finitely determined or both not finitely determined. The determinacy of germs of functions of one variable is now easily characterized by the following theorem.

Theorem (13). A germ $[f] \in m_1$ has determinacy k if and only if k is the smallest positive integer satisfying $f^{(k)}(0) \neq 0$.

Proof. Let $[f]$ have determinacy k. Since $T_f^k \neq 0$, a smallest integer $l \leq k$ with $f^{(l)}(0) \neq 0$ exists. If ε is the sign of $f^{(l)}(0)$, then $[f] \sim [\varepsilon^{l-1}x^l]$ and $\det[\varepsilon^{l-1}x^l] = l$ by (1.34). Using (12), one obtains $l = k$. To prove the converse, let k be the smallest positive integer satisfying $f^{(k)}(0) \neq 0$. Then $[f]$ is k-determined by (1.34). Since $T_f^{k-1} = 0$ is true, $\det[f] = k$ results.

For smooth functions of more than one variable, an immediate characterization of 1-determinacy can be given.

Theorem (14). A germ $[f] \in m$ is 1-determined if and only if the origin is not a critical point of f.

Proof. If $[f] \in m$ and $\det[f] = 1$, then $T_f^1 \neq 0$, implying $Df(0) \neq 0$. The converse follows from (1.35) using (12).

When the origin is a critical point of $[f] \in m$, then 2-determinacy of $[f]$ is equivalent to the nondegeneracy of the origin. The Morse Lemma (1.37), (12), and (14) prove that nondegeneracy is sufficient, and its necessity will be proved in (46).

The goal now is to show that a certain algebraic condition implies that a germ is k-determined. This condition is stated in terms of the following notion.

Definition. The **Jacobi ideal** $\mathcal{J}[f]$ of a germ $[f] \in \mathcal{E}$ is the ideal of $\mathcal{E}$ generated by the germs of the partial derivatives $D_i f, i = 1, \ldots, n$, i.e.,

$$\mathcal{J}[f] = \langle D_1 f, \ldots, D_n f \rangle_{\mathcal{E}}.$$

It will be proved next that if m^k is contained in the ideal $\langle m \mathcal{J}[f] \rangle$, then $[f]$ is k-determined. The proof employs the Lemma of Nakayama concerning finitely generated modules. The few notions from module theory needed in this context will be recalled first.

Let $\mathcal{R}$ be a commutative ring with identity 1. A **module** M over $\mathcal{R}$ is an abelian group $(M, +)$ together with an **outer product** $\mathcal{R} \times M \to M$,

$(r, m) \mapsto rm$, such that for all $r, s \in \mathcal{R}$ and $m, n \in M$,

$$r(m + n) = rm + rn$$
$$(r + s)m = rm + sm$$
$$r(sm) = (rs)m$$
$$1m = m.$$

Note that $\mathcal{R}$ is a module over itself and that the ideals of $\mathcal{R}$ are exactly those subrings of $\mathcal{R}$ that are also modules over $\mathcal{R}$.

A **submodule** N of M is a subgroup of $(M, +)$ such that $rn \in N$ for all $r \in \mathcal{R}$ and $n \in N$. Clearly, N is a module over $\mathcal{R}$. If A is a subset of M and $\mathcal{A}$ is an ideal of $\mathcal{R}$, then

$$\langle A \rangle_{\mathcal{A}} := \left\{ \sum_{j=1}^{n} r_j a_j : r_j \in \mathcal{A}, a_j \in A, n \in \mathbf{N} \right\}$$

is a submodule of M. The module $\langle A \rangle_{\mathcal{R}}$ is called the **submodule of M generated by** A. A module M is **finitely generated** if there is a finite subset A of M with $M = \langle A \rangle_{\mathcal{R}}$. Let N be a submodule of a module M over $\mathcal{R}$. With the natural outer product $\mathcal{R} \times M/N \to M/N$ given by $(r, m + N) \mapsto rm + N$, the factor group M/N becomes a module over $\mathcal{R}$ called the **factor module** of M by N.

> **Nakayama's Lemma** (15). Let M be a module over $\mathcal{R}$. Suppose that $\mathcal{M}$ is an ideal of $\mathcal{R}$ with $\mathcal{M} \neq \mathcal{R}$ such that every element in $\mathcal{R}$ lying outside of $\mathcal{M}$ is invertible. Let A be a nonempty finite subset of M. If $A \subset \langle A \rangle_{\mathcal{M}}$, then $A = \{0\}$.

> **Proof.** Let $B \subset A$ be such that A is contained in $\langle B \cup \{0\} \rangle_{\mathcal{M}}$ and $k := \operatorname{card}(B)$ is minimal. The assertion is that $k = 0$, i.e., $B = \emptyset$. Assume the contrary, and let $B = \{b_1, \ldots, b_k\}$ with $k \geq 1$. Then $0 \notin B$ and $b_1 = \sum_{\kappa=1}^{k} r_\kappa b_\kappa$ with $r_\kappa \in \mathcal{M}$. Hence

$$(1 - r_1)b_1 = \sum_{\kappa=2}^{k} r_\kappa b_\kappa,$$

> where the sum is zero if $k = 1$. Now $1 - r_1$ is not in $\mathcal{M}$, because otherwise $\mathcal{M}$ would contain the identity, implying the contradiction $\mathcal{M} = \mathcal{R}$. Thus, $1 - r_1$ is invertible and

$$b_1 = \sum_{\kappa=2}^{k} (1 - r_1)^{-1} r_\kappa b_\kappa \in \langle \{b_2, \ldots, b_k, 0\} \rangle_{\mathcal{M}},$$

> which contradicts the minimality of k.

Nakayama's Lemma will frequently be applied throughout the development of Catastrophe Theory. Often, the following version will be needed.

Corollary (16). Let K and L be submodules of M, and let K be finitely generated. Then $K \subset L + \langle K \rangle_{\mathcal{M}}$ implies $K \subset L$.

Proof. Let $\emptyset \neq A \subset K$ be finite such that $K = \langle A \rangle_{\mathcal{R}}$ is true. Then obviously $\langle A \rangle_{\mathcal{M}} = \langle K \rangle_{\mathcal{M}}$ holds. Therefore, $A \subset L + \langle A \rangle_{\mathcal{M}}$ follows, which implies

$$\tilde{A} := \{a + L : a \in A\} \subset (\langle A \rangle_{\mathcal{M}} + L)/L,$$

where $\tilde{A}$ is a nonvoid finite subset of the factor module M/L. An element of $(\langle A \rangle_{\mathcal{M}} + L)/L$ is of the form

$$\left(\sum_j r_j a_j \right) + L = \sum_j r_j (a_j + L)$$

with $r_j \in \mathcal{M}$, $a_j \in A$ by definition of the outer product of a factor module. This verifies

$$(\langle A \rangle_{\mathcal{M}} + L)/L = \langle \tilde{A} \rangle_{\mathcal{M}},$$

and the assertion results from (15).

In particular, if K is a finitely generated submodule of M such that $K \subset \langle K \rangle_{\mathcal{M}}$ is valid, then $K = \{0\}$ holds.

In the following, the ring $\mathcal{E}$ will be considered as a module over itself. Notice that the set of germs in $\mathcal{E}$ that are not in m is *exactly* the set of invertible elements of $\mathcal{E}$. The submodules of $\mathcal{E}$ coincide with the ideals of $\mathcal{E}$, and modules over $\mathcal{E}$ are real vector spaces, since the germs of the constant functions can be identified with $\mathbf{R}$. Note that (10) can be written as

$$(17) \qquad\qquad m^{k+l} = \langle m^k \rangle_{m^l}.$$

In particular, $m^{k+1} = \langle m^k \rangle_m$ holds. Furthermore, the submodules m^k of $\mathcal{E}$ are finitely generated, as expressed by (9). Therefore, if $\mathcal{A}$ is an ideal of $\mathcal{E}$, then (16) implies

$$(18) \qquad\qquad m^k \subset \mathcal{A} \quad \text{if and only if} \quad m^k \subset \mathcal{A} + m^{k+1}.$$

Corollary (19). A subset A of m^k generates the ideal m^k if and only if $\{[f] + m^{k+1} : [f] \in A\}$ spans the vector space m^k/m^{k+1}.

Proof. Suppose that $m^k = \langle A \rangle_{\mathcal{E}}$ holds. Then

$$\langle \{[f] + m^{k+1} : [f] \in A\} \rangle \subset m^k/m^{k+1}$$

is trivially true. In order to show that this inclusion is an equality, let $\nu \in \mathbf{N}_0$ with $|\nu| = k$. Then $[x^\nu] = \Sigma_j [g_j][f_j]$ with $[g_j] \in \mathcal{E}$ and $[f_j] \in A$.

Put $a_j := g_j(0)$ and $\tilde{g}_j := g_j - a_j$. Then $[\tilde{g}_j]$ is in m. Due to (10), the germ $[h] := \Sigma_j [\tilde{g}_j][f_j]$ lies in m^{k+1}. If $[f] := \Sigma_j a_j [f_j]$, then $[x^\nu] = [f] + [h]$, and therefore $[x^\nu] + m^{k+1} = [f] + m^{k+1}$. Since $[f] \in \langle A \rangle$, the first part of the assertion follows by (9).

Now let $\{[f] + m^{k+1} : [f] \in A\}$ span m^k / m^{k+1}. This implies $m^k \subset \langle A \rangle_{\mathcal{E}} + m^{k+1}$, so that $m^k \subset \langle A \rangle_{\mathcal{E}}$ follows from (18). Since $\langle A \rangle_{\mathcal{E}} \subset m^k$ holds trivially, $\langle A \rangle_{\mathcal{E}} = m^k$ is valid.

The vector space m^k / m^{k+1} is isomorphic to the vector space of all homogeneous polynomials of degree k in n variables, since the latter is the image of the linear map $[f] \mapsto T_f^k$ defined on m^k with kernel m^{k+1}. Therefore

$$(20) \qquad \dim(m^k / m^{k+1}) = \binom{n + k - 1}{k}$$

holds (see Exercise 4.1). The simple rule

$$\langle \langle A \rangle_{\mathcal{A}} \langle B \rangle_{\mathcal{E}} \rangle = \langle AB \rangle_{\mathcal{A}}$$

for subsets A and B of $\mathcal{E}$ and $\mathcal{A}$ an ideal of $\mathcal{E}$, which is proved in Exercise 4.2, is very useful for explicitly calculating a generating set of an ideal. Nakayama's Lemma will often be used to verify the following type of statements.

Example (21). $\langle x^2, y^2 \rangle_{\mathcal{E}} = \langle x^2 + y^3, x^3 + y^2 \rangle_{\mathcal{E}}$.

Proof. Let $\mathcal{A}$ be the ideal on the left of the equality and $\mathcal{B}$ be the one on the right. Obviously, $\mathcal{B} \subset \mathcal{A}$ holds. For the opposite inclusion, it suffices by (16) to show $\mathcal{A} \subset \mathcal{B} + \langle \mathcal{A} \rangle_m$. Now,

$$\langle \mathcal{A} \rangle_m = \langle \langle x, y \rangle_{\mathcal{E}} \langle x^2, y^2 \rangle_{\mathcal{E}} \rangle = \langle x^3, xy^2, x^2 y, y^3 \rangle_{\mathcal{E}}$$

holds. Therefore, the germs of $x^2 = (x^2 + y^3) - y^3$ and $y^2 = (y^2 + x^3) - x^3$ lie in $\mathcal{B} + \langle \mathcal{A} \rangle_m$, which proves the assertion.

There is also a direct proof of (21). Indeed,

$$x^2 = (x^2 + y^3) \frac{1}{1 - xy} + (x^3 + y^2) \frac{-y}{1 - xy}$$

shows that $[x^2]$ is in $\mathcal{B}$. Analogously, it is proved that $[y^2]$ lies in $\mathcal{B}$. However, finding these representations for x^2 and y^2 is not obvious.

The next lemma, whose proof relies on the Existence Theorem for Solutions of Differential Equations (see Appendix A.3), is a basic tool to find a sufficient condition for finite determinacy. In addition, it can be used to calculate the determinacy of germs (see Exercise 4.13). It will also enter the proof of the uniqueness up to equivalence of the residual singularity (48). The natural embedding of $\mathcal{E}_n$ in $\mathcal{E}_{n+1}$ will be used frequently. Consider a function of n variables $x_1, \cdots, x_n$ as being a function of $n + 1$ variables $x_1, \cdots, x_n, t$ that does not depend on the last variable t.

Sufficient Criterion for Equivalence of Germs (22). Two germs $[f]$ and $[g]$ in $\mathcal{E}_n$ are equivalent if for every $s \in [0,1]$ there are germs $[\psi_i] \in \mathcal{E}_{n+1}, i = 1, \ldots, n$, such that

$$\psi_i(0,t) = 0 \quad \text{and}$$

(23)
$$g(x) - f(x) = \sum_{i=1}^{n} \psi_i(x,t) \frac{\partial}{\partial x_i}[f(x) + (s+t)(g(x) - f(x))]$$

hold for all (x,t) in an open neighborhood of the origin in $\mathbf{R}^{n+1}$.

Proof. Let f and g be given on the same open set U in $\mathbf{R}^n$. Define $F: U \times \mathbf{R} \to \mathbf{R}$ by $F(x,t) := (1-t)f(x) + tg(x)$ and $F_t: U \to \mathbf{R}$ by $F_t(x) := F(x,t)$. Then F and F_t are smooth, and $F_0 = f, F_1 = g$. The proof of Lemma (22) will be broken down into five parts, reducing the assertion step by step to an initial value problem for an ordinary differential equation.

(i) $[f] \sim [g]$ if for every $s \in [0,1]$ there is an open interval I_s in $\mathbf{R}$ containing s such that $[F_t] \sim [F_s]$ for all $t \in I_s$.

To see this, note that $A := \{s \in [0,1]: [F_s] \sim [f]\}$ is open in $[0,1]$, since $I_s \cap [0,1] \subset A$ for every $s \in A$. But A is also closed, because for every adherent point s of A one has $I_s \cap A \neq \emptyset$, and hence $[F_s] \sim [f]$, giving $s \in A$. Since A is a nonempty subset of the connected unit interval $[0,1]$, it is equal to it, and $1 \in A$ follows.

From now on fix an $s \in [0,1]$.

(ii) A sufficient condition for the existence of an interval I_s as in (i) is the following one: There is an open set V in U containing the origin, an open interval $I \subset \mathbf{R}$ with $0 \in I$, and a smooth map $\Phi: V \times I \to \mathbf{R}^n$ satisfying

$$(\alpha) \quad \Phi(x,0) = x \quad \text{for all} \quad x \in V,$$
$$(\beta) \quad \Phi(0,t) = 0 \quad \text{for all} \quad t \in I,$$
$$(\gamma) \quad \Phi(x,t) \in U \quad \text{and} \quad F(\Phi(x,t), s+t) = F_s(x)$$
$$\text{for all} \quad (x,t) \in V \times I.$$

In order to verify this statement, consider $\Phi_t: V \to \mathbf{R}^n$ defined by $\Phi_t(x) := \Phi(x,t)$ for $t \in I$. The map Φ_t is smooth and preserves the origin. Moreover, $\det(D\Phi_t(0)) \neq 0$ for all t in some open interval $\tilde{I} \subset I$ with $0 \in \tilde{I}$. This follows from $D\Phi_0(0) = E_n$, since $t \mapsto \det(D\Phi_t(0))$ is a continuous function. Hence, Φ_t is a local diffeomorphism at 0 for all $t \in \tilde{I}$. By (γ), one obtains $[F_{s+t}][\Phi_t] = [F_s]$. Hence $[F_t] \sim [F_s]$ for $t \in I_s := s + \tilde{I}$.

(iii) Using the notation in (ii), let (α), (β), and $\Phi(V \times I) \subset U$ be satisfied. Then (γ) follows from

$$\frac{d}{dt}F(\Phi(x,t), s+t) = 0,$$

because of the initial condition $F(\Phi(x,0),s) = F_s(x)$. Hence condition (γ) can be replaced by

$$\Phi(x,t) \in U \quad \text{and}$$

$$(\gamma') \quad \frac{\partial}{\partial t}F(\Phi(x,t),s+t) + \sum_{i=1}^{n}D_iF(\Phi(x,t),s+t)\frac{\partial}{\partial t}\Phi_i(x,t) = 0$$

for all $(x,t) \in V \times I$,

where Φ_i is the ith component of Φ.

(iv) Here it will be shown that (α), (β), and (γ') can be fulfilled if one finds a smooth map $\psi \colon U \times J \to \mathbf{R}^n$, where J is an open interval $J \subset \mathbf{R}$ around the origin, satisfying

$$(\delta) \quad \psi(0,t) = 0 \quad \text{for all} \quad t \in J,$$

$$(\varepsilon) \quad \frac{\partial}{\partial t}F(x,s+t) + \sum_{i=1}^{n}\psi_i(x,t)D_iF(x,s+t) = 0$$

for all $(x,t) \in U \times J$,

where ψ_i is the ith component of ψ.
    Indeed, for x in U consider the differential equation

$$(*) \qquad\qquad \frac{\partial}{\partial t}\Phi(x,t) = \psi(\Phi(x,t),t)$$

with the initial condition $\Phi(x,0) = x$. According to the existence theorem in Appendix A.3, there is an open set V in U containing the origin and an open interval $I \subset J$ with $0 \in I$, such that for every $x \in V$ there is a unique solution $I \to \mathbf{R}^n, t \mapsto \Phi(x,t)$, of the differential equation $(*)$ with $\Phi(x,t) \in U$ and $\Phi(x,0) = x$, such that $\Phi \colon V \times I \to \mathbf{R}^n$ is smooth.
    Obviously Φ satisfies (α). Condition (γ') follows from (ε). Moreover (β) is valid due to (δ) and the uniqueness of the solution of $(*)$.

(v) By hypothesis (23), a smooth map $\psi = (\psi_1,\ldots,\psi_n)$ exists satisfying (δ) and (ε), which is evident when the definition of F is inserted in (ε). This completes the proof.

Sufficient Criterion for Determinacy (24).   A germ $[f] \in \mathcal{E}$ is k-determined if

$$m^{k+1} \subset \langle m^2 \mathcal{J}[f]\rangle$$

holds.

Proof. Consider a germ $[g]$ in $\mathcal{E}_n$ with $j^k[f] = j^k[g]$. Then $[g]-[f] \in m_n^{k+1}$. In order to show that $[f]$ and $[g]$ are equivalent, Lemma (22) will be applied. Toward that end, fix $s \in [0,1]$.

Let $[h] \in \mathcal{E}_{n+1}$ be given by $h(x,t) := f(x) + (s+t)(g(x) - f(x))$ on an open set of $\mathbf{R}^{n+1}$ containing the origin. Then

$$[D_i f] = [D_i h] + [s+t]([D_i f] - [D_i g]) \quad \text{for} \quad i = 1, \ldots, n,$$

where $[s+t]$ denotes the germ of the function $(x,t) \mapsto s+t$. Since $[D_i f] - [D_i g]$ is in m_n^k, it follows that

$$\langle m_n^2 \mathcal{J}[f] \rangle_{\mathcal{E}_{n+1}} \subset \langle m_n \{[D_1 h], \ldots, [D_n h]\} \rangle_{\mathcal{E}_{n+1}} + \langle m_n^{k+1} \rangle_{m_{n+1}}.$$

The last term results from

$$m_n^2 \{[s+t]\} m_n^k \subset m_n m_{n+1} \mathcal{E}_{n+1} m_n^k \subset m_{n+1} m_n^{k+1}.$$

By hypothesis, $\langle m_n^{k+1} \rangle_{\mathcal{E}_{n+1}}$ is contained in the left side. The Corollary (16) to Nakayama's Lemma yields

$$m_n^{k+1} \subset \langle m_n \{[D_1 h], \ldots, [D_n h]\} \rangle_{\mathcal{E}_{n+1}}.$$

This indeed implies (23).

The inclusion $m^{k+1} \subset \langle m^2 \mathcal{J}[f] \rangle$ obviously follows from $m^k \subset \langle m \mathcal{J}[f] \rangle$. Consequently,

(25) $m^k \subset \langle m \mathcal{J}[f] \rangle$ implies that $[f] \in \mathcal{E}$ is k-determined.

In general, the latter inclusion is easier to check than the former. However, Example (27) shows that the condition given in (25) really is weaker than that given in (24). A germ $[f]$ satisfying $m^k \subset \langle m \mathcal{J}[f] \rangle$ is called **k-complete**.

The next example is particularly important for Catastrophe Theory, since it treats germs appearing in Thom's first classification theorem; see (1.12).

Example (26). The germ of $f \colon \mathbf{R}^2 \to \mathbf{R}$, $f(x,y) := x^3 + y^3$, as well as the germ of $f \colon \mathbf{R}^2 \to \mathbf{R}$, $f(x,y) := x^3 - xy^2$ has determinacy 3.

Proof. Since $T_f^2 = 0$, $[f]$ is not 2-determined. Obviously, $\mathcal{J}[f] = \langle x^2, y^2 \rangle_{\mathcal{E}}$, and therefore

$$\langle m \mathcal{J}[f] \rangle = \langle \langle x, y \rangle \langle x^2, y^2 \rangle \rangle_{\mathcal{E}} = \langle x^3, x^2 y, xy^2, y^3 \rangle_{\mathcal{E}} = m^3$$

by (9). Hence $\det[f] = 3$ follows.

The same line of argument shows that the germ of $f \colon \mathbf{R}^2 \to \mathbf{R}$, $f(x,y) := x^3 - xy^2$ has determinacy 3; see Exercise 4.4.

It is interesting to note that for k greater than 2 the determinacy of the germ of $f \colon \mathbf{R}^2 \to \mathbf{R}$, $f(x,y) := x^3 \pm xy^k$, is $2k - 2$; see Exercise 4.13.

Example (27). If $f \colon \mathbf{R}^2 \to \mathbf{R}$ is defined by $f(x,y) := x^4 + y^4$, then $\det[f] = 4$. Nevertheless, m^4 is not contained in $\langle m \mathcal{J}[f] \rangle$.

Proof. Clearly, $[f]$ is not 3-determined due to $T_f^3 = 0$. Now $\mathcal{J}[f] = \langle x^3, y^3 \rangle_\varepsilon$, and thus

$$\langle m\mathcal{J}[f] \rangle = \langle \langle x, y \rangle \langle x^3, y^3 \rangle \rangle_\varepsilon = \langle x^4, x^3 y, xy^3, y^4 \rangle_\varepsilon.$$

Since this ideal cannot contain the germ of $x^2 y^2$, it follows that $m^4 \not\subset \langle m\mathcal{J}[f] \rangle$. However, (9) implies

$$\langle m^2 \mathcal{J}[f] \rangle = \langle \langle x, y \rangle \langle x^4, x^3 y, xy^3, y^4 \rangle \rangle_\varepsilon$$
$$= \langle x^5, x^4 y, x^2 y^3, x^3 y^2, xy^4, y^5 \rangle_\varepsilon = m^5.$$

Thus, $\det[f] = 4$ results from (24).

The criterion given in (24) for k-determinacy is not necessary, as seen by the following example, due to D. Siersma [Sie].

Example (28). The germ of $f \colon \mathbf{R}^2 \to \mathbf{R}$, $f(x, y) := x^3 + xy^3$ has determinacy 4, although m^5 is not contained in $\langle m^2 \mathcal{J}[f] \rangle$. Furthermore, $m^5 \subset \langle m\mathcal{J}[f] \rangle$ holds.

Proof.

(i) First of all it will be shown that $m^5 \subset \langle m\mathcal{J}[f] \rangle$ is valid, which implies that $[f]$ is 5-determined by (25). Due to $\mathcal{J}[f] = \langle 3x^2 + y^3, xy^2 \rangle_\varepsilon$, one has

$$\langle m\mathcal{J}[f] \rangle = \langle 3x^3 + xy^3, x^2 y^2, 3x^2 y + y^4, xy^3 \rangle_\varepsilon$$
$$= \langle x^3, x^2 y^2, 3x^2 y + y^4, xy^3 \rangle_\varepsilon.$$

It is immediate that the germs of $x^5, x^4 y, x^3 y^2, x^2 y^3$, and xy^4 lie in $\langle m\mathcal{J}[f] \rangle$. The same is true for the germ of $y^5 = y(3x^2 y + y^4) - 3x^2 y^2$. Hence, by (9), m^5 is contained in $\langle m\mathcal{J}[f] \rangle$.

(ii) Now it will be proved that $[f]$ is 4-determined by applying the following Lemma (29). Consider a homogeneous polynomial of degree 5:

$$p(x, y) = c_0 x^5 + c_1 x^4 y + c_2 x^3 y^2 + c_3 x^2 y^3 + c_4 xy^4 + c_5 y^5.$$

The objective is to find a local diffeomorphism $\varphi = (\varphi_1, \varphi_2)$ at the origin in $\mathbf{R}^2$ that vanishes there, such that the germ of $f \circ \varphi - (f + p)$ is in m^6, or loosely speaking, such that

$$f \circ \varphi = f + p + \text{terms of order at least 6}.$$

This means that the following equation has to be solved in a neighborhood of the origin:

$$(*) \qquad \varphi_1(x, y)^3 + \varphi_1(x, y)\varphi_2(x, y)^3 = x^3 + xy^3 + p(x, y)$$
$$+ \text{ terms of order at least 6}.$$

It suffices to consider for φ_1 and φ_2 just polynomials of degree at most 3 and 2, respectively. To see this, consider, for example, φ_2. If $h := \varphi_2 - T_{\varphi_2}^2$, then $\varphi_2 = T_{\varphi_2}^2 + h$ with $[h] \in m^3$. Of course, the germs of φ_1 and φ_2 are in m. Then the germ of every term in $\varphi_1 \varphi_2^3$ containing h lies in m^6 and can be neglected. The argument concerning φ_1 is similar. Therefore, it may be supposed that

$$\varphi_1(x,y) = a_{10}x + a_{01}y + a_{20}x^2 + a_{11}xy + a_{02}y^2$$
$$+ a_{30}x^3 + a_{21}x^2y + a_{12}xy^2 + a_{03}y^3$$
$$\varphi_2(x,y) = b_{10}x + b_{01}y + b_{20}x^2 + b_{11}xy + b_{02}y^2.$$

The task now is to choose the coefficients a_ν and b_ν in such a manner that the equation (*) is satisfied up to powers of order at most 5. Comparing coefficients on both sides of (*) implies that the following conditions are necessary and sufficient for (*) to hold:

$$a_{10} = 1, \quad a_{01} = 0, \quad a_{20} = \tfrac{1}{3}c_5^3, \quad a_{11} = -c_5^2, \quad a_{02} = c_5,$$
$$b_{10} = -c_5, \quad b_{01} = 1, \quad b_{02} = \tfrac{1}{3}(c_4 + c_5^2),$$

as well as

$$a_{30} + c_5^2 b_{20} = \tfrac{1}{3}c_0,$$
$$a_{21} + c_5^2 b_{11} - 2c_5 b_{20} = \tfrac{1}{3}c_1,$$
$$a_{12} + b_{20} - 2c_5 b_{11} = \tfrac{1}{3}(c_2 - c_4 c_5^2 - c_5^4),$$
$$a_{03} + b_{11} = \tfrac{1}{3}(c_3 + 2c_4 c_5 + \tfrac{5}{3}c_5^3).$$

Choosing $b_{20} := b_{11} := 0$, one easily finds a solution to the above equations in the coefficients. Hence the corresponding polynomials φ_1 and φ_2 satisfy (*). Since $a_{10} = b_{01} = 1$ and $a_{01} = 0$ holds, φ is a local diffeomorphism by (1.29).

(iii) The determinacy of $[f]$ is 4, because if $[f]$ were 3-determined, then by (12) the germ of $T_f^3(x,y) = x^3$ would be 3-determined as a germ in $\mathcal{E}_2$. This would yield the following contradiction. Since $[x^3] \sim [x^3 + y^4]$, a $[\varphi] \in \mathcal{G}_2$ would exist with $\varphi_1(x,y)^3 = x^3 + y^4$ around the origin. Hence $\varphi_1(0,y) = |y|^{4/3}$, which is impossible because this function is not smooth at the origin.

Another way to show that $[f]$ is not 3-determined is to apply Corollary (45).

(iv) Here the claim $m^5 \not\subset \langle m^2 \mathcal{J}[f] \rangle$ will be justified. The ideal $\langle m^2 \mathcal{J}[f] \rangle$ is generated by

$$\{x,y\}\{x^3, x^2y^2, 3x^2y + y^4, xy^3\}$$
$$= \{x^4, x^3y^2, 3x^3y + xy^4, x^2y^3, x^3y, x^2y^3, 3x^2y^2 + y^5, xy^4\}.$$

Hence,

$$\langle m^2 \mathcal{J}[f] \rangle = \langle x^4, x^3 y, x^2 y^3, xy^4, 3x^2 y^2 + y^5 \rangle_{\mathcal{E}}$$

holds, and all generators of m^5 excluding the germ of y^5 are in $\langle m^2 \mathcal{J}[f] \rangle$. To see that $[y^5]$ cannot lie in $\langle m^2 \mathcal{J}[f] \rangle$, suppose the contrary. Then for $[h_i]$ in $\mathcal{E}_2$, $i = 1, \ldots, 5$, one has the equation

$$(1 - h_5(x,y))y^5 = h_1(x,y)x^4 + h^2(x,y)x^3 y + h_3(x,y)x^2 y^3$$
$$+ h_4(x,y)xy^4 + h_5(x,y)3x^2 y^2$$

for small x and y. Setting $x = 0$, it follows that $h_5(0,0) = 1$. But the above equation also implies $h_5(0,0) = 0$, as is easily verified by differentiating this equation twice with respect to x and also twice with respect to y and then evaluating at the origin. Thus, $m^5 \not\subset \langle m^2 \mathcal{J}[f] \rangle$ follows.

The tedious computations in part (ii) are not necessary if one proves (ii) using a method based on (22). This method even allows one to prove $\det[x^3 \pm xy^k] = 2k - 2$ for $k \geq 3$; see Exercise 4.13.

Lemma (29). Let k and l be positive integers with $l < k$. Suppose that $[f] \in \mathcal{E}$ is k-determined. Then $[f]$ is l-determined if and only if for any polynomial

$$p(x) := \sum_{l < |\nu| \leq k} a_\nu x^\nu$$

in n variables, $j^k[T_f^l + p] = j^k[T_f^k \circ \varphi]$ is valid for some $[\varphi]$ in $\mathcal{G}$.

Proof. Let $[f]$ be l-determined. Then the germ of T_f^k is l-determined, too, because the l-jets of $[f]$ and $[T_f^k]$ coincide and $[f] \sim [T_f^k]$ holds. For p as above, $[T_f^l + p] \sim [T_f^k]$ follows. Thus, a $[\varphi] \in \mathcal{G}$ exists with $[T_f^l + p] = [T_f^k \circ \varphi]$, giving $j^k[T_f^l + p] = j^k[T_f^k \circ \varphi]$.

To prove the converse, consider a $[g] \in \mathcal{E}$ with $j^l[g] = j^l[f]$. Then $T_g^k = T_f^l + p$ holds for a polynomial p as above. By assumption a germ $[\varphi] \in \mathcal{G}$ can be found such that $j^k[g] = j^k[T_g^k] = j^k[T_f^k \circ \varphi]$ results. An application of Lemma (11) gives $j^k[g \circ \varphi^{-1}] = j^k[f]$, and the k-determinacy implies $[f] \sim [g]$.

What is missing now is a necessary condition for a germ $[f]$ in $\mathcal{E}$ to be k-determined. The rest of this chapter is devoted to proving that

(30) $$m^{k+1} \subset \langle m \mathcal{J}[f] \rangle$$

is such a condition. Using a plausible assumption, it is easy to see how one arrives at this inclusion. The heuristic argument goes as follows. Let $[f] \in \mathcal{E}$ be k-determined. Suppose that $[g]$ is in m^{k+1}, and let t be a real number. By hypothesis, $[f] + t[g]$ is equivalent to $[f]$. Hence the germ $[\varphi_t] \in \mathcal{G}$ of a smooth coordinate transformation exists such that $[f][\varphi_t] = [f] + t[g]$ holds. Without restriction, φ_0 is the identity. Now assume that the map $(x,t) \mapsto \Phi(x,t) := \varphi_t(x)$ depends *smoothly* not only on x but also on t and that

$$(31) \qquad f(\Phi(x,t)) = f(x) + tg(x)$$

is valid in a neighborhood of the origin in $\mathbf{R}^{n+1}$. Differentiating (31) with respect to t and evaluating at $(x,0)$ gives

$$\sum_{i=1}^{n} D_i f(x) \frac{\partial \Phi_i}{\partial t}(x,0) = g(x)$$

for $\Phi = (\Phi_1, \ldots, \Phi_n)$, due to $\Phi(x,0) = x$. Because of $\Phi(0,t) = 0$, clearly $(\partial \Phi_i/\partial t)(0,0) = 0$ is true, and (30) results.

It is important to emphasize that condition (30) concerns *infinite* dimensional vector spaces, although it follows from

$$(32) \qquad m^{k+1}/m^{k+2} \subset (\langle m\mathcal{J}[f]\rangle + m^{k+2})/m^{k+2}$$

by the implication (18) of Nakayama's Lemma (15). Inclusion (32) is a statement about subspaces of the vector space $\mathcal{E}/m^{k+2}$, which by (8) is isomorphic to the *finite* dimensional space J^{k+1}. The plausible assumption mentioned above can be circumvented by replacing $\mathcal{E}$ with the jet space J^{k+1} and the group $\mathcal{G}$ by a finite dimensional Lie group $\mathcal{G}^{k+1}$ and then showing that the analogous assumption is really fulfilled. The major tool for an elementary and direct proof of this is the following Linearization Lemma (37).

Remark. The Linearization Lemma (37) is formulated in a manner that avoids concepts such as submanifolds and tangent spaces to manifolds used in Mather's original proof as well as in all other known proofs (see [M III, 8.1; Tro, 2.11; BL, 11.9; W, 2.9]). In particular, the theorem stating that the orbits of a Lie group action are immersed submanifolds and thus have a tangent space (see, e.g., [Gib, Appendix]) is not needed. Using the language of this theorem, (37) proves that the tangent vector to a curve whose image lies completely in an orbit is itself an element of the tangent space to that orbit. Figures 1 and 2, which appear later in this chapter, illustrate the significance of this nontrivial fact, which has not been treated in the standard proofs of Mather's Necessary Condition for Determinacy (38). At any rate, it is not sufficient just to cite the theorem, according to which the tangent space to the orbit is the image of the tangent space to the group under the differential of the orbital map, without showing that the tangents to the curve are in the tangent spaces to the orbit.

It should be mentioned, however, that the action of the Lie group $\mathcal{G}^k$ on J^k is **semialgebraic**, as is easily verified; see (36). A deep result involving the Tarski–Seidenberg theorem states that under a semialgebraic Lie group action orbits are **regular** submanifolds, i.e., they carry the induced topology; see [Gib, Appendix B; BCR, BM, BR, Co, L]. Of course, for orbits that are regular submanifolds it is immediate that (37) holds (see part (i) of the proof of (37)).

The advantage of (37) is in enabling a complete, elementary, and direct proof of Mather's theorem (38) to be given. A consequence of (37) is that a curve lying in an orbit can be locally lifted to a curve lying in the group (see Exercise 4.7). Moreover, as shown in [CH], (37) and Exercise 4.7 can be rather straightforwardly generalized to smooth actions of an arbitrary σ-compact Lie group G. This is indeed more than just semialgebraic actions. For example, it includes the action of $\mathbf{R}$ on the double torus by the irrational flow

$$(t; z_1, z_2) \mapsto (z_1 \exp(it), z_2 \exp(2\pi it))$$

whose orbits are dense. Obviously, the lifting of curves in an orbit (see [CH,3]) implies that G-orbits are **weakly embedded submanifolds** (a notion due to [P]) and therefore, in turn, satisfy the Linearization Lemma. Actually G-orbits are even **initial submanifolds**. This results from the first chapter of [KMS,5.14 Theorem and 2.17 Lemma], which can be consulted for further information.

Definition. Let k be a positive integer. Then

$$j^k[\varphi] := (j^k[\varphi_1], \ldots, j^k[\varphi_n])$$

is called the **k-jet of the germ $[\varphi]$ in $\mathcal{G}$ of the local diffeomorphism $\varphi = (\varphi_1, \ldots, \varphi_n)$ at the origin** with $\varphi(0) = 0$. The set of all k-jets of germs in $\mathcal{G}$ is denoted by $\mathcal{G}^k$ or, more explicitly, by $\mathcal{G}_n^k$.

The map

$$\mathcal{G}^k \times \mathcal{G}^k \to \mathcal{G}^k, \quad (j^k[\varphi], j^k[\psi]) \mapsto j^k([\varphi][\psi])$$

is well defined by Lemma (11) and is a natural group multiplication on $\mathcal{G}^k$; the identity is $j^k[id]$, and the inverse of $j^k[\varphi]$ is $j^k([\varphi]^{-1})$ for $[\varphi] \in \mathcal{G}$. Obviously, the jet map

$$j^k : \mathcal{G} \to \mathcal{G}^k, \quad [\varphi] \mapsto j^k[\varphi]$$

is a group epimorphism. In what follows $\mathcal{G}^k$ is identified with an open subset U^k of $\mathbf{R}^{n(N-1)}$ by taking coefficients, just as the jet space J^k is identified with $\mathbf{R}^N$ after (6). Thus, explicit formulae for the group operations of $\mathcal{G}^k$ and for the natural action of $\mathcal{G}^k$ on J^k are obtained showing the smoothness of these maps. Since that is straightforward, it may suffice to glance over the computations in (33) – (36).

Lemma (33). Let N be the binomial coefficient $\binom{n+k}{k}$, let ν denote a multi-index in $\mathbf{N}_0^n$, and consider the subset

$$U^k := \{\xi = (\xi_{i\nu})_{\substack{1 \le i \le n \\ 0 < |\nu| \le k}} \in \mathbf{R}^{n(N-1)} : \det(\xi_{i\nu})_{\substack{1 \le i \le n \\ |\nu| = 1}} \ne 0\}$$

of $\mathbf{R}^{n(N-1)}$. Then U^k is open in $\mathbf{R}^{n(N-1)}$, and $\mathcal{G}^k$ is naturally identified with U^k by means of the bijection

$$U^k \to \mathcal{G}^k, \quad \xi \mapsto j^k[p]$$

for $p := (p_i)_{1 \leq i \leq n}$ with $p_i(x) := \sum_{0 < |\nu| \leq k} \xi_{i\nu} x^\nu$, $x \in \mathbf{R}^n$.

Proof. Consider the set $Q^k := \{p \in (P^k)^n : p(0) = 0\}$ of polynomials in $(\mathbf{R}[x_1, \ldots, x_n])^n$. Since Q^k has dimension $n(N-1)$ (see (6)), it can be identified with $\mathbf{R}^{n(N-1)}$ via the map that assigns to a polynomial its coefficients

$$p(x) = \left(\sum_{0 < |\nu| \leq k} \xi_{i\nu} x^\nu \right)_{1 \leq i \leq n} \longmapsto \xi := (\xi_{i\nu})_{\substack{1 \leq i \leq n \\ 0 < |\nu| \leq k}}.$$

Those polynomials in Q^k defining a local diffeomorphism at the origin will be denoted by G^k. Taking coefficients, this set corresponds to the subset U^k of $\mathbf{R}^{n(N-1)}$ by (1.29), since the Jacobi matrix $D\varphi(0)$ of $\varphi \in G^k$ at the origin is

$$(\xi_{i\nu})_{\substack{1 \leq i \leq n \\ |\nu|=1}}.$$

The continuity of the determinant function yields that U^k is open in $\mathbf{R}^{n(N-1)}$. Now G^k is bijectively mapped onto $\mathcal{G}^k$ by $\varphi \mapsto j^k[\varphi]$. The map is well defined, since $[\varphi] \in \mathcal{G}$ for $\varphi \in G^k$. Its injectivity is obvious due to $\varphi = T_\varphi^k = (T_{\varphi_1}^k, \ldots, T_{\varphi_n}^k)$ for $\varphi \in G^k$. The inverse map is $j^k[\varphi] \mapsto T_\varphi^k$ for $[\varphi] \in \mathcal{G}$, since $j^k[T_\varphi^k] = j^k[\varphi]$, and because

$$T_{\varphi_i}^k(x) =: \sum_{0 < |\nu| \leq k} \xi_{i\nu} x^\nu$$

implies $D\varphi_i(0) = (\xi_{i\nu})_{|\nu|=1}$, so that $\det(\xi_{i\nu})_{\substack{1 \leq i \leq n \\ |\nu|=1}} = \det D\varphi(0) \neq 0$ and hence $T_\varphi^k \in G^k$.

Lemma (34). The group multiplication

$$\mathcal{G}^k \times \mathcal{G}^k \to \mathcal{G}^k, \quad (j^k[\varphi], j^k[\psi]) \mapsto j^k([\varphi][\psi])$$

and inversion

$$\mathcal{G}^k \to \mathcal{G}^k, \quad j^k[\varphi] \mapsto j^k([\varphi]^{-1})$$

are smooth maps when $\mathcal{G}^k$ is identified with U^k as in (33).

Proof. Let G^k be defined as in the proof of (33). First of all, note that the group multiplication on G^k induced by that on $\mathcal{G}^k$ is

$$G^k \times G^k \to G^k, \quad (\varphi, \psi) \mapsto T_{\varphi \circ \psi}^k,$$

i.e., it is given by truncating the composition of two polynomial maps. The identity is the map id on $\mathbf{R}^n$, and the inverse of $\varphi \in G^k$ is T_ψ^k, where ψ is a local inverse of φ at the origin. The group structure on G^k is transferred to U^k by taking coefficients. The notation

$$U^k \times U^k \to U^k, \quad (\xi, \eta) \mapsto \xi \cdot \eta,$$
$$U^k \to U^k, \quad \xi \mapsto \xi^{-1},$$

will be used, and ε will refer to the identity in U^k. Since

$$x_i = \sum_{0 < |\nu| \le k} \varepsilon_{i\nu} x^\nu \quad \text{for} \quad i = 1, \dots, n,$$

one has

$$\varepsilon_{i\nu} = 1 \quad \text{if} \quad \nu_j = \delta_{ij}, \quad \text{for} \quad j = 1, \dots, n, \quad \text{and} \quad \varepsilon_{i\nu} = 0 \quad \text{otherwise.}$$

In order to calculate the group multiplication on U^k more explicitly, let $\varphi = (\varphi_1, \dots, \varphi_n)$ and $\psi = (\psi_1, \dots, \psi_n)$ be in G^k, i.e.,

$$\varphi_i(x) = \sum_{0 < |\mu| \le k} \xi_{i\mu} x^\mu \quad \text{and} \quad \psi_j(x) = \sum_{0 < |\lambda| \le k} \eta_{j\lambda} x^\lambda.$$

Then

$$\varphi_i(\psi(x)) = \sum_{0 < |\mu| \le k} \xi_{i\mu} \prod_{j=1}^n \left(\sum_{0 < |\lambda| \le k} \eta_{j\lambda} x^\lambda \right)^{\mu_j}$$

holds for $i = 1, \dots, n$. After rearranging the terms of

$$\varphi_i(\psi(x)) =: \sum_{0 < |\nu| \le k} \tau_{i\nu} x + \text{higher order terms}$$

according to powers of x, one obtains for the coefficients

$$\tau_{i\nu} = \sum_{0 < |\mu| \le k} \xi_{i\mu} M_{\mu\nu}(\eta),$$

where $M_{\mu\nu}(\eta)$ is a polynomial in η. For more details about $M_{\mu\nu}$ see Exercise 4.5. Since τ is the product $\xi \cdot \eta$, this shows that each component $(\xi, \eta) \mapsto (\xi \cdot \eta)_{i\nu}$ of the group multiplication is a smooth function. The product can be written more succinctly in matrix notation as

$$\xi \cdot \eta = \xi L(\eta),$$

where $L(\eta)$ is the n-fold direct sum of the $(N-1) \times (N-1)$ matrix

$$M(\eta) := (M_{\mu\nu}(\eta))_{0 < |\mu|, |\nu| \le k}.$$

Clearly, the map $\eta \mapsto L(\eta)$ is smooth. It is also injective, since $L(\eta) = L(\xi)$ implies $\varepsilon \cdot \eta = \varepsilon L(\eta) = \varepsilon L(\xi) = \varepsilon \cdot \xi$ and therefore $\eta = \xi$.

Now it will be shown that inversion is also a smooth operation. This will follow from the formula

(35) $$\xi^{-1} = \varepsilon L(\xi)^{-1}$$

because the smoothness of the map $\xi \mapsto L(\xi)$ implies that of $\xi \mapsto L(\xi)^{-1}$, as an application of Cramer's rule shows. Of course, the invertibility of $L(\xi)$ for every ξ in U^k has to be verified first. Note that

$$L(\xi)L(\eta) = L(\xi \cdot \eta)$$

holds for all $\xi, \eta \in U^k$. This results from the associativity of the group operation on U^k—i.e., $(\zeta \cdot \xi) \cdot \eta = \zeta \cdot (\xi \cdot \eta)$—which implies

$$\zeta L(\xi)L(\eta) = \zeta L(\xi \cdot \eta) \text{ for } \zeta, \xi, \eta \in U^k.$$

Keeping ξ and η fixed, ζ runs through U^k, which as an open subset spans all of $\mathbf{R}^{n(N-1)}$. Thus the matrices $L(\xi)L(\eta)$ and $L(\xi \cdot \eta)$ coincide. Starting from $\xi E_{n(N-1)} = \xi = \xi \cdot \varepsilon = \xi L(\varepsilon)$, a similar argument yields

$$L(\varepsilon) = E_{n(N-1)}.$$

Finally, $\xi^{-1} \cdot \xi = \varepsilon$ implies $L(\xi^{-1})L(\xi) = E_{n(N-1)}$, showing the invertibility of $L(\xi)$ as well as the fact that $L(\xi)^{-1} = L(\xi^{-1})$ holds. Now from $\xi^{-1} \cdot \xi = \varepsilon$, assertion (35) also follows.

Another way to express the two results (33) and (34) is to say that $\mathcal{G}^k$ is a **finite dimensional Lie group** admitting a global coordinate system. Moreover, it follows that the map

$$\mathcal{G}^k \to GL(n(N-1), \mathbf{R}), \quad j^k[\varphi] \mapsto L(\xi)$$

with

$$\xi_{i\nu} := \frac{1}{\nu!} D_i^\nu \varphi(0), \quad 1 \le i \le n, \quad 0 < |\nu| \le k,$$

is a faithful $n(N-1)$-dimensional matrix representation of $\mathcal{G}^k$. Notice that it is the n-fold direct sum of the faithful $(N-1)$-dimensional matrix representation $\mathcal{G}^k \to GL(N-1, \mathbf{R}), j^k[\varphi] \mapsto M(\xi)$, of $\mathcal{G}^k$. By taking k-jets, the action of the group $\mathcal{G}$ on the set of germs $\mathcal{E}$ induces an action of $\mathcal{G}^k$ on the jet space J^k that is smooth:

Lemma (36). Let N be the binomial coefficient $\binom{n+k}{k}$. The action

$$J^k \times \mathcal{G}^k \to J^k, \quad (j^k[f], j^k[\varphi]) \mapsto j^k[f]j^k[\varphi] := j^k([f][\varphi])$$

of $\mathcal{G}^k$ on J^k is a smooth map when $\mathcal{G}^k$ is identified with the open set U^k of $\mathbf{R}^{n(N-1)}$ as in (33) and J^k is identified with $\mathbf{R}^N$ as usual by taking coefficients, i.e.,

$$j^k[f] \mapsto \left(\frac{1}{\nu!} D^\nu f(0) \right)_{0 \le |\nu| \le k} \quad \text{for } \nu \in \mathbf{N}_0^n.$$

Proof. The action of U^k on $\mathbf{R}^N$ is given specifically by

$$\mathbf{R}^N \times U^k \to \mathbf{R}^N, \quad (z, \xi) \mapsto z \cdot \xi,$$

where $z \in \mathbf{R}^N$ has components $(z_\nu)_{0 \leq |\nu| \leq k}$, and $z \cdot \xi$ is defined by

$$(z \cdot \xi)_0 := z_0, \quad (z \cdot \xi)_\nu := \sum_{0 < |\mu| \leq k} z_\mu M_{\mu\nu}(\xi) \quad \text{for} \quad |\nu| > 0.$$

Since $M_{\mu\nu}(\xi)$ is a polynomial in ξ, the assertion follows.

Obviously, $\mathcal{G}$ is a group of algebra isomorphisms of $\mathcal{E}$, i.e., if $[\varphi]$ is in $\mathcal{G}$, the induced map

$$\mathcal{E} \to \mathcal{E}, \quad [f] \mapsto [f][\varphi]$$

is an algebra isomorphism (see Exercise 4.6). Lemma (11) shows that $[f] \in \mathcal{E}$ is in m^k if and only if $[f][\varphi]$ is in m^k for all $[\varphi] \in \mathcal{G}$, i.e., the above map leaves the ideals m^k invariant. This implies that the map $J^k \to J^k$, $j^k[f] \mapsto j^k([f][\varphi])$, is well defined. $\mathcal{G}^k$ is then clearly a group of algebra isomorphisms of J^k leaving invariant the subalgebra

$$j^k(m) := \{j^k[f] : [f] \in m\}$$

as well as the subalgebra of the constants.

To set the terminology, call a smooth map $c \colon I \to U$ of an open interval I containing the origin into an open subset U of $\mathbf{R}^n$ a **curve in** U **based at the point** $p \in U$ if $c(0) = p$ holds. Then $c'(0) \in \mathbf{R}^n$ is the **tangent to the curve** c **at** p. Identifying $\mathcal{G}^k$ and J^k with U^k and $\mathbf{R}^N$, respectively, one may also speak about curves in $\mathcal{G}^k$ and J^k.

Linearization Lemma (37). Let c be a curve in J^k based at the k-jet $z := j^k[f]$ of a germ $[f]$ in $\mathcal{E}$ whose image lies in the orbit $z\mathcal{G}^k$. Then there is a curve C in $\mathcal{G}^k$ based at the identity with

$$c'(0) = \frac{d}{dt}[zC(t)]_{t=0}.$$

Proof. Using the above notation, it is more convenient to prove the analogous assertion for U^k acting on $\mathbf{R}^N$ instead of $\mathcal{G}^k$ acting on J^k. Thus z will be considered as a point in $\mathbf{R}^N$. Furthermore the **orbital map** $\varrho \colon U^k \to \mathbf{R}^N, \varrho(\xi) := z \cdot \xi$, will be considered as a map into $\mathbf{R}^{N-1}$ given by $\varrho(\xi) := zM(\xi)$, omitting the 0th component of z, since this component is left invariant under the group action. Notice that the rank r of the Jacobi matrix $D\varrho(\xi)$ is constant. This follows because $M(\xi) = M(\xi \cdot \eta)M(\eta^{-1})$ holds for ξ and η in U^k, implying $DM(\xi) = L(\eta) \cdot DM(\xi \cdot \eta)M(\eta^{-1})$, which means

$$D_{ik}M_{\mu\nu}(\xi) = \sum_{\alpha, j, \lambda} L_{ik,j\lambda}(\eta)D_{j\lambda}M_{\mu\alpha}(\xi \cdot \eta)M_{\alpha\nu}(\eta^{-1}).$$

Hence, by choosing $\eta = \xi^{-1}$, one obtains

$$D\varrho(\xi) = L(\xi)^{-1} D\varrho(\varepsilon) M(\xi).$$

This proves that r is equal to the rank of $D\varrho(\varepsilon)$. Therefore, the Rank Theorem (see Appendix A.4) can be applied, yielding the existence of

a diffeomorphism u of an open neighborhood W of the identity ε in U^k onto the open cube

$$K := \{x \in \mathbf{R}^{n(N-1)} : |x_l| < 1, 1 \leq l \leq n(N-1)\}$$

with $u(\varepsilon) = 0$, and

a diffeomorphism v of an open neighborhood V of z in $\mathbf{R}^{N-1}$ onto the open cube

$$L := \{x \in \mathbf{R}^{N-1} : |x_l| < 1, 1 \leq l \leq N - 1\}$$

with $v(z) = 0$

such that $v(\varrho(\xi)) = \tilde{\varrho}(u(\xi))$ holds for $\xi \in W$, where $\tilde{\varrho}: K \to L$ is defined by

$$\tilde{\varrho}(x_1, \ldots, x_{n(N-1)}) := (x_1, \ldots, x_r, 0, \ldots, 0).$$

Now let $c: I \to \mathbf{R}^{N-1}$ be a curve based at z with $c(I) \subset \varrho(U^k)$. It is no restriction to assume $c(I) \subset V$. In the following sections (i) and (ii), the Linearization Lemma will be proved under certain additional assumptions. These sections serve as a motivation for the general proof of the lemma, which is then given in (iii), (iv), and (v).

(i) It will be shown here that the assertion follows immediately if an open interval $J \subset I$ containing 0 exists with $c(J) \subset \varrho(W)$.

In this case, for any $t \in J$, let ξ_t be in W such that $c(t) = \varrho(\xi_t)$. Since $v(c(t)) = \tilde{\varrho}(u(\xi_t))$, the vector $v(c(t))$ lies in $L \subset \mathbf{R}^{N-1}$ and is of the form $(\gamma_1(t), \ldots, \gamma_r(t), 0, \ldots, 0)$. Add $(n-1)(N-1)$ zeros to make $\gamma(t) := (\gamma_1(t), \ldots, \gamma_r(t), 0, \ldots, 0)$ be in $\mathbf{R}^{n(N-1)}$ and hence in K. Obviously, $\gamma: J \to K$ is smooth and $\gamma(0) = 0$ holds. Define $C := u^{-1} \circ \gamma$. Then $v(\varrho(C(t))) = \tilde{\varrho}(\gamma(t)) = v(c(t))$ implies $\varrho \circ C = c$ on all of J and hence $c'(0) = (\varrho \circ C)'(0)$.

(ii) Actually, to prove the assertion of (37), just sequences (t_n) in $I \backslash \{0\}$ and (ξ_n) in U^k suffice for which $c(t_n) = \varrho(\xi_n), t_n \to 0$, and $\xi_n \to \xi$ for some ξ in U^k hold.

In order to verify this, note that it is no restriction to assume $\xi = \varepsilon$, and therefore that (ξ_n) is in W. Indeed,

$$z = \lim_{n \to \infty} c(t_n) = \varrho(\xi)$$

means that ξ, and hence ξ^{-1}, leaves z invariant so that (ξ_n) can be replaced by $(\xi^{-1} \cdot \xi_n)$. As in (i), note that $\upsilon(c(t_n))$ lies in L and is of the form $(\gamma_1(t_n), \ldots, \gamma_r(t_n), 0, \ldots, 0)$. Again, add enough zeros to make $\gamma(t_n) := (\gamma_1(t_n), \ldots, \gamma_r(t_n), 0, \ldots, 0)$ lie in K. Since $\upsilon \circ c$ is differentiable,

$$\lim_{n \to \infty} \frac{\gamma_i(t_n)}{t_n}$$

exists for $i = 1, \ldots, r$. Hence, there is a vector $\tilde{\zeta}$ in $\mathbf{R}^{n(N-1)}$ satisfying

$$\tilde{\zeta} = \lim_{n \to \infty} \frac{\gamma(t_n)}{t_n}.$$

Let J be an open interval containing 0 such that $t\tilde{\zeta} \in K$ for $t \in J$. Then $C(t) := u^{-1}(t\tilde{\zeta})$, $t \in J$, is a curve in W based at the identity. Now,

$$\frac{d}{dt}\upsilon(\varrho(C(t)))\Big|_{t=0} = \frac{d}{dt}\tilde{\varrho}(t\tilde{\zeta})\Big|_{t=0} = \tilde{\varrho}(\tilde{\zeta}) = \frac{d}{dt}\upsilon(c(t))\Big|_{t=0}.$$

Therefore, $(d/dt)\varrho(C(t))|_{t=0} = c'(0)$, since υ is a diffeomorphism.

Note that this result implies $c'(0) = D\varrho(\varepsilon)\zeta$ for $\zeta := C'(0) \in \mathbf{R}^{n(N-1)}$.

(iii) Not even the assumption stated in (ii) concerning (t_n) and (ξ_n) will be necessary to prove (37). It turns out that the following is all that is needed: For any $\varepsilon > 0$ there is an $s \in I$ with $|s| \le \varepsilon$ and sequences (t_n) in $I \setminus \{s\}$ and (ξ_n) in U^k for which $c(t_n) = \varrho(\xi_n)$, $t_n \to s$, and $\xi_n \to \xi$ hold for some ξ in U^k.

First the existence of such sequences will be proved. It is no restriction to assume that $I_\varepsilon := [-\varepsilon, \varepsilon] \subset I$. By hypothesis, for $t \in I_\varepsilon$ there is a ξ_t in U^k with $c(t) = \varrho(\xi_t)$. Since U^k is open in $\mathbf{R}^{n(N-1)}$, it is the union of countably many compact subsets (for instance, take the closed balls in U^k with rational centers and radii). Since I_ε is uncountably infinite, it contains uncountably many elements t whose corresponding ξ_t all lie in the same compact subset of U^k. Consequently, there is a sequence (t_l) in I_ε with $t_l \neq t_m$ for $l \neq m$ such that (ξ_{t_l}) has a limit in U^k. The compactness of I_ε implies the existence of a convergent subsequence of (t_l) in I_ε, proving the existence of sequences (t_n) and (ξ_n) as in (iii).

(iv) Now the lemma will be proved under the assumption that a certain map is continuous; this continuity will then be verified in part (v).

Denote by H the set of all $s \in I$ for which there are sequences (t_n) in $I \setminus \{s\}$ and (ξ_n) in U^k satisfying $c(t_n) = \varrho(\xi_n)$, $t_n \to s$, and $\xi_n \to \xi$ for some ξ in U^k. According to (iii) the origin $0 \in I$ is an adherent point of H. Define ϱ_s on U^k by $\varrho_s(\xi) := c(s) \cdot \xi$. By an argument analogous to that in (ii)—replacing ϱ by ϱ_s and the origin in $\mathbf{R}$ by s—it follows that for $s \in H$

$$(*)\qquad\qquad\qquad c'(s) = D\varrho_s(\varepsilon)\zeta$$

holds, where $\zeta \in \mathbf{R}^{n(N-1)}$ depends on s. More explicitly, due to $\varrho_s(\xi) = c(s)M(\xi)$, Equation (*) reads

$$c'(s)_\nu = \sum_\mu c(s)_\mu \sum_{j,\kappa} D_{j\kappa} M_{\mu\nu}(\varepsilon)\zeta_{j\kappa} \quad \text{with}$$

$$0 < |\nu|, |\mu|, |\kappa| \leq k, \quad \text{and} \quad 1 \leq j \leq n.$$

This means that $c'(s)$ is in

$$T_{c(s)} := \langle c(s)D_{j\kappa}M(\varepsilon): 1 \leq j \leq n, \kappa \in \mathbf{N}_0^n \quad \text{with} \quad 0 < |\kappa| \leq k\rangle,$$

which is the so-called **tangent space at the point** $c(s)$ **to the orbit** $\varrho_s(U^k)$. Notice that it is a linear subspace of $\mathbf{R}^{N-1}$.

For s in I let P_s denote the orthogonal projection of $\mathbf{R}^{N-1}$ onto $T_{c(s)}$. Orthogonality is defined with respect to the Euclidean scalar product on $\mathbf{R}^{N-1}$. As a linear map, P_s can be regarded as a real $(N-1)\times(N-1)$ matrix acting on $\mathbf{R}^{N-1}$ by matrix multiplication. In particular, $P_s(c'(s)) = c'(s)$ holds for s in H, since $c'(s)$ is in $T_{c(s)}$.

The proof of the lemma is accomplished by showing that the matrix elements of P_s are continuous functions of s. The reason is that in this case, the equation $P_s(c'(s)) = c'(s)$ holds even when s is the origin, i.e., it holds for the adherent point $0 \in I$ of H. This in turn means $c'(0) \in T_z$, so that there is a $\zeta \in \mathbf{R}^{n(N-1)}$ with

$$c'(0) = \sum_{j,\kappa} \zeta_{j\kappa} z D_{j\kappa} M(\varepsilon) = D\varrho(\varepsilon)\zeta.$$

Then $c'(0)$ is trivially equal to $(\varrho \circ C)'(0)$, where for small $t, C(t) := \varepsilon + t\zeta$ is a curve in U^k based at ε.

(v) It remains to verify the continuity of the map $s \mapsto P_s$.

First it will be shown that the dimension of the tangent space $T_{c(s)} = P_s(\mathbf{R}^{N-1})$ is the same for all $s \in I$. The reason for this is that the entire curve c lies in the orbit through z, i.e., $c(I) \subset \varrho(U^k)$. More generally, if $a := \varrho(\xi) = zM(\xi)$ for $\xi \in U^k$, then the invertible matrix $M(\xi)$ maps T_z onto T_a. To see this, take a generator $aD_{j\kappa}M(\varepsilon)$ of T_a. Due to

$$M(\xi)M(\eta) = M(\xi \cdot \eta \cdot \xi^{-1})M(\xi) = M(\xi L(\eta)L(\xi^{-1}))M(\xi),$$

this generator is equal to

$$zD_{j\kappa}M(\xi)M(\eta)|_{\eta=\varepsilon} = \left(\sum_{i,\delta} \zeta_{i\delta} z D_{i\delta} M(\varepsilon)\right) M(\xi)$$

with $\zeta := \xi(D_{j\kappa}L(\varepsilon))L(\xi^{-1})$. Hence, it is the image under $M(\xi)$ of an element in T_z. Thus $\dim T_a \leq \dim T_z$ follows, and by interchanging the roles of a and z, equality is obtained.

Let $r := \dim T_z$. Fix t in I and choose r linearly independent vectors $b_1(t), \ldots, b_r(t)$ in $T_{c(t)}$ of the form $c(t)D_{j\kappa}M(\varepsilon)$. Since c is continuous, the maps $s \mapsto b_i(s)$ are continuous, and there is a neighborhood I_t of t in I such that $b_1(s), \ldots b_r(s)$ form a basis of $T_{c(s)}$ for all s in I_t. By means of the Gram–Schmidt orthogonalization process, change to an orthonormal basis $\tilde{b}_1(s), \ldots, \tilde{b}_r(s)$. Note that the maps $s \mapsto \tilde{b}_i(s)$ are also continuous on I_t. Now the formula

$$P_s = \langle \tilde{b}_1(s), \cdot \rangle \tilde{b}_1(s) + \cdots + \langle \tilde{b}_r(s), \cdot \rangle \tilde{b}_r(s)$$

proves the asserted continuity at t.

It should be remarked that the Linearization Lemma can be used to show that C may actually be chosen to fulfill $zC(t) = c(t)$ for t in a small interval J around 0; see Exercise 4.7. In other words, a curve c whose image lies in the orbit $z\mathcal{G}^k$ can be lifted locally to a curve C in the group $\mathcal{G}^k$, i.e., the diagram

commutes. The significance of the Linearization Lemma can be illustrated by considering Fig. 1, which shows the image of the map

Figure 1   **Figure 2**

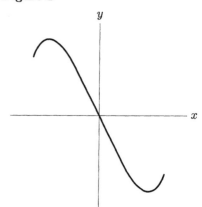

$$\varrho \colon \mathbf{R} \to \mathbf{R}^2, \quad \varrho(t) := (\sin(2\arctan t), \sin(4\arctan t)).$$

As $t \to -\infty (t \to +\infty)$, $\varrho(t)$ approaches the origin along that part of the curve in the second (fourth) quadrant. The tangent vector to ϱ at the origin has the coordinates $(2, 4)$. Figure 2 shows a curve in $\mathbf{R}^2$ whose image is contained in $\varrho(\mathbf{R})$ but whose tangent at the origin is obviously *not* tangent to ϱ there. The Linearization Lemma (37) implies that curves like those in Fig.1 cannot be orbits.

Using (37), it is now easy to prove the following

Necessary Condition for Determinacy (38). Suppose that a germ $[f] \in \mathcal{E}$ is k-determined. Then

$$m^{k+1} \subset \langle m \mathcal{J}[f] \rangle$$

holds.

Proof. Let $[g]$ be in m^{k+1}. Recall that it suffices to show that $j^{k+1}[g]$ is in $j^{k+1}(\langle m \mathcal{J}[f] \rangle)$, since this verifies (32), from which the assertion follows.

Let t be real. Because $[f] + t[g]$ is k-equivalent to $[f]$, it lies in the orbit through $[f]$ under $\mathcal{G}$. Consequently, its jet

$$c(t) := j^{k+1}([f] + t[g]) = j^{k+1}[f] + t j^{k+1}[g]$$

lies in the orbit through $z := j^{k+1}[f]$ under $\mathcal{G}^{k+1}$. Then (37) yields a smooth map $C \colon J \to \mathcal{G}^{k+1}$ on an open interval J containing the origin such that $C(0) = j^{k+1}[id]$ and

$$\frac{d}{dt}[zC(t)]_{t=0} = c'(0) = j^{k+1}[g].$$

To evaluate the derivative of $zC(t)$, identify J^{k+1} with P^{k+1} and $\mathcal{G}^{k+1}$ with G^{k+1} as before (6) and in the proof of (33). Then z is equal to T_f^{k+1} and $C(t)$ is $(C_i(t))_{1 \leq i \leq n}$ with polynomials

$$C_i(t)(x) =: \sum_{0 < |\nu| \leq k+1} \xi_{i\nu}(t) x^\nu$$

whose coefficients $\xi_{i\nu} \colon J \to \mathbf{R}$ are smooth. Moreover, $zC(t) = T_{f \circ C(t)}^{k+1}$ holds. Obviously,

$$\frac{d}{dt} f(C(t)(x)) \bigg|_{t=0} = \sum_{i=1}^{n} D_i f(x) C_i'(0)(x)$$

lies in $\langle m \mathcal{J}[f] \rangle$. By interchanging the order of the derivatives,

$$\frac{d}{dt} T_{f \circ C(t)}^{k+1} \bigg|_{t=0} = T_{(d/dt) f \circ C(t)|_{t=0}}^{k+1}$$

follows, proving $j^{k+1}[g] \in j^{k+1}(\langle m \mathcal{J}[f] \rangle)$.

Observe that $C(t)(x) =: C(x,t)$ depends smoothly on x and t. Therefore, the heuristic differentiability assumption of the map Φ mentioned after (30) is verified for its counterpart C in the jet space.

The condition given in (38) is not sufficient for k-determinacy, as Example (26) shows. A simpler example to illustrate this fact is given by $f(x) := x^k$, since $m^k = \langle m \mathcal{J}[f] \rangle$ holds but obviously $[f]$ is not $(k-1)$-determined.

The determinacy of a germ can now be narrowed down to two possible values.

Theorem (39). A germ $[f] \in \mathcal{E}$ is finitely determined if and only if there is a positive integer k for which

$$m^k \subset \langle m\mathcal{J}[f] \rangle$$

holds. If k is the smallest integer with this property, then $\det[f]$ is either k or $k-1$.

Proof. The first assertion results from (25) and (38). Put $d := \det[f]$. Then $m^{d+1} \subset \langle m\mathcal{J}[f] \rangle$, due to (38), and (25) implies $m^{d-1} \not\subset \langle m\mathcal{J}[f] \rangle$.

As illustrated in Exercise 4.13, Lemma (22) can be applied to find out whether the determinacy of a germ is either k or $k-1$. If $d := \det[f]$, then m^d may or may not be contained in the ideal $\langle m\mathcal{J}[f] \rangle$. This is illustrated by $f(x) := x^d$, for which $m^d \subset \langle m\mathcal{J}[f] \rangle$ holds, and Example (28), for which $m^d \not\subset \langle m\mathcal{J}[f] \rangle$ is true. Notice that Example (28) shows that $m^{d+1} \not\subset \langle m^2\mathcal{J}[f] \rangle$ can happen.

Corollary (40). The finite determinacy of a germ $[f] \in \mathcal{E}$ is equivalent to

$$m^k \subset \mathcal{J}[f]$$

for some positive integer k.

Proof. The inclusion $m^k \subset \mathcal{J}[f]$ implies $m^{k+1} \subset \langle m\mathcal{J}[f] \rangle$. On the other hand, $\langle m\mathcal{J}[f] \rangle \subset \mathcal{J}[f]$ holds. Therefore, the assertion follows from (39).

Only functions with isolated critical points can be finitely determined.

Corollary (41). If a germ $[f] \in m^2$ is finitely determined, then the origin is an isolated critical point.

Proof. After (40), there is a $k \in \mathbf{N}$ with $m^k \subset \mathcal{J}[f]$. Thus, for $\nu \in \mathbf{N}_0^n$, $|\nu| = k$, there is an open set U in $\mathbf{R}^n$ containing 0 and functions $h_{\nu,i} \colon U \to \mathbf{R}$, $1 \leq i \leq n$, so that

$$x^\nu = \sum_{i=1}^n h_{\nu,i}(x) D_i f(x) \quad \text{for} \quad x \in U.$$

If $p = (p_1, \ldots, p_n) \in U$ is a critical point of f, then $D_i f(p) = 0$ and hence $(p_i)^k = 0$ for $1 \leq i \leq n$. Consequently, $p = 0$.

Clearly, the converse of (41) is not true, as $f(x) := \exp(-1/|x|)$ for nonzero $x \in \mathbf{R}$ and $f(0) := 0$ shows. For a nonflat counterexample, take $f \colon \mathbf{R}^2 \to \mathbf{R}$, $f(x,y) := (x^2 + y^2)^2$; see Exercise 4.8.

Example (42). The germ of $f \colon \mathbf{R}^2 \to \mathbf{R}$, $f(x,y) := x^2 y$, is not finitely determined.

Proof. The set of all critical points is the y-axis, and the assertion follows from (41).

Another function appearing in Thom's first classification theorem (see (1.12)) can now be treated.

Example (43). The germ of $f: \mathbf{R}^2 \to \mathbf{R}$, $f(x,y) := x^2y + y^4$, has determinacy 4.

Proof. Due to Example (42), $[f]$ is not 3-determined. Moreover,

$$\mathcal{J}[f] = \langle xy, x^2 + 4y^3 \rangle \varepsilon$$

implies

$$\langle m\mathcal{J}[f] \rangle = \langle x^2y, x^3 + 4xy^3, xy^2, x^2y + 4y^4 \rangle \varepsilon.$$

This ideal contains the germs of x^2y, y^4, xy^2, xy^3, and x^3 and hence also the germs of x^4, x^3y, and x^2y^2. Therefore, it encompasses the ideal m^4. An application of (25) proves $\det[f] = 4$.

Example (44). The germ of $f: \mathbf{R}^2 \to \mathbf{R}$, $f(x,y) := -x^2y - y^4$, has determinacy 4 and is not equivalent to $[-f]$.

Proof. Obviously $\det[f] = \det[-f]$ always holds, and $\det[f] = 4$ follows from (43). Assume that $[f]$ is equivalent to $[-f]$. Then there is a local diffeomorphism $\varphi = (\varphi_1, \varphi_2)$ at $0 \in \mathbf{R}^2$ with $\varphi(0) = 0$ satisfying $f(\varphi(x,y)) = x^2y + y^4$ on a neighborhood of the origin in $\mathbf{R}^2$. By (1.25), there are real numbers a_i, b_i and germs $[g_i], [h_i]$ in $\mathcal{E}_1$, $i = 1, 2$, such that $\varphi_i(x,0) = a_ix + x^2g_i(x)$ and $\varphi_i(0,y) = b_iy + y^2h_i(y)$ holds for small x and y. Hence, for $x = 0$ the equation $f(\varphi(x,y)) = x^2y + y^4$ yields

$$-(b_1 + yh_1(y))^2(b_2 + yh_2(y)) - y(b_2 + yh_2(y))^4 = y$$

after both sides are divided by y^3. Putting $y = 0$, it follows that $b_1^2b_2 = 0$ is true. This implies $b_2 = 0$, since $b_1 = 0$ would lead to the contradiction $-b_2^4 = 1$. Evaluating the equation $f(\varphi(x,y)) = x^2y + y^4$ for $y = 0$ yields, similarly, $a_2 = 0$. However, $a_2 = b_2 = 0$ implies $\det D\varphi(0) = a_1b_2 - a_2b_1 = 0$, which is impossible by (1.29).

The following is an immediate consequence of (41).

Corollary (45). If f is independent of one of its variables, and $[f]$ is in m^2, then $[f]$ is not finitely determined.

The hypothesis $[f] \in m^2$ in (45) is necessary, as the projection $f(x_1, \ldots, x_n) := x_1$ shows. Note that functions g of l variables with $g(0) = 0$ that have a critical point at the origin and whose germs are finitely determined induce naturally functions $f(x_1, \ldots, x_n) := g(x_1, \ldots, x_l)$ of n variables for $n > l$ whose germs are not finitely determined by (45).

The last corollary can be used to characterize 2-determinacy.

Corollary (46). A germ $[f] \in m^2$ has determinacy 2 if and only if the origin is a nondegenerate critical point of f.

Proof. As already mentioned after the proof of (14), nondegeneracy implies 2-determinacy by the Morse Lemma (1.37) and by (12).

Now let $[f] \in m^2$ be 2-determined, but assume that 0 is a degenerate critical point of f. According to the Reduction Lemma (3.2) and by (12), it is no restriction to suppose

$$f(x_1, \ldots, x_n) = q_{sr}(x_1, \ldots, x_r) + g(x_{r+1}, \ldots, x_n)$$

for small x where $r < n$, q_{sr} is a normal quadratic form on $\mathbf{R}^r$, and $[g]$ is in $\mathcal{E}_{n-r}$ such that g has rank 0 at the origin. Therefore $T_f^2 = q_{sr}$ holds and $[f] \sim [q_{sr}]$ follows. This is a contradiction by (12), since $[q_{sr}]$ is not finitely determined as a germ in $\mathcal{E}_n$ after (45).

In view of (46), Corollary (41) generalizes statement (1.30).

The remainder of this chapter is devoted to the question of the uniqueness of the **residual singularity**. Recall that according to the Reduction Lemma (3.2) a germ in n variables with a critical point of rank r at the origin is equivalent to the sum of the germ of a nondegenerate quadratic form in r variables and a residual singularity in $n - r$ other variables with a totally degenerate critical point at the origin. It will be shown now that the residual singularity is unique up to equivalence of germs.

Recently, Frank Hofmaier [H] gave an elementary proof of this important result. Previously, only a rough sketch by Thom himself was available in the literature (see [T1]), which uses a result from an extensive theory developed by J.C. Tougeron [Tou]. From the criterion for equivalence (22) one derives first a

Sufficient Criterion for Equivalence in m^3 (47). Let $[f]$ and $[g]$ be two germs in m^3 such that the difference $[f] - [g]$ lies in the ideal $\langle \mathcal{J}[f] \mathcal{J}[f] \rangle$. Then $[f]$ and $[g]$ are equivalent.

Proof. It suffices to show that $[f]$ and $[g]$ fulfill the premises of the criterion (22). By assumption the difference $[h] := [f] - [g]$ has a representation

$$[h] = \sum_{i,j=1}^{n} [h_{ij}][D_i f][D_j f]$$

with some $[h_{ij}] \in \mathcal{E}$. Partial differentiation yields

$$[D_i h] = \sum_{j=1}^{n} [c_{ij}][D_j f]$$

with some $[c_{ij}] \in m$ for $[f]$ is in m^3. In matrix notation, these equations read

$$h(x) = Df(x) H(x) (Df(x))^t \text{ and } (Dh(x))^t = C(x) (Df(x))^t$$

for small x, where gradients are row vectors.

Now fix $s \in [0,1]$ and let E be the identity matrix. Due to $[c_{ij}] \in m$, the matrix $E + (s+t)C(x)$ is invertible for small (x,t). Set

$$\psi(x,t) := Df(x)H(x)\,(E + (s+t)C(x))^{-1}.$$

Then $\psi(0,t) = 0$ and

$$\sum_{i=1}^{n} \psi_i(x,t)\frac{\partial}{\partial x_i}\left(f(x) + (s+t)(g(x) - f(x))\right) =$$

$$\psi(x,t)\left((Df(x))^t + (s+t)C(x)\,(Df(x))^t\right) =$$

$$\psi(x,t)\,(E + (s+t)C(x))\,(Df(x))^t =$$

$$Df(x)H(x)\,(Df(x))^t = h(x) = g(x) - f(x)$$

completes the proof.

The main result follows.

Uniqueness of the Residual Singularity up to Equivalence (48). Let $[f]$ and $[g]$ be in m_n^3 and let $q : \mathbf{R}^m \to \mathbf{R}$, $q(y) = -y_1^2 - \cdots - y_s^2 + y_{s+1}^2 + \cdots + y_m^2$, be the nondegenerate normal quadratic form with index s in m new variables (1.21). Suppose that $[f]+[q]$ and $[g]+[q]$ are equivalent in $\mathcal{E}_{n+m}$. Then $[f]$ and $[g]$ are equivalent in $\mathcal{E}_n$.

Proof. By assumption there is a $[\tau] \in \mathcal{G}_{n+m}$ satisfying

$$g(x) + q(x) = f(\alpha(x,y)) + q(\beta(x,y))$$

for small (x,y), where $\tau(x,y) =: (\alpha(x,y),\beta(x,y))$. Differentiation with respect to y_k yields

$$(*) \quad 2\varepsilon_k y_k = \sum_{i=1}^{n} D_i f(\alpha(x,y))\frac{\partial \alpha_i}{\partial y_k}(x,y) + 2\sum_{j=1}^{m} \varepsilon_j \beta_j(x,y)\frac{\partial \beta_j}{\partial y_k}(x,y)$$

with $\varepsilon_j \in \{-1,1\}$ the signs of q.

By the Implicit Function Theorem (see Appendix A.1) there is a unique smooth map β_0 on an open neighborhood of $0 \in \mathbf{R}^n$ into $\mathbf{R}^n$ such that $\beta_0(0) = 0$ and $\beta(x,\beta_0(x)) = 0$ hold. Indeed, β vanishes at $(0,0)$ and the functional matrix $B := (\beta_{lj})_{1\leq l,j \leq m}$ with $\beta_{lj} := (\partial\beta_j/\partial y_l)(0,0)$ is invertible, since differentiating the equation $(*)$ with respect to y_l yields Kronecker's symbol

$$\delta_{lk} = \sum_{j=1}^{m} \frac{\varepsilon_j}{\varepsilon_k}\beta_{lj}\beta_{kj},$$

whence $B^{-1} = ((\varepsilon_j/\varepsilon_k)\beta_{kj})_{1\leq j,k \leq m}$ for $E = BB^{-1}$.

Now $\tau^{-1}(z.0) =: (\gamma(z), \delta(z))$ is defined on an open neighborhood of $O \in$ $\mathbf{R}^n$. Note that γ and δ are smooth satisfying $\gamma(0) = 0$, $\alpha(\gamma(z), \delta(z)) = z$, and $\beta(\gamma(z), \delta(z)) = 0$. Hence, $\delta(z) = \beta_0(\gamma(z))$. Therefore, $x \mapsto \alpha(x, \beta_0(x))$ is a local inverse of γ, and $[\gamma] \in \mathcal{G}_n$ holds.

The proof is accomplished by showing that the germ $[\tilde{g}] := [g][\gamma]$ is equivalent to $[f]$. The initial equation and (*) yield at $(x, y) = (\gamma(z), \delta(z))$ for small z the equations

$$\tilde{g}(z) + q(\delta(z)) = f(z)$$

and

$$\delta_k(z) = \frac{1}{2\varepsilon_k} \sum_{i=1}^{n} D_i f(z) \frac{\partial \alpha_i}{\partial y_k} (\gamma(z), \delta(z)).$$

Since $q(\delta(z)) = \sum_{j=1}^{m} \varepsilon_j (\delta_j(z))^2$, these equations imply that $[\tilde{g}] - [f]$ lies in $\langle \mathcal{J}[f] \mathcal{J}[f] \rangle$. Thus $[\tilde{g}] \sim [f]$ holds by (47).

According to the Reduction Lemma, a germ with a degenerate critical point at the origin reduces to a germ with rank zero at the origin. Now by (48) the latter is unique up to equivalence of germs. For a slightly more general consequence of (48) see Exercise 4.19.

Exercises

4.1.   Show that the set P^k of all polynomials in $\mathbf{R}[x_1, \ldots, x_n]$ of degree at most k is a real vector space of dimension $N = \binom{n+k}{k}$, and conclude that

$$\dim(m^k/m^{k+1}) = \binom{n+k-1}{k}$$

holds (see (6), (8), (20), (33), and (36)).

4.2.   Show that

$$\langle \langle A \rangle_{\mathcal{A}} \langle B \rangle_{\mathcal{E}} \rangle = \langle AB \rangle_{\mathcal{A}}$$

holds for subsets A and B of $\mathcal{E}$ and an ideal $\mathcal{A}$ of $\mathcal{E}$.

4.3.   Show the following generalization of Exercise 4.2. Let $\mathcal{R}$ be a commutative ring with identity, and let $\mathcal{A}$ be an ideal of $\mathcal{R}$. Then

$$\langle \langle A \rangle_{\mathcal{A}} \langle B \rangle_{\mathcal{R}} \rangle_{\mathcal{R}} = \langle AB \rangle_{\mathcal{A}}$$

holds for subsets A and B of $\mathcal{R}$.

4.4.   Show that $\det[f] = 3$ holds for the function $f: \mathbf{R}^2 \to \mathbf{R}$, $f(x, y) := x^3 - xy^2$, which appears in Thom's first classification theorem.

4.5. According to Lemma (34), the group multiplication $(\xi, \eta) \mapsto \tau := \xi \cdot \eta$ on U^k is given by

$$\tau_{i\nu} := \sum_{0 < |\mu| \le k} \xi_{i\mu} M_{\mu\nu}(\eta),$$

where $M_{\mu\nu}(\eta)$ is a polynomial in η.

(i) Prove the formula

$$M_{\mu\nu}(\eta) = \sum_{S \in \boldsymbol{S}_{\mu\nu}} \frac{\mu!}{S!} \eta^S$$

with

$$\boldsymbol{S}_{\mu\nu} := \left\{ S = (S_{i\lambda})_{\substack{1 \le i \le n \\ 0 < |\lambda| \le k}} : S_{i\lambda} \in \mathbf{N}_0 \text{ with} \right.$$

$$\left. \sum_\lambda S_{i\lambda} = \mu_i, \ \sum_{i,\lambda} S_{i\lambda}\lambda_j = \nu_j \text{ for } 1 \le i, j \le n \right\}.$$

Here, as usual,

$$S! = \prod_{i,\lambda} S_{i\lambda}! \quad \text{and} \quad \eta^S = \prod_{i,\lambda} \eta_{i\lambda}^{S_{i\lambda}}.$$

(ii) Show that the matrices $M(\eta)$ have the following triangular structure:

$$M(\eta) = \begin{pmatrix} M^1(\eta) & & & \\ & M^2(\eta) & & * \\ & & \ddots & \\ 0 & & & M^k(\eta) \end{pmatrix}$$

where $M^l(\eta) := (M_{\mu\nu}(\eta))_{|\mu|=|\nu|=l}$ and where $*$ represents unspecified entries.

(iii) Prove that the submatrices $M^l(\eta)$ depend only on $\eta_{i\lambda}$ for $1 \le i \le n$ and $|\lambda| = 1$ and are homogeneous, i.e., $M^l(\alpha\eta) = \alpha^l M^l(\eta)$ for all real α.

(iv) Verify that the following relations are valid:

$$M^l(\varepsilon) = E_d \quad \text{for} \quad d := \binom{n-1+l}{n-1}$$

and

$$M^l(\eta)M^l(\xi) = M^l(\eta \cdot \xi) \quad \text{for} \quad \eta, \xi \in U^k.$$

4.6.

(a)   Let $[\varphi]$ be in $\mathcal{E}_{n,n}$ with $\varphi(0) = 0$. Show that its **pullback**,

$$\varphi^* : \mathcal{E} \to \mathcal{E}, \quad \varphi^*[f] := [f \circ \varphi],$$

is an algebra homomorphism with $\varphi^*[1] = [1]$. Verify that $[\varphi]$ is in $\mathcal{G}$ if and only if φ^* is surjective.

(b)   Show that $\mathcal{G}$ is a group of algebra isomorphisms of $\mathcal{E}$ leaving the identity invariant, i.e., the pullback of $[\varphi] \in \mathcal{G}$ is an algebra isomorphism and $[1][\varphi] = [1]$ holds. Conclude that

$$m^k[\varphi] = m^k$$

is true for $[\varphi] \in \mathcal{G}$ and all positive integers k.

4.7.   Let c be a curve in J^k based at the k-jet $z := j^k[f]$ of a germ $[f]$ in $\mathcal{E}$ whose image lies in the orbit $z\mathcal{G}^k$. Show that there is a curve C in $\mathcal{G}^k$ based at the identity with $c(t) = zC(t)$ for t sufficiently small.
Hint: Use the results in part (iv) of the proof of the Linearization Lemma (37). In particular, consider the surjective linear map

$$D\varrho_t(\varepsilon) : \mathbf{R}^{n(N-1)} \to T_{c(t)}$$

onto the tangent space to the orbit at the point $c(t)$, and recall that $c'(t)$ is in $T_{c(t)}$, and ε is the identity in U^k.

Show first that a right inverse $r(t)$ of $D\varrho_t(\varepsilon)$ exists depending smoothly on t near the origin. Denote by $l(\xi)$ the derivative at ε of the left translation $\eta \mapsto \xi \cdot \eta$ by ξ. Since this translation is a diffeomorphism on U^k, $l(\xi)$ is a linear bijection of $\mathbf{R}^{n(N-1)}$ onto itself.

Now define C as the solution of the differential equation

$$C'(t) = l(C(t))r(t)c'(t)$$

satisfying the initial condition $C(0) = \varepsilon$. Show that the above equation implies $(c(t) \cdot C(t)^{-1})' = 0$.

4.8.   Show that the germ of $f : \mathbf{R}^2 \to \mathbf{R}$, $f(x,y) := (x^2 + y^2)^2$, is not finitely determined, although the origin is a nonflat, isolated critical point of f. Recall that a germ is finitely determined neither at a flat point nor, by (41), at a nonisolated critical point.

4.9.   Determine all $a \in \mathbf{R}$ for which the germ of $f : \mathbf{R}^2 \to \mathbf{R}$, $f(x,y) := x^4 + ax^2y^2 + y^4$, is finitely determined. Compute $\det[f]$ in the case of finite determinacy.

4.10.   Show for $f : \mathbf{R}^2 \to \mathbf{R}$, $f(x,y) := x^2y + y^2$, that $m^4 \subset \langle m\mathcal{J}[f] \rangle$ holds but also $m^4 \not\subset \langle m^2 with \mathcal{J}[f] \rangle$. Prove $\det[f] = 4$.

4.11. Let k be a nonnegative integer, let a be real, and consider the germ of $f\colon \mathbf{R}^2 \to \mathbf{R}$, $f(x,y) := x^2 y + ay^k$. Find the determinacy of $[f]$ depending on k and a.

4.12. Find the determinacy of the germs $[f]$ in $\mathcal{E}_2$ if

(a) $f(x,y) := x^2 + y^4$.

(b) $f(x,y) := x^3 + y^k$ or $x^3 - y^k$ for $k \in \mathbf{N}$ greater than 2.

(c) $f(x,y) := x^5 + y^5$.

4.13. Let k be an integer greater than 2. Show that the germs $[x^3 + xy^k]$ and $[x^3 - xy^k]$ have determinacy $2k - 2$.
Hint: First show that $\det[x^3 \pm xy^k] \in \{2k-2, 2k-1\}$ by applying (39). Next find a local diffeomorphism at the origin transforming a germ in $[x^3 \pm xy^k] + m^{2k-1}$ to a germ in $[x^3 \pm xy^k + ay^{2k-1}] + m^{2k}$. Finally, apply (22).

4.14. Show that the ideal m^∞ of $\mathcal{E}$ is not finitely generated.

4.15. Let $[g]$ be a germ in m_r^2 and $[h]$ a germ in m_s^2. Consider the germ $[f]$ in m_{r+s}^2 given by $f(x,y) := g(x) + h(y)$ for small x and y. If $[f]$ is k-determined, prove that $[g]$ and $[h]$ are $(k+1)$-determined.

4.16. Let $z := j^k[f]$ be the k-jet of a germ $[f]$ in $\mathcal{E}$ and consider the smooth orbital map

$$\mathcal{G}^k \to J^k, \quad j^k[\varphi] \mapsto z j^k[\varphi];$$

see (36). Its derivative at the identity is a linear map from $\mathbf{R}^{n(N-1)}$ into J^k, and its image $T_z(z\mathcal{G}^k)$ is a linear subspace of J^k, the **tangent space at the point z to the orbit** $z\mathcal{G}^k$. Show that

$$T_z(z\mathcal{G}^k) = j^k(\langle m\mathcal{J}[f]\rangle)$$

holds.

4.17. Let k be a positive integer and let $[f]$ in $\mathcal{E}$ be k-determined. Suppose that $[g]$ in $\mathcal{E}$ and $[\varphi]$ in $\mathcal{G}$ satisfy

$$j^k[g] = j^k[f]j^k[\varphi],$$

which means that $j^k[g]$ lies in the orbit $j^k[f]\mathcal{G}^k$ or, in other words, that $j^k[g]$ is **equivalent to $j^k[f]$ under $\mathcal{G}^k$**. Show that $[f]$ and $[g]$ are equivalent, i.e., that $[g] \in [f]\mathcal{G}$.

4.18. Let $\mathcal{E}_n$ and $\mathcal{E}_m$ be embedded in $\mathcal{E}_{n+m}$ as usual, and denote the first n variables by $x = (x_1, \ldots, x_n)$ and the last m variables by $y = (y_1, \ldots, y_m)$.

Prove that the direct sum

$$\mathcal{E}_{n+m} = \langle m_n^k \rangle \mathcal{E}_{n+m} \oplus \langle x^{\nu} : 0 \le |\nu| < k \rangle \mathcal{E}_m$$

holds for every positive integer k. Note the special case $k = 1$:

$$\mathcal{E}_{n+m} = \langle m_n \rangle \mathcal{E}_{n+m} \oplus \mathcal{E}_m.$$

Hint: Use (1.28).

4.19. Let $[f]$ and $[g]$ be two germs in m_n^3 and let q_1 and q_2 be two nondegenerate (not necessarily normal) quadratic forms in other variables. Then $[f] + [q_1]$ and $[g] + [q_2]$ are equivalent if and only if $[f]$ is equivalent to $[g]$ and $[q_1]$ is equivalent to $[q_2]$. Recall that the latter means that q_1 and q_2 have the same number of variables as well as the same rank and index. Generally speaking, two nonequivalent germs in m_n^3 cannot be made equivalent by adding nondegenerate quadratic forms in new variables.

4.20. Let q be a nondegenerate quadratic form in r variables, and let $[h]$ be a germ in m_s^2. Consider the germ $[f]$ in m_{r+s}^2 given by $f(x, y) := q(x) + h(y)$ for small x and y. Show that

$$\det[f] = \det[h] \ (\le \infty)$$

holds. In particular, the determinacy of a germ in m^2 is equal to the determinacy of any representative of its residual singularity.

Solutions

4.1. Clearly, the dimension N of P^k is equal to the number of elements of the set $\{\nu \in \mathbb{N}_0^n : |\nu| \le k\}$. This number is determined by the following combinatorial consideration: Mark $n + k$ points on a line, and then discard n of these points. A bijective correspondence can be defined as follows between the remaining points and the indicated set of multi-indices ν: Denote the number of points up to the first discarded point by ν_1, and let ν_l be the number of points lying between the $(l-1)$st and the lth discarded point for $l = 2, \ldots, n$.

Since there are exactly $\binom{n+k}{n} = \binom{n+k}{k}$ different ways to choose n out of $n + k$ points, this proves that $N = \binom{n+k}{n}$.

Since m^k/m^{k+1} is isomorphic to the vector space of all homogeneous polynomials of degree k in n variables, its dimension is equal to $\dim P^k - \dim P^{k-1}$, which is

$$\binom{n+k}{k} - \binom{n+k-1}{k-1} = \binom{n+k-1}{n-1} = \binom{n+k-1}{k}.$$

4.2. An element of $\langle A \rangle_{\mathcal{A}} \langle B \rangle_{\mathcal{E}}$ can be written as

$$\left(\sum_i r_i a_i \right) \left(\sum_j s_j b_j \right) = \sum_{i,j} r_i s_j a_i b_j$$

for $r_i \in \mathcal{A}, s_j \in \mathcal{E}, a_i \in A$, and $b_j \in B$. It has the form $\Sigma_k t_k c_k$ with $t_k \in \mathcal{A}$ and $c_k \in AB$. Since $\mathcal{A}$ is also a real vector space, linear combinations of elements $\Sigma_k t_k c_k$ are of the same form. Therefore, the left side of the asserted equation is contained in the right side.

The reverse inclusion holds, since an element $\Sigma_k r_k a_k b_k$ is a sum of terms $(r_k a_k)(1 b_k)$, which obviously lie in $\langle A \rangle_{\mathcal{A}} \langle B \rangle_{\mathcal{E}}$.

4.3. The solution follows quite analogously to that of Exercise 4.2 using the fact that $\mathcal{A}\mathcal{R} = \mathcal{A}$.

4.4. Since $T_f^2 = 0$, $[f]$ is not 2-determined. Obviously, $\mathcal{J}[f] = \langle 3x^2 - y^2, xy \rangle_{\mathcal{E}}$, and therefore

$$\langle m \mathcal{J}[f] \rangle = \langle \{x,y\}\{3x^2 - y^2, xy\} \rangle_{\mathcal{E}}$$
$$= \langle 3x^3 - xy^2, x^2 y, 3x^2 y - y^3, xy^2 \rangle_{\mathcal{E}} = \langle x^3, x^2 y, xy^2, y^3 \rangle_{\mathcal{E}} = m^3$$

by (9). Hence $\det[f] = 3$ follows by (25).

4.5. In order to calculate the group multiplication $\tau = \xi \cdot \eta$, consider

$$\varphi_i(x) := \sum_{0 < |\mu| \leq k} \xi_{i\mu} x^\mu, \qquad \psi_j(x) := \sum_{0 < |\lambda| \leq k} \eta_{j\lambda} x^\lambda.$$

Then

$$\varphi_i(\psi(x)) = \sum_{0 < |\mu| \leq k} \xi_{i\mu} (\psi(x))^\mu \quad \text{and}$$

$$(\psi(x))^\mu = \prod_{j=1}^n \left(\sum_{0 < |\lambda| \leq k} \eta_{j\lambda} x^\lambda \right)^{\mu_j}.$$

Now apply the general formula

$$\left(\sum_{r \in R} \alpha_r \right)^m = \sum_{|\varrho| = m} \frac{m!}{\varrho!} \alpha^\varrho$$

for a finite set R and $\varrho \in \mathbf{N}_0^R$. This yields

$$(\psi(x))^\mu = \prod_{j=1}^n \left(\sum_{|S_j| = \mu_j} \frac{\mu_j!}{S_j!} \eta_j^{S_j} x^{\kappa_j} \right)$$

with $S_j = (S_{j\lambda})_{0<|\lambda|\leq k}$, $S_{j\lambda} \in \mathbf{N}_0$, $|S_j| = \Sigma_\lambda S_{j\lambda}$, $S_j! = \prod_\lambda S_{j\lambda}!$, $\eta_j^{S_j} = \prod_\lambda \eta_{j\lambda}^{S_{j\lambda}}$, and $\kappa_{ji} := \Sigma_\lambda S_{j\lambda}\lambda_i$. Therefore, after rearranging the terms according to powers of x one obtains

$$(\psi(x))^\mu = \sum_{0<|\nu|\leq k} M_{\mu\nu}(\eta)x^\nu$$

with $M_{\mu\nu}(\eta)$ as in (i).

In order to verify (ii), note that for $S \in \boldsymbol{S}_{\mu\nu}$ the relations

$$|\mu| = \sum_{i,\lambda} S_{i\lambda} \quad \text{and} \quad |\nu| = \sum_{i,\lambda} S_{i\lambda}|\lambda|$$

hold. Hence, only in the case $|\nu| \geq |\mu|$ the index set $\boldsymbol{S}_{\mu\nu}$ is not void.

The homogeneity (iii) follows immediately from (i), since $\Sigma_{i,\lambda}S_{i\lambda} = l$. For $|\mu| = |\nu|$ one has $|\lambda| = 1$ if $S_{i\lambda} \neq 0$. Therefore $M^l(\eta)$ depends only on $\eta_{i\lambda}$ for $|\lambda| = 1$.

The relations (iv) are valid, since the analogous relations hold for the matrices M themselves, as shown in the proof of (34). Note that d is the number of elements of the set $\{\mu \in \mathbf{N}_0^n : |\mu| = l\}$. This number is determined by the following combinatorial consideration. Mark $n - 1 + l$ points on a line and then discard $n - 1$ of these points. In this way every possible $\mu \in \mathbf{N}_0^n$ with $|\mu| = l$ is uniquely represented. There are exactly $d = \dbinom{n-1+l}{n-1}$ different ways to choose $n - 1$ out of $n - 1 + l$ points.

4.6.

(a) For $[f], [g] \in \boldsymbol{\mathcal{E}}$, $\alpha \in \mathbf{R}$, and small x, one has $(f+\alpha g)(\varphi(x)) = f(\varphi(x)) + \alpha g(\varphi(x))$, $(fg)(\varphi(x)) = f(\varphi(x))g(\varphi(x))$, and $1(\varphi(x)) = 1$. Let φ^* be surjective. Then there is an $[f_i] \in \boldsymbol{\mathcal{E}}_n$ with $\varphi^*[f_i] = [x_i]$ for $i = 1, \ldots, n$. The map $\psi := (f_1, \ldots, f_n)$ is defined for small x. Clearly, $[\psi]$ lies in $\boldsymbol{\mathcal{E}}_{n,n}$ and satisfies $\psi(\varphi(x)) = x$ for small x. Hence $D\psi(0) \circ D\varphi(0) = id_{\mathbf{R}^n}$ by the chain rule. This implies that $D\varphi(0)$ is invertible, and $[\varphi] \in \boldsymbol{\mathcal{G}}$ follows from (1.29). Conversely, if $[\varphi] \in \boldsymbol{\mathcal{G}}$, then obviously φ^* is surjective.

(b) If ψ is a local inverse of φ at the origin, $f(\varphi(\psi(x))) = f(\psi(\varphi(x))) = f(x)$ is valid for small x. This shows that $[\varphi] \in \boldsymbol{\mathcal{G}}$ acts as an algebra isomorphism on $\boldsymbol{\mathcal{E}}$. Clearly, $[\varphi]$ leaves the identity invariant.

Plainly, $\boldsymbol{\mathcal{E}}[\varphi] = \boldsymbol{\mathcal{E}}$ and $m[\varphi] = m$, since $f(0) = 0$ if and only if $f(\varphi(0)) = 0$. Therefore, $m^{k+1}[\varphi] = \langle mm^k\rangle[\varphi] = \langle m[\varphi]m^k[\varphi]\rangle = \langle mm^k\rangle = m^{k+1}$ by induction on k. Notice that the second equality holds, because φ^* is an algebra isomorphism.

4.7.   In the following, the notations of (37) are used. Since $D\varrho_t(\varepsilon)$ is a surjective linear map onto the r-dimensional space $T_{c(t)}$, there are r linearly independent vectors $a_1, \ldots, a_r$ in $\mathbf{R}^{n(N-1)}$ whose images $D\varrho_t(\varepsilon)a_1, \ldots, D\varrho_t(\varepsilon)a_r$ form a basis of $T_{c(t)}$ for $t = 0$. This they do for sufficiently small t, since $D\varrho_t(\varepsilon)$ depends continuously, even smoothly, on t. Hence, the linear map $r(t) : T_{c(t)} \to \mathbf{R}^{n(N-1)}$ given by $D\varrho_t(\varepsilon)a_i \mapsto a_i$, $i = 1, \dots, r$, is a right inverse of $D\varrho_t(\varepsilon)$ depending smoothly on t near the origin.

Let $\Lambda(\xi): U^k \to U^k$, $\Lambda(\xi)\eta := \xi \cdot \eta$, be the left translation by ξ. Because this is a diffeomorphism, $l(\xi) := D\Lambda(\xi)(\varepsilon)$ is a linear bijection of $\mathbf{R}^{n(N-1)}$ onto itself depending smoothly on $\xi \in U^k$. Its inverse is given by $D\Lambda(\xi^{-1})(\xi)$, which follows immediately from $\Lambda(\xi^{-1})\Lambda(\xi)\eta = \eta$ for all $\eta \in U^k$ by the chain rule.

By (37) the tangent vector $c'(t)$ lies in $T_{c(t)}$. Thus one can consider its image under $r(t)$. According to Appendix A.3, the differential equation

(*) $$C'(t) = l(C(t))r(t)c'(t)$$

has a unique smooth solution C for small t satisfying $C(0) = \varepsilon$.

Now it will be shown that $c(t) = c(0) \cdot C(t)$ for small t. From (*) one obtains

$$c'(t) = D\varrho_t(\varepsilon)D\Lambda(C(t)^{-1})(C(t))C'(t)$$

$$= \frac{d}{ds}\varrho_t\left(\Lambda(C(t)^{-1})C(s)\right)\Big|_{s=t}$$

$$= \frac{d}{ds}\varrho_t(C(t)^{-1} \cdot C(s))\Big|_{s=t}$$

$$= \frac{d}{ds}c(t) \cdot (C(t)^{-1} \cdot C(s))\Big|_{s=t}.$$

The last term is equal to $c(t)M(C(t)^{-1})M(C(t))'$. From

$$0 = \left(M(C(t)^{-1})M(C(t))\right)'$$
$$= M(C(t)^{-1})'M(C(t)) + M(C(t)^{-1})M(C(t))',$$

it follows that $c'(t) = -c(t)M(C(t)^{-1})'M(C(t))$, implying

$$0 = c'(t)M(C(t)^{-1}) + c(t)M(C(t)^{-1})' = \left(c(t)M(C(t)^{-1})\right)'$$

and proving the assertion.

4.8. Since

$$\left(\frac{\partial}{\partial x}\right)^4 f(x, y) = 24 \quad \text{and} \quad Df(x, y) = 4(x^2 + y^2)(x, y) = (0, 0)$$

only for $x = y = 0$, the origin is a nonflat, isolated critical point of f.

Assume that $[f]$ were finitely determined. Then, according to (40), a positive integer k would exist for which $m^k \subset \mathcal{J}[f]$ holds. Since

$$\mathcal{J}[f] = \langle x(x^2 + y^2), y(x^2 + y^2)\rangle_{\mathcal{E}} \subset [x^2 + y^2]\mathcal{E},$$

this would imply $x^k = (x^2 + y^2)g(x, y)$ near the origin in $\mathbf{R}^2$ for some $[g]$ in $\mathcal{E}$. Thus, $g(0,0) = 0$ and $g(x, y) = x^k/(x^2 + y^2)$ for $(x, y) \neq (0, 0)$, which is impossible, because g is smooth.

4.9.   According to Exercise 4.8, $[f]$ is not finitely determined for $a = 2$. The same is true for $a = -2$, which is proved analogously. The latter also follows from (41), since the points (x, x) are all critical.

Now let $a = 2b$ and $|b| \neq 1$. Clearly, $[f]$ is not 3-determined, since $T_f^3 = 0$. As suggested by the proof of (27), the sufficient criterion for determinacy (24) can be applied to prove that $\det[f] = 4$ holds. Toward this purpose, one finds

$$\langle m^2 \mathcal{J}[f] \rangle = \langle \{x^2, xy, y^2\}\{x^3 + bxy^2, bx^2y + y^3\} \rangle_{\mathcal{E}}$$
$$= \langle x^5 + bx^3y^2, bx^4y + x^2y^3, x^4y + bx^2y^3, bx^3y^2 + xy^4,$$
$$x^3y^2 + bxy^4, bx^2y^3 + y^5 \rangle_{\mathcal{E}}.$$

Since $|b| \neq 1$ is true,

$$x^2y^3 = \frac{1}{1 - b^2}(bx^4y + x^2y^3) - \frac{1}{1 - b^2}(x^4y + bx^2y^3)$$

and

$$x^4y = \frac{1}{1 - b^2}(x^4y + bx^2y^3) - \frac{1}{1 - b^2}(bx^4y + x^2y^3)$$

hold. Analogous expressions are obtained for xy^4 and x^3y^2. Therefore,

$$\langle m^2 \mathcal{J}[f] \rangle = \langle x^5, x^4y, x^2y^3, x^3y^2, xy^4, y^5 \rangle_{\mathcal{E}}$$

follows. This is exactly m^5 by (9), and hence (24) is satisfied for $k = 4$.

Summarizing the results, $\det[f] = \infty$ for $|a| = 2$ and $\det[f] = 4$ otherwise.

4.10.   Since $\mathcal{J}[f] = \langle xy, x^2 + 2y \rangle_{\mathcal{E}}$, one obtains

$$\langle m\mathcal{J}[f] \rangle = \langle \{x, y\}\{xy, x^2 + 2y\} \rangle = \langle x^2y, x^3 + 2xy, xy^2, x^2y + 2y^2 \rangle_{\mathcal{E}}$$
$$= \langle x^2y, x^3 + 2xy, y^2 \rangle_{\mathcal{E}}.$$

Clearly, $\langle m\mathcal{J}[f] \rangle$ contains the germs $[y^4]$, $[xy^3]$, $[x^2y^2]$, and $[x^3y]$. The germ of x^4 also lies in $\langle m\mathcal{J}[f] \rangle$, due to $x^4 = (x^3 + 2xy)x - 2x^2y$. Therefore $m^4 \subset \langle m\mathcal{J}[f] \rangle$ follows by (9).

Now it will be shown that m^4 is not contained in $\langle m^2 \mathcal{J}[f] \rangle$. From

$$\langle m^2 \mathcal{J}[f] \rangle = \langle \{x, y\}\{x^2y, x^3 + 2xy, y^2\} \rangle_{\mathcal{E}}$$
$$= \langle x^3y, x^4 + 2x^2y, xy^2, x^2y^2, x^3y + 2xy^2, y^3 \rangle_{\mathcal{E}}$$
$$= \langle x^3y, x^4 + 2x^2y, xy^2, y^3 \rangle_{\mathcal{E}},$$

it is obvious that the germs of y^4, xy^3, x^2y^2, and x^3y lie in $\langle m^2 \mathcal{J}[f] \rangle$. However, it will be shown that $[x^4]$ is not contained in $\langle m^2 \mathcal{J}[f] \rangle$. Assume the contrary. Then there are germs $[g], [h], [j]$, and $[k]$ in $\mathcal{E}_2$ such that

$$x^4 = (x^4 + 2x^2y)g(x, y) + x^3yh(x, y) + xy^2j(x, y) + y^3k(x, y)$$

holds for small x and y. Differentiating this equation at the origin, it follows for $(\partial/\partial x)^4$ that $4! = 4!g(0)$ and for $(\partial/\partial x)^2(\partial/\partial y)$ that $0 = 4g(0)$, which is a contradiction. This proves that $[x^4]$, and hence m^4, is not contained in $\langle m^2 \mathcal{J}[f] \rangle$.

Since $m^4 \not\subset \langle m^2 \mathcal{J}[f] \rangle$ implies $m^3 \not\subset \langle m \mathcal{J}[f] \rangle$, it follows $\det[f] \in \{3, 4\}$ by (39). Which one of these two values actually holds can not be decided by the sufficient criterion for determinacy (24) nor by the necessary condition for determinacy (38), because $\det[f]$ equal to 3 cannot be confirmed by (24) and cannot be discarded by (38). Using a trick, it will be shown that $\det[f] = 4$ is true. Assume that $\det[f]$ were 3. Consider the local diffeomorphism $\varphi = (\varphi_1, \varphi_2)$ at the origin defined by $\varphi_1(x, y) := x, \varphi_2(x, y) := y - \frac{1}{2}x^2$; see (1.29). By means of φ, the germ of f is equivalent to the germ of $\varphi_1(x, y)^2 \varphi_2(x, y) + \varphi_2(x, y)^2 = -\frac{1}{4}x^4 + y^2$, whose determinacy would also be 3 by (12). This, however, implies that $[y^2]$ would be 3-determined in $\mathcal{E}_2$, in contradiction to (45). Consequently, $\det[f] = 4$ follows.

4.11. By Example (42) the germ of f is not finitely determined if $a = 0$. Of course, the same is true for $k = 0$. Let $a \neq 0$ and $k \geq 1$.

(i) For $k = 1$, the origin is not a critical point of f and $\det[f] = 1$ follows from (14).

(ii) Consider the case $k = 2$. Using the linear coordinate transformation $u := |a|^{-1/4}x, v := \varepsilon|a|^{1/2}y$, with ε denoting the sign of a, $[f]$ is equivalent to the germ $[\varepsilon(u^2v + v^2)]$. Since its determinacy obviously does not depend on ε, $\det[f] = 4$ follows from (12) and Exercise 4.10.

(iii) For $k \geq 3$ it will be shown that $\det[f] = k$. The determinacy of $[f]$ is at least k, since otherwise in the case $k = 3$ the contradiction $T_f^2 \neq 0$ would follow, and in the case $k > 3$ the germ $[x^2y]$ would have finite determinacy, contradicting (42). On the other hand, the criterion (25) shows that $[f]$ actually is k-determined, since

$$\langle m \mathcal{J}[f] \rangle = \langle \{x, y\}\{xy, x^2 + aky^{k-1}\} \rangle_{\mathcal{E}}$$
$$= \langle x^2y, x^3 + akxy^{k-1}, xy^2, x^2y + aky^k \rangle_{\mathcal{E}} = \langle x^2y, xy^2, x^3, y^k \rangle_{\mathcal{E}},$$

which obviously contains m^k by (9).

4.12.

(a) Obviously, $\det[f] > 3$, since otherwise $[x^2] \in \mathcal{E}_2$ would be finitely determined, contradicting (45). Due to $\boldsymbol{J}[f] = \langle x, y^3 \rangle_{\mathcal{E}}$, it follows that

$$\langle m \boldsymbol{J}[f] \rangle = \langle x^2, xy^3, xy, y^4 \rangle_{\mathcal{E}} \supset m^4$$

holds, which implies that $[f]$ is 4-determined by (25). Thus, $\det[f] = 4$.

(b)   It will be shown that $\det[f] = k$. Since $[x^3 - y^k] \sim -[x^3 + y^k]$ it suffices to consider $f(x, y) := x^3 + y^k$. One finds

$$\langle m \boldsymbol{J}[f] \rangle = \langle x^3, x^2 y, xy^{k-1}, y^k \rangle_{\mathcal{E}},$$

which contains m^k due to (9). Since $\det[f]$ must be greater than $k - 1$ because of (45), this implies that $\det[f] = k$ by (25).

(c)   Obviously, $[x^3 y^3]$ is not contained in

$$\langle m \boldsymbol{J}[f] \rangle = \langle x^5, xy^4, x^4 y, y^5 \rangle_{\mathcal{E}}.$$

Hence $m^6 \not\subset \langle m \boldsymbol{J}[f] \rangle$, and $[f]$ is not 5-determined by (38). On the other hand,

$$\langle m^2 \boldsymbol{J}[f] \rangle = \langle x^6, x^2 y^4, x^5 y, xy^5, x^4 y^2, y^6 \rangle_{\mathcal{E}}$$

contains m^7, and the sufficient criterion for determinacy (24) yields that $[f]$ is 6-determined. Hence, $\det[f] = 6$.

4.13.

(i) Using (39), it will be shown that $\det[x^3 \pm xy^k]$ is either $2k - 1$ or $2k - 2$. After (9),

$$\boldsymbol{A} := \langle m \boldsymbol{J}[x^3 \pm xy^k] \rangle = \langle x^3, x^2 y^{k-1}, xy^k, 3x^2 y \pm y^{k+1} \rangle$$

contains m^{2k-1}, since the germs of

$$y^{2k-1} = (3x^2 y \pm y^{k+1})(\mp y^{k-2}) \pm 3(x^2 y^{k-1}),$$

$xy^{2k-2} = xy^k(y^{k-2})$, $x^2 y^{2k-3} = x^2 y^{k-1}(y^{k-2})$, and $x^{3+a} y^b = x^3(x^a y^b)$ for $a, b \geq 0$ lie in m^{2k-1}. But m^{2k-2} is not contained in $\boldsymbol{A}$, because obviously $[y^{2k-2}]$ is not in $\boldsymbol{A}$.

(ii) Now it will be shown that $[x^3 \pm xy^k] + [j]$ with $[j]$ in m^{2k-1} is equivalent to $[x^3 \pm xy^k] + a[y^{2k-1}]$ for some real a. Since $[x^3 \pm xy^k]$ is $(2k-1)$-determined after (i), it suffices to suppose that j is a homogeneous polynomial of degree $2k - 1$. Let $j(x, y) = ay^{2k-1} + bxy^{2k-2} + x^2 p(x, y)$, where p is a homogeneous polynomial of degree $2k - 3$ and a, b are real. Then

$$\varphi(x, y) := \left(x - \frac{1}{3} p(x, y), y \mp \frac{b}{k} y^{k-1} \right)$$

is a local diffeomorphism at the origin that transforms $[x^3 \pm xy^k] + [j]$ to $[x^3 \pm xy^k + ay^{2k-1}]$ plus a germ in m^{2k}. This proves (ii), since $[x^3 \pm xy^k]$ is $(2k - 1)$-determined.

(iii) The proof is accomplished by showing that $[x^3 \pm xy^k]$ and $[x^3 \pm xy^k + ay^{2k-1}]$ are equivalent. This will be done with the aid of (22). Indeed it is not too hard to guess that (23) is satisfied for

$$\psi_1(x,y,t) = \frac{ay^{k-1}}{\pm 1 + 3\tau^2 y^{k-2}}, \qquad \psi_2(x,y,t) := \frac{3a}{k} \frac{\tau y^{k-1} \mp x}{\pm 1 + 3\tau^2 y^{k-2}}$$

with

$$\tau := \frac{2k-1}{k} a(s+t).$$

4.14. First it will be shown that $\langle m^\infty \rangle_m = m^\infty$. Clearly, $\langle m^\infty \rangle_m = \langle mm^\infty \rangle$ is contained in m^∞. To verify the reverse inclusion, take $[f] \in m^\infty$. Let $f_1(x) := f(x)/x_1$ for $x_1 \neq 0$ and $f_1(x) := 0$ for $x = 0$. Taylor's formula (1.27) yields $[f_1] \in m^k$ for every $k = 1, 2, \ldots$. Hence, $[f_1] \in m^\infty$ and $[f] \in mm^\infty$.

Now, if m^∞ were finitely generated, Nakayama's Lemma would imply the contradiction that m^∞ is the null ideal; see the remark after the proof of (16).

4.15. According to the necessary condition for determinacy (38), the ideal m_{r+s}^{k+1} is contained in $\langle m_{r+s} with \mathcal{J}[f] \rangle$. Obviously, the ideal $with \mathcal{J}[f]$ of $\mathcal{E}_{r+s}$ is generated by the union of the ideal $with \mathcal{J}[g]$ of m_r and the ideal $with \mathcal{J}[h]$ of m_s. Hence, $[x^\mu]$ for $|\mu| = k+1$ is a sum of terms $[a][j] + [b][k]$ with $[a]$, $[b]$ in m_{r+s}, $[j]$ in $with \mathcal{J}[g]$ and $[k]$ in $with \mathcal{J}[h]$. For $y = 0$ and small x, one obtains x^μ as a sum of functions of the kind $a(x,0)j(x)$ whose germs lie in $m_r with \mathcal{J}[g]$. This proves $m_r^{k+1} \subset \langle m_r with \mathcal{J}[g] \rangle$ by (9), and (25) implies that $[g]$ is $(k+1)$-determined. Analogously, $[h]$ is $(k+1)$-determined.

4.16. Let $j^k[\varphi]$ be an element in $\mathcal{G}^k$. Without restriction, the components of φ can be assumed to be polynomials, i.e.,

$$\varphi_i(x) = \sum_{0 < |\lambda| \leq k} \xi_{i\lambda} x^\lambda.$$

For small x

$$\frac{\partial}{\partial \xi_{j\kappa}} f(\varphi(x)) \bigg|_{\xi=\varepsilon} = \sum_i D_i f(x) \delta_{ij} x^\kappa = x^\kappa D_j f(x).$$

Hence, $T_z(z\mathcal{G}^k)$ is the linear span of $j^k[x^\kappa D_j f]$ for $1 \leq j \leq n$ and $0 < |\kappa| \leq k$. Clearly, this is equal to $j^k(\langle m \mathcal{J}[f] \rangle)$.

4.17. Note that $j^k[f] = j^k[g] \left(j^k[\varphi] \right)^{-1} = j^k[g \circ \psi]$, where ψ denotes a local inverse of $[\varphi]$. Hence, $[f]$ and $[g \circ \psi]$ are k-equivalent. Since $[f]$ is k-determined, $[f]$ and $[g \circ \psi] = [g][\psi]$ are eqivalent. This implies the assertion.

4.18. For $[f] \in \mathcal{E}_{n+m}$ set

$$S_f^k(x,y) := \sum_{0 \leq |\nu| \leq k} \frac{1}{\nu!} D_1^\nu f(0,y) x^\nu$$

with

$$D_1^\nu := \left(\frac{\partial}{\partial x_1}\right)^{\nu_1} \cdots \left(\frac{\partial}{\partial x_n}\right)^{\nu_n}.$$

Then $R(x,y) := f(x,y) - S_f^{k-1}(x,y)$ satisfies $D_1^\mu R(0,y) = 0$ for small y if $0 \le |\mu| \le k-1$. Hence (1.28) yields

$$R(x,y) = \sum_{i_1,\ldots,i_k=1}^{n} g_{i_1\ldots i_k}(x,y)x_{i_1}\ldots x_{i_k}$$

for some $[g_{i_1\ldots i_k}] \in \mathcal{E}_{n+m}$, and $[R]$ lies in $\langle m_n^k\rangle\mathcal{E}_{n+m}$. Since $[S_f^{k-1}]$ is in $\langle x^\nu : 0 \le |\nu| < k\rangle\mathcal{E}_m$, the decomposition

$$\mathcal{E}_{n+m} = \langle m_n^k\rangle\mathcal{E}_{n+m} + \langle x^\nu : 0 \le |\nu| < k\rangle\mathcal{E}_m$$

follows. Due to $\langle m_n^k\rangle\mathcal{E}_{n+m} \cap \langle x^\nu : 0 \le |\nu| < k\rangle\mathcal{E}_n = \{0\}$ the proof is completed.

4.19. For the nontrivial implication of the assertion, assume that $[f] + [q_1]$ and $[f] + [q_2]$ are equivalent. By linear equivalence transformations of these germs affecting only q_1 and q_2, the latter become normal quadratic forms. Since equivalent germs have the same number of variables as well as the same rank and index, it follows from the remark after (3.1) that the normal forms of q_1 and q_2 are the same. Now the assertion holds by (48).

4.20. Note that it suffices to show that $[h]$ is k-determined if $[f]$ is k-determined and vice versa.

First the special case is treated that $[h]$ is in m_s^3.

Let $[f]$ be k-determined and let a germ $[\tilde{h}]$ in $\mathcal{E}_s$ be k-equivalent to $[h]$. Then $[\tilde{h}]$ is in m_s^3. Since the germs $[f]$ and $[q]+[\tilde{h}]$ are k-equivalent, they are equivalent. Exercise 4.19 yields the equivalence of $[h]$ and $[\tilde{h}]$. Thus $[h]$ is k-determined.

Conversely, if $[h]$ is k-determined and $[\tilde{f}]$ in $\mathcal{E}_{r+s}$ is k-equivalent to $[f]$, then $[\tilde{f}]$ is equal to $[q] + [\tilde{h}]$ where $[\tilde{h}]$ in $\mathcal{E}_s$ is k-equivalent to $[h]$. It follows that $[\tilde{h}]$ is equivalent to $[h]$ and hence $[\tilde{f}]$ equivalent to $[f]$. Thus $[f]$ is k-determined.

In the general case, due to the Reduction Lemma (3.2) it is no restriction to suppose that $[h]$ is the sum $[p]+[g]$ of the germ of a nondegenerate quadratic form p and a germ $[g]$ of rank zero in the remaining variables. Then $[f] = [q]+[p]+[g]$ holds and—applying the previous result twice—one obtains $\det[f] = \det[g] = \det([p] + [g]) = \det[h]$.

Chapter 5

Codimension

The codimension of a germ of a smooth function is a notion that plays a central role in Catastrophe Theory. Thom's first theorem (6.1) classifies the degenerate critical points of functions whose germs are at most 4-codimensional. The codimension gives exactly the number of parameters of the so-called universal folding of such germs—a concept treated in Chapter 7. Here it is shown that equivalent germs have the same codimension and that finite determinacy means having a finite codimension. Then two lower bounds for the codimension are proved. One is with respect to the determinacy and the other is in terms of the corank. The lower bound in terms of the corank is sharpened in Exercise 5.10.

Recall that ideals of $\mathcal{E}$ are vector spaces and that the origin is a critical point of a germ $[f]$ in $\mathcal{E}$ if and only if the Jacobi ideal $\mathcal{J}[f]$ of $[f]$ lies in the maximal ideal m of $\mathcal{E}$.

Definition. Let $[f]$ be a germ in m^2. Then

$$\mathrm{cod}[f] := \dim m/\mathcal{J}[f]$$

is called the **codimension** of $[f]$.

Thus, $\mathrm{cod}[f]$ is a nonnegative integer or infinity. Attention should be drawn to the fact that the use of the term codimension is not unified. Some authors refer to $\dim \mathcal{E}/\mathcal{J}[f]$ as the codimension of $[f]$. In the literature, for $[f] \in \mathcal{E}$, $\dim \mathcal{E}/\mathcal{J}[f]$ is also called the **Milnor number** of $[f]$. Note that

(1) $$\mathrm{cod}[f] = \dim \mathcal{E}/\mathcal{J}[f] - 1$$

holds, because $\dim \mathcal{E}/m = 1$. Some examples illustrating this notion will be given now.

Example (2). Let k be an integer with $k \geq 2$ and let a be a nonzero real number. Then the germ of ax^k is in m_1^2, and $\mathrm{cod}[ax^k] = k - 2$ holds. A basis of $m_1/\mathcal{J}[ax^k]$ is given by

$$[x] + \mathcal{J}[ax^k], \quad [x^2] + \mathcal{J}[ax^k], \ldots, [x^{k-2}] + \mathcal{J}[ax^k].$$

Proof. $\mathcal{J}[ax^k] = \langle x^{k-1}\rangle_{\mathcal{E}_1} = m_1^{k-1}$ by (4.9). An application of (4.8) and (4.6) yields $\dim \mathcal{E}_1/m_1^{k-1} = k - 1$ and the linear independence of the indicated cosets in $m_1/\mathcal{J}[ax^k]$ from which the assertion follows.

An example of a function whose germ has infinite codimension is the function $f: \mathbf{R}^2 \to \mathbf{R}$, $f(x,y) := x^2$. This is proved by applying the following lemma.

Lemma (3). Let $n \geq 2$. If f is independent of one of its variables and if $[f]$ is in m^2, then $[f]$ has an infinite codimension.

Proof. Let f be independent of the nth variable x_n. It suffices to show that the set
$$\{[x_n^k] + \mathcal{J}[f]: k \in \mathbf{N}\}$$
is linearly independent in $m/\mathcal{J}[f]$. Notice that $\mathcal{J}[f] \subset \langle m_{n-1}\rangle_{\mathcal{E}_n}$ holds. Therefore, the only polynomial $p \in \mathbf{R}[x_n]$ whose germ lies in $\mathcal{J}[f]$ is the null polynomial. This proves the assertion.

Before we show that equivalent germs have the same codimension, note that if $[\varphi] \in \mathcal{G}$ is the germ of a local diffeomorphism at the origin with $\varphi(0) = 0$, then

(4) $\mathcal{J}([f][\varphi]) = (\mathcal{J}[f])[\varphi]$

holds for $[f] \in \mathcal{E}$, where
$$(\mathcal{J}[f])[\varphi] := \{[h][\varphi]: [h] \in \mathcal{J}[f]\}.$$
Indeed, for $[g] := [f][\varphi]$ the chain rule implies
$$D_i g = \sum_{j=1}^{n}(D_j f) \circ \varphi \, D_i \varphi_j.$$

Let $[\psi] := [\varphi]^{-1}$. Then $D_i\varphi_j = (D_i\varphi_j) \circ \psi \circ \varphi$ yields $\mathcal{J}[g] \subset (\mathcal{J}[f])[\varphi]$. Analogously, $\mathcal{J}[f] \subset (\mathcal{J}[g])[\psi]$ holds and $(\mathcal{J}[f])[\varphi] \subset \mathcal{J}[g]$ follows.

Theorem (5). Suppose that $[f]$ is a germ in m^2. If $[g] \in \mathcal{E}$ is equivalent to $[f]$, then $\mathrm{cod}[f] = \mathrm{cod}[g]$.

Proof. Let $[\varphi]$ be in $\mathcal{G}$ such that $[g] = [f][\varphi]$ holds. The algebra isomorphism $\mathcal{E} \to \mathcal{E}$, $[h] \mapsto [h][\varphi]$, obviously leaves m invariant, inducing an algebra isomorphism $m \to m$. Because of (4), the ideal $\mathcal{J}[f]$ is mapped onto $\mathcal{J}[g]$. Hence $m/\mathcal{J}[f] \to m/\mathcal{J}[g]$, $[h] + \mathcal{J}[f] \mapsto [h][\varphi] + \mathcal{J}[g]$, is an algebra isomorphism.

Note that (5) includes the case of infinite codimension.

Now, if q_{sr}, $0 \leq s \leq r \leq n$, denotes the normal quadratic form on $\mathbf{R}^n$ defined in (1.21), then its germ is clearly in m^2. When $r < n$ holds, then $\mathrm{cod}[q_{sr}] = \infty$ follows from (3). The converse is also true, because if $r = n$, then $\mathcal{J}[q_{sn}] = m$, and hence, $\mathrm{cod}[q_{sn}] = 0$. In particular, $\mathrm{cod}[q_{sr}] = 0$ is valid if and only if q_{sr} is nondegenerate. This statement will now be generalized.

Theorem (6). Let $[f]$ be a germ in m^2. Then $\mathrm{cod}[f] = 0$ holds if and only if the origin is a nondegenerate critical point of f and hence, by the Morse Lemma if and only if f is equivalent to the germ of a nondegenerate quadratic form.

Proof. Obviously, $\mathrm{cod}[f] = 0$ is equivalent to $m = \mathcal{J}[f]$. Now if $\mathrm{cod}[f] = 0$ is valid, then m^3 can be written as $\langle m^2 \mathcal{J}[f] \rangle$, and $[f]$ is 2-determined by (4.24). Thus 0 is nondegenerate by (4.46). Conversely, the Morse Lemma (1.37) implies that $[f]$ is equivalent to the germ of a nondegenerate normal quadratic form q_{sn}. Then the assertion follows from Theorem (5) and the remark preceding this theorem.

Notice that a germ $[f]$ in m^2 is 0-codimensional precisely when it is 2-determined according to Theorem (6) and (4.46). A more general relationship between the codimension and the determinacy of a germ can be concluded from the following theorem.

Theorem (7). Let $\mathcal{A}$ be an ideal of $\mathcal{E}$. Then the factor space $\mathcal{E}/\mathcal{A}$ is finite dimensional if and only if there is a nonnegative integer k satisfying $m^k \subset \mathcal{A}$. If k is the smallest nonnegative integer with this property, then

$$\infty > c_0 > c_1 > \cdots > c_{k-1} > c_k = 0$$

holds for

$$c_j := \dim(\mathcal{A} + m^j)/\mathcal{A}.$$

Proof. The inclusion $m^k \subset \mathcal{A}$ implies $\dim \mathcal{E}/\mathcal{A} \leq \dim \mathcal{E}/m^k < \infty$ by (4.8). To prove the converse, let $c_0 := \dim \mathcal{E}/\mathcal{A}$ be finite. The decreasing sequence of subspaces of $\mathcal{E}$,

$$\mathcal{E} \supset \mathcal{A} + m \supset \mathcal{A} + m^2 \supset \cdots \supset \mathcal{A} + m^{j-1} \supset \mathcal{A} + m^j \supset \cdots \supset \mathcal{A},$$

yields

$$\infty > c_0 \geq c_1 \geq c_2 \geq \cdots \geq c_{j-1} \geq c_j \geq \cdots.$$

This implies that there is a $k \in \mathbf{N}_0$ with $\mathcal{A} + m^k = \mathcal{A} + m^{k+1}$, because otherwise (c_j) would be an infinite strictly decreasing sequence of nonnegative integers, which is absurd. But this equation is equivalent to $m^k \subset \mathcal{A}$ by (4.18), a consequence of Nakayama's Lemma, and $m^k \subset \mathcal{A}$ holds if and only if $c_k = 0$.

Notice that $c_0 = \dim \mathcal{E}/\mathcal{A}$ in Theorem (7). If $\mathcal{A}$ is the Jacobi ideal $\mathcal{J}[f]$ of a germ $[f]$ in m^2, then $c_0 - 1 = c_1 = \mathrm{cod}[f]$. The following corollary is an immediate consequence of (7) and (4.40).

Corollary (8). A germ $[f] \in m^2$ has finite codimension if and only if it is finitely determined.

Corollary (9). Let $[f]$ be a germ in m^2. Then $\det[f] \le \operatorname{cod}[f] + 2$ holds.

Proof. Let $c_1 := \operatorname{cod}[f]$ be finite. Due to (7), there is a smallest positive integer k satisfying $m^k \subset \mathcal{J}[f]$, and $c_1 \ge k - 1$ holds. Hence $m^{k+1} \subset \langle m\mathcal{J}[f]\rangle$, which by (4.25) implies $\det[f] \le k + 1$, showing the assertion.

Note that for functions f of one variable, equality in (9) holds, i.e., $\det[f] = \operatorname{cod}[f] + 2$ for $[f] \in m_1^2$. This follows from (4.13) and (1.34), using (2) and (5). For functions f of n variables with a nondegenerate critical point at the origin and $f(0) = 0$, one also has $\det[f] = \operatorname{cod}[f] + 2$. This follows because (4.46) implies $\det[f] = 2$, and (6) yields $\operatorname{cod}[f] = 0$. Notice that for f as in (3), $\det[f] = \operatorname{cod}[f] = \infty$ is valid for (3) and (8).

In the next three examples, those polynomials in two variables appearing in Thom's first classification theorem (6.1) (see also (1.12)) are treated. It turns out that their codimension and determinacy coincide; recall that the determinacy was calculated in (4.26) and (4.43). The following remark is used in these examples to calculate the codimension: If k and l are integers with $0 \le k \le l$, then

$$(10) \qquad m^k = \langle x^\nu : k \le |\nu| \le l \rangle \oplus m^{l+1}$$

holds. To see this, observe that the right-hand side of this equation is contained in the left-hand side by (4.9). To verify the opposite inclusion, let $[f]$ be in m^k. Then the germ of the remainder $R := f - T_f^l$ is in m^{l+1} by Taylor's formula (1.27), whereas the germ of the lth Taylor polynomial T_f^l of f at the origin is a linear combination of monomials $[x^\nu]$, $k \le |\nu| \le l$. Clearly, the sum is direct.

Example (11). The germ of $f \colon \mathbf{R}^2 \to \mathbf{R}$, $f(x,y) := x^3 + y^3$, has codimension 3. A basis of $m/\mathcal{J}[f]$ is given by the cosets represented by $[x]$, $[y]$, and $[xy]$.

Proof. Clearly, $\mathcal{J}[f] = \langle x^2, y^2 \rangle_{\mathcal{E}}$ holds. Then $m = \langle x, y, xy \rangle + \mathcal{J}[f]$ is valid, since $m = \langle x, y, x^2, xy, y^2 \rangle + m^3$ follows from (10) and $\langle x^2, y^2 \rangle + m^3 \subset \mathcal{J}[f]$ from (4.9). This implies that the cosets determined by $[x]$, $[y]$, and $[xy]$ span the vector space $m/\mathcal{J}[f]$. To see their linear independence, consider the equation

$$\alpha x + \beta y + \gamma xy + g(x,y)x^2 + h(x,y)y^2 = 0$$

around the origin in $\mathbf{R}^2$ for real α, β, γ, and $[g]$, $[h]$ in $\mathcal{E}$. The germ of $\alpha x + \beta y$ has to lie in m^2, since the germs of the other summands do. Thus, α and β are zero. After evaluating the partial derivative $\partial^2/\partial x \partial y$ at the origin, $\gamma = 0$ follows.

Example (12). The germ of $f \colon \mathbf{R}^2 \to \mathbf{R}$, $f(x,y) := x^3 - xy^2$, is 3-codimensional. A basis of $m/\mathcal{J}[f]$ is given by the cosets represented by $[x]$, $[y]$, and $[x^2 + y^2]$.

Proof. Since $\mathcal{J}[f] = \langle 3x^2 - y^2, xy \rangle_{\mathcal{E}}$, the equality $m = \langle x, y, x^2 + y^2 \rangle + \mathcal{J}[f]$ holds. This follows from $m = \langle x, y, x^2, xy, y^2 \rangle + m^3$ by (10), from the obvious inclusion $m^3 \subset \mathcal{J}[f] \subset m$, and because $\langle x^2, y^2 \rangle = \langle x^2 + y^2, 3x^2 - y^2 \rangle$ is true. Therefore the cosets represented by $[x]$, $[y]$, and $[x^2 + y^2]$ span $m/\mathcal{J}[f]$. They also form a basis, as can be seen by considering the equation

$$\alpha x + \beta y + \gamma(x^2 + y^2) + g(x, y)(3x^2 - y^2) + h(x, y)xy = 0$$

around the origin in $\mathbf{R}^2$, where α, β, and γ are real and $[g]$, $[h]$ are in $\mathcal{E}$. As in the proof of (11), $\alpha = \beta = 0$ follows. Now differentiating this simplified equation twice with respect to x at the origin gives $2\gamma + 6g(0, 0) = 0$ and with respect to y gives $2\gamma - 2g(0, 0) = 0$, implying $\gamma = 0$.

Example (13). The germ of $f \colon \mathbf{R}^2 \to \mathbf{R}$, $f(x, y) := x^2 y + y^4$, has $\mathrm{cod}[f] = 4$. A basis of $m/\mathcal{J}[f]$ is given by the cosets of the germs of x, y, x^2, and y^2.

Proof. Obviously, $\mathcal{J}[f] = \langle xy, x^2 + 4y^3 \rangle_{\mathcal{E}}$ so that $m = \langle x, y, x^2, y^2 \rangle + \mathcal{J}[f]$ is valid. To see this, notice that (10) implies

$$m = \langle x, y, x^2, xy, y^2, x^3, x^2 y, xy^2, y^3 \rangle + m^4.$$

Moreover, $m^4 \subset \mathcal{J}[f]$ is true, since, e.g., the germs of $x^4 = (x^2 + 4y^3)x^2 - xy(4xy^2)$ and $y^4 = (x^2 + 4y^3)(y/4) - xy(x/4)$ lie in $\mathcal{J}[f]$. Finally, it is not hard to verify $\langle x^3, x^2 y, xy^2, y^3 \rangle \subset \langle x^2 \rangle + \mathcal{J}[f]$. From this the claim involving m follows, which shows that the cosets of $[x]$, $[y]$, $[x^2]$, and $[y^2]$ span the quotient space $m/\mathcal{J}[f]$. Their linear independence is shown by considering the equation

$$\alpha x + \beta y + \gamma x^2 + \delta y^2 + g(x, y)xy + h(x, y)(x^2 + 4y^3) = 0$$

for real $\alpha, \beta, \gamma, \delta$, and $[g]$, $[h]$ in $\mathcal{E}$, and arguing similarly as in the previous two examples.

In the remaining part of this chapter a lower bound for the codimension of a germ $[f]$ in m^2 will be proved employing the **corank of f at the origin**, denoted by $\mathrm{cor}[f]$. This will be a marked improvement in comparison to the estimation given in (9) using the determinacy of $[f]$, because $\det[f]$ is considerably more difficult to calculate than $\mathrm{cor}[f]$.

Recall that $\mathrm{cor}[f]$ is the dimension of the null space of the Hessian matrix $D^2 f(0)$ of f at the origin; see Chapter 3. Note also that when $[f]$ is in m^2, then by Theorem (6), $\mathrm{cor}[f] = 0$ if and only if $\mathrm{cod}[f] = 0$. The following three lemmas will be used to prove a general relationship between the corank and the codimension of a germ $[f] \in m^2$. The proof of the first lemma is another application of the Existence Theorem of Solutions of a Differential Equation quoted in Appendix A.3.

Lemma (14).   Let $[h_i]$ be in $\mathcal{E}$ for $i = 1, \ldots, n$, and consider the differential operator $X : \mathcal{E} \to \mathcal{E}$ defined by

$$X[f] := \sum_{i=1}^{n} [h_i][D_i f].$$

If $(h_1(0), \ldots, h_n(0)) \neq 0 \in \mathbf{R}^n$, then a $[\varphi]$ in $\mathcal{G}$ exists such that $(X[f])[\varphi] = [D_1(f \circ \varphi)]$ holds for all $[f] \in \mathcal{E}$.

Proof. By applying the chain rule to $D_1(f \circ \varphi)$ and specializing $[f]$ to $[x_j]$ for $j = 1, \ldots, n$, the assertion is seen to be equivalent to finding a local diffeomorphism $\varphi = (\varphi_1, \ldots, \varphi_n)$ at 0 with $\varphi(0) = 0$ satisfying

$$D_1 \varphi_i = h_i \circ \varphi$$

for $i = 1, \ldots, n$ on an open neighborhood of the origin. Let $h := (h_1, \ldots, h_n)$ and let each h_i be defined on an open neighborhood U of the origin in $\mathbf{R}^n$. Consider the differential equation

(*) $$\frac{\partial}{\partial t} \Phi(x, t) = h(\Phi(x, t))$$

with the initial condition $\Phi(x, 0) = x$. The existence theorem in Appendix A.3 ensures the existence of an open set V in U with $0 \in V$ and an open interval I in $\mathbf{R}$ containing 0 such that for every $x \in V$ there is a unique solution $I \to \mathbf{R}^n$, $t \mapsto \Phi(x, t)$, of (*) with $\Phi(x, t) \in U$ as well as $\Phi(x, 0) = x$ and such that $\Phi : V \times I \to \mathbf{R}^n$ is smooth.

Let k satisfy $h_k(0) \neq 0$ and define

$$\varphi(x_1, \ldots, x_n) := \Phi((x_k, x_2, \ldots, x_{k-1}, 0, x_{k+1}, \ldots, x_n), x_1),$$

where for $k = 1$ one has $\varphi(x_1, \ldots, x_n) = \Phi((0, x_2, \ldots, x_n), x_1)$. Evidently, $\varphi(0) = 0$ holds, and $D_1 \varphi = h \circ \varphi$ follows from (*). It remains to verify that $D\varphi(0)$ is invertible. By permuting the first and the kth row, $D\varphi(0)$ becomes a matrix with nonzero entries only in the first column and in the diagonal. The diagonal elements are $h_k(0), 1, \ldots, 1$, implying $\det D\varphi(0) = (-1)^{k+1} h_k(0) \neq 0$.

Lemma (15).   Suppose that $[f]$ is a germ in m^2. Then

(16) $$\mathcal{J}[f] = \langle m\mathcal{J}[f] \rangle + \mathbf{R}[D_1 f] + \cdots + \mathbf{R}[D_n f],$$

and this sum is direct when $\mathrm{cod}[f] < \infty$ holds.

Proof. The right side of the equation (16) is obviously contained in the left side. To verify the converse, let $[g]$ be in $\mathcal{J}[f]$, which means that

$$[g] = \sum_{i=1}^{n} [h_i][D_i f] \quad \text{with } [h_i] \in \mathcal{E}.$$

Then

$$[g] = \sum_{i=1}^{n} [\tilde{h}_i][D_i f] + \sum_{i=1}^{n} h_i(0)[D_i f]$$

for $\tilde{h}_i := h_i - h_i(0)$, proving Equation (16).

Now let $\mathrm{cod}[f] < \infty$. The sum in Equation (16) is direct, if

$$\sum_{i=1}^{n} [h_i][D_i f] = 0, \quad [h_i] \in \mathcal{E},$$

implies $h_i(0) = 0$ for all $i = 1, \ldots, n$. Assuming the contrary, Lemma (14) shows that a $[\varphi]$ in $\mathcal{G}$ exists such that

$$0 = \sum_{i=1}^{n} [h_i \circ \varphi][(D_i f) \circ \varphi] = [D_1(f \circ \varphi)]$$

holds. But then $f \circ \varphi$ does not depend on x_1. The contradiction $\mathrm{cod}[f] = \infty$ follows from (3) and (5).

An immediate result of Lemma (15) for $[f] \in m^2$ using (10) is

(17) $\mathrm{cod}[f] = \dim(m/\langle m\boldsymbol{\mathcal{J}}[f]\rangle) - n = \dim(m^2/\langle m\boldsymbol{\mathcal{J}}[f]\rangle).$

Recall that when $[f]$ is in m^2 and $k := \mathrm{cor}[f] \geq 1$, then the origin is a degenerate critical point of f, and f is reducible there to a function g of just k variables for which the origin is a totally degenerate critical point by the Reduction Lemma (3.2). In the following, $\mathcal{E}_{n-k}$ (respectively $\mathcal{E}_k$) is naturally embedded in $\mathcal{E}_n$ by considering a function of $n - k$ (respectively k) variables as being a function of n variables that does not depend on the last k (first $n - k$) variables.

Lemma (18). Let $[f]$ be a germ in m_n^2 such that $k := \mathrm{cor}[f]$ is not zero. Denote by $[g] \in m_k^3$ the germ of the function of k variables to which f can be reduced by (3.2). Then $\mathrm{cod}[f] = \mathrm{cod}[g]$ holds.

Proof. Let q be a normal quadratic form on $\mathbf{R}^{n-k}$ with the property that $[q] + [g] \sim [f]$; see (3.1) and (3.2). It suffices to prove $\mathrm{cod}([q] + [g]) = \mathrm{cod}[g]$ due to (5). In $\boldsymbol{\mathcal{J}}([q] + [g]) = \langle \boldsymbol{\mathcal{J}}[q]\rangle_{\mathcal{E}_n} + \langle \boldsymbol{\mathcal{J}}[g]\rangle_{\mathcal{E}_n}$ the first summand is equal to $\langle m_{n-k}\rangle_{\mathcal{E}_n}$. The second summand can be replaced by the ideal $\boldsymbol{\mathcal{J}}[g]$ in $\mathcal{E}_k$, due to $\mathcal{E}_n = \langle m_{n-k}\rangle_{\mathcal{E}_n} + \mathcal{E}_k$. To verify this equation, let $[h]$ be in $\mathcal{E}_n$ and define $\tilde{h}(x,y) := h(x,y) - h(0,y)$ for sufficiently small (x,y) in $\mathbf{R}^{n-k} \times \mathbf{R}^k$. Then $[h - \tilde{h}]$ lies in $\mathcal{E}_k$. Since $\tilde{h}(0,y) = 0$, (1.28) can be applied, yielding $[\tilde{h}] \in \langle m_{n-k}\rangle_{\mathcal{E}_n}$ (see also Exercise 4.18). Thus,

$$\boldsymbol{\mathcal{J}}([q] + [g]) = \langle m_{n-k}\rangle_{\mathcal{E}_n} + \boldsymbol{\mathcal{J}}[g]$$

follows.

The linear map

$$m_n \to m_k/\mathcal{J}[g], \quad [h] \mapsto [h - \tilde{h}] + \mathcal{J}[g]$$

is surjective. Its kernel is $\mathcal{J}([q] + [g])$, which follows from the above considerations. This implies $\mathrm{cod}([q] + [g]) = \mathrm{cod}[g]$.

In other words, the codimension of a germ in m^2 is equal to the codimension of any representative of its residual singularity. A lower bound for the codimension of a germ in m^2 using its corank can now be shown.

Theorem (19). If $[f]$ is a germ in m^2 and $k := \mathrm{cor}[f]$, then

$$\mathrm{cod}[f] \geq \tfrac{1}{2}k(k+1)$$

holds.

Proof. The cases $k = 0$ and $\mathrm{cod}[f] = \infty$ are trivial. Therefore, suppose $k \geq 1$ and $\mathrm{cod}[f] < \infty$. By (18), there is a $[g] \in m_k^3$ with $\mathrm{cod}[f] = \mathrm{cod}[g]$. Let $\mathcal{A}$ be the ideal $\langle m_k \mathcal{J}[g] \rangle$ in $\mathcal{E}_k$. Then $\mathrm{cod}[g] = \dim \mathcal{E}_k/\mathcal{A} - k - 1$ after (17). Now $\mathcal{J}[g] \subset m_k^2$ implies $\mathcal{A} \subset m_k^3$, yielding $\dim \mathcal{E}_k/\mathcal{A} \geq \dim \mathcal{E}_k/m_k^3$. Due to (4.8), $\dim \mathcal{E}_k/m_k^3 = \binom{k+2}{2}$ holds, and the assertion follows.

Exercise 5.10 brings an improvement of (19). In the next chapter, germs whose codimension is at most 4 will be classified. Because of (19), this means that only germs of corank 1 or 2 must be considered.

Exercises

5.1. Compute the codimension of the germ of $f \colon \mathbf{R}^2 \to \mathbf{R}$, $f(x,y) := x^2 y + ay^k$, where a is a real number and $k \geq 2$ is an integer. Find a basis of $m/\mathcal{J}[f]$ when a is not zero.

5.2. Find a basis of the subspace $\{[p] + \mathcal{J}[f] \colon p \in \mathbf{R}[x,y]\}$ of $\mathcal{E}/\mathcal{J}[f]$ for $f \colon \mathbf{R}^2 \to \mathbf{R}$, $f(x,y) := (x^2 + y^2)^2$.

5.3. Compute the codimension of the germ of $f \colon \mathbf{R}^2 \to \mathbf{R}$, $f(x,y) := x^4 + ax^2 y^2 + y^4$, for real a. Find a basis of $m/\mathcal{J}[f]$ when $|a|$ is not 2.

5.4. Find a basis of $m/\mathcal{J}[f]$ for $f \colon \mathbf{R}^2 \to \mathbf{R}$ with

(a) $f(x,y) := x^2 + y^4$ (b) $f(x,y) := x^3 + y^k$ or $x^3 - y^k$ and $k \geq 2$
(c) $f(x,y) := x^5 + y^5$ (d) $f(x,y) := x^3 + xy^k$ or $x^3 - xy^k$ and $k \geq 2$.

In (d) conclude that $\mathrm{cod}[f + g] \geq 2k - 1$ holds if $[g]$ is a germ in m^{k+2}.

5.5. Let $\mathcal{A}$ be an ideal in $\mathcal{E}$ with $0 < \dim(\mathcal{E}/\mathcal{A}) < \infty$. Show that there are integers k, $l \geq 1$ satisfying

$$m^k \subset \mathcal{A} \subset m^l \quad \text{and} \quad m^{k-1} \not\subset \mathcal{A} \not\subset m^{l+1}.$$

5.6. Let l be a nonnegative integer. Show that m_1^l is the unique ideal $\mathcal{A}$ in $\mathcal{E}_1$ satisfying $\dim(\mathcal{E}_1/\mathcal{A}) = l$.

5.7. Two ideals $\mathcal{A}$ and $\mathcal{B}$ of $\mathcal{E}$ will be defined to be equivalent if there is a $[\varphi] \in \mathcal{G}$ such that $\mathcal{A} = \mathcal{B}[\varphi]$ holds; see also Exercise 4.6. Up to equivalence find all ideals $\mathcal{A}$ of $\mathcal{E}_2$ satisfying $\dim(\mathcal{E}_2/\mathcal{A}) = 3$.

5.8. Determine the codimension of the germ of $f \colon \mathbf{R}^4 \to \mathbf{R}$, $f(x,y,z,w) := \sin y \sin(x-z) + \sin x \sin(z^2 + w^2)$.

5.9. Let $[f]$ be a finitely determined germ in m^2. Show that

$$\mathrm{cod}[f] \leq \binom{n+d}{n} - 1$$

if $d := \det[f]$.

5.10. Let $[f]$ be a germ in m_n^k for $k \geq 3$. Prove that

$$\mathrm{cod}[f] \geq \binom{n+k+l-1}{k+l-1} - n\binom{n+l}{l} - 1$$

holds for all positive integers l. Show that $(k-1)^n - 1$ is an upper bound for the right-hand side of this inequality by computing the codimension of $[x_1^k + x_2^k + \cdots + x_n^k]$. Conclude from this that $\mathrm{cod}[f] \geq 7$ for $n \geq 3$ and also that the bounds for the codimension in the following table cannot be improved:

$[f]$ in	cod $[f]$ not less than
m_1^k	$k - 2$
m_2^k	$k(k-2)$
m_3^3	7

Finally, for $[f]$ in m_n^2 show that

$$\mathrm{cod}[f] \geq \frac{s}{6}(s^2 + 5) \geq \frac{s}{2}(s+1)$$

holds where s denotes the corank of $[f]$, thereby improving the estimate in (19).

Hint: Let $[g_1], \ldots, [g_r]$ be germs in $\mathcal{E}$ and denote the ideal they generate by $\mathcal{A}$. Show that

$$\mathcal{A} = \langle m^{l+1} \mathcal{A} \rangle + L,$$

where L is the linear space spanned by $[x^\nu g_j]$ for $0 \leq |\nu| \leq l$ and $1 \leq j \leq r$. Deduce from this that $(\mathcal{J}[f] + m^{k+l})/m^{k+l}$ is at most $n\binom{n+l}{l}$-dimensional.

5.11.   Compute the codimension and determinacy of the germ of the cubic form

$$x^3 + y^3 + z^3 + 3cxyz,$$

on $\mathbf{R}^3$, where c is a real parameter.

Solutions

5.1.   If a is zero, then obviously $\mathcal{J}[f] = [x]m$, and the cosets represented by the germs of y^l for $l \in \mathbf{N}$ are linearly independent in $m/\mathcal{J}[f]$. Therefore, $\text{cod}[f] = \infty$. Now let a be nonzero. Then $\mathcal{J}[f] = \langle xy, x^2 + aky^{k-1} \rangle_{\mathcal{E}}$ holds. It will be shown that the cosets represented by $[x], [y], [y^2], \ldots, [y^{k-1}]$ form a basis of $m/\mathcal{J}[f]$, implying $\text{cod}[f] = k$.

To verify the linear independence, consider the equation

$$(*) \qquad \alpha x + \sum_{i=1}^{k-1} \alpha_i y^i = g(x,y)xy + h(x,y)(x^2 + aky^{k-1})$$

for small x and y, where $\alpha, \alpha_1, \ldots, \alpha_{k-1}$ are real and $[g],[h]$ lie in $\mathcal{E}$. An immediate consequence of $(*)$ is $\alpha = 0$ and $h(x,y) = yh_1(x,y)$ for some $[h_1] \in \mathcal{E}$. Thus,

$$\sum_{i=0}^{k-2} \alpha_{i+1}y^i - aky^{k-1}h_1(x,y) = x(g(x,y) + xh_1(x,y))$$

follows, which implies $\alpha_i = 0$ for $i = 1, \ldots, k-1$.

Now it will be shown that the linearly independent cosets above span $m/\mathcal{J}[f]$. If $k = 2$, then $m^3 \subset \mathcal{J}[f]$. According to (10), it suffices to note that $[x^2], [y^2] \in \langle x, y \rangle + \mathcal{J}[f]$. In the case $k \geq 3$, it is obvious that $m^k \subset \mathcal{J}[f]$ holds; see also the solution of Exercise 4.11. Hence, according to (10), it suffices to show $[x^b y^c] \in \langle x, y^{k-1} \rangle + \mathcal{J}[f]$ for $b \geq 1$, $c \geq 0$, and $b + c \leq k - 1$. If $c \geq 1$, this is due to $\langle xy \rangle_{\mathcal{E}} \in \mathcal{J}[f]$. The cases $[x]$ and $[x^2]$ are also obvious. For $b \geq 3$, one finds $x^b = (x^2 + aky^{k-1})x^{b-2} - (xy)akx^{b-3}y^{k-2}$.

5.2.   It is immediate that $\mathcal{J}[f] = [x^2 + y^2]m$ holds. Now it will be shown that the cosets represented by the germs of $x^l, x^l y, y^2$ for $l \in \mathbf{N}_0$ form a basis of $\{[p] + \mathcal{J}[f] : p \in \mathbf{R}[x,y]\}$.

In order to see that they span this space, it suffices to show that the cosets of $[x^a y^b]$ for $a, b \in \mathbf{N}_0$ are linear combinations of these germs. Clearly, this is true for $b = 0, 1$ with a arbitrary and for $a = 0$, $b = 2$. Consider the remaining case $[x^a y^b]$, where $b \geq 2$ and $a + b > 2$. Then $[x^a y^b] + \mathcal{J}[f] = [x^a y^b] + [x^2 + y^2] \cdot [-x^a y^{b-2}] + \mathcal{J}[f] = -[x^{a+2} y^{b-2}] + \mathcal{J}[f]$. By repeating the argument finitely many times, a reduction to the previous cases is achieved.

Obviously, the cosets considered above are linearly independent if the equation

(*) $$p(x) + yq(x) + \alpha y^2 = (x^2 + y^2)h(x, y),$$

where $p, q \in \mathbf{R}[x]$, $\alpha \in \mathbf{R}$, and $[h] \in m$, implies $p = q = 0$ and $\alpha = 0$. Indeed, $\alpha = 0$ is an easy consequence of (*). Moreover, for $y = 0$ one has $p(x) = x^2 h(x, 0)$, so that $p(x) = x^2 p_1(x)$ with $p_1 \in \mathbf{R}[x]$ and $h(x, y) = p_1(x) + y h_1(x, y)$ by (1.28) for some $[h_1] \in \mathcal{E}$. Inserting this in (*), the new equation $q(x) - y p_1(x) = (x^2 + y^2) h_1(x, y)$ follows. By repeating the argument, one finds $p_1(x) + y q_1(x) = (x^2 + y^2) h_2(x, y)$ with $q(x) = x^2 q_1(x)$ and $[h_2] \in \mathcal{E}$. Clearly, after finitely many steps the left side vanishes, implying $p = 0$ and $q = 0$.

5.3. The cases $|a| = 2$ and $|a| \neq 2$ turn out to be essentially different, since $\mathrm{cod}[f] = \infty$ for $|a| = 2$ and $\mathrm{cod}[f] = 8$ otherwise.

(i) For $a = 2$, one has $\mathcal{J}[f] = [x^2 + y^2]m$. Then the cosets represented by $[x^l]$, $l \in \mathbf{N}$, are linearly independent in $m/\mathcal{J}[f]$. To see this, consider a polynomial $p \in \mathbf{R}[x]$ contained in $\mathcal{J}[f]$. Then the function given by $p(x)/(x^2 + y^2)$ for $x^2 + y^2 \neq 0$ and vanishing at the origin is smooth. This is only true if $p = 0$, since the $2n$-th derivative with respect to y at $y = 0$ of $(x^2 + y^2)^{-1}$ is equal to $(-1)^n (2n)! x^{-2n-2}$. Consequently, $\mathrm{codim}[f] = \infty$. Note that this result also follows from Exercise 5.2, or from (8) using Exercise 4.8.

The case $a = -2$ is treated similarly, yielding the same result. It also follows from (8) using Exercise 4.9.

(ii) For $a = 2b$ and $|b| \neq 1$ one has $\mathcal{J}[f] = \langle x^3 + bxy^2, bx^2 y + y^3 \rangle_\mathcal{E}$. The claim is that the cosets of $[x], [y], [x^2], [xy], [y^2], [xy^2], [x^2 y]$, and $[x^2 y^2]$ form a basis of $m/\mathcal{J}[f]$.

Observe first that these elements span $m/\mathcal{J}[f]$. Since $m^5 \subset \mathcal{J}[f]$ holds, as is shown in the solution of Exercise 4.9, it suffices by (10) to verify that $[x^4], [x^3 y], [xy^3]$, and $[y^4]$ lie in $\langle x^2 y^2 \rangle + \mathcal{J}[f]$. This is true, since $x^4 = -bx^2 y^2 + (x^3 + bxy^2)x$ and

$$x^3 y = \frac{1}{1 - b^2}(x^3 + bxy^2)y - \frac{b}{1 - b^2}(bx^2 y + y^3)x.$$

The terms xy^3 and y^4 are given by interchanging x and y in the expression for $x^3 y$ and x^4, respectively.

Now the linear independence will be proved. Due to $\mathcal{J}[f] \subset m^3$, it suffices to consider the equation

$$\alpha xy^2 + \beta x^2 y + \gamma x^2 y^2 = (x^3 + bxy^2)g(x, y) + (bx^2 y + y^3)k(x, y)$$

for small x and y, where α, β, and γ are real and $[g], [k]$ lie in $\mathcal{E}$. When $x = 0$, this yields $k(0, y) = 0$ for small y. Hence $k(x, y) = xk_1(x, y)$

for some $[k_1] \in \mathcal{E}$ by (1.28). Analogously, $g(x,y) = yg_1(x,y)$ follows for some $[g_1] \in \mathcal{E}$. Dividing by xy, a new equation,

$$\alpha y + \beta x + \gamma xy = (x^2 + by^2)g_1(x,y) + (bx^2 + y^2)k_1(x,y),$$

is obtained, obviously implying $\alpha = \beta = \gamma = 0$. In particular, $\mathrm{cod}[f] = 8$ follows.

5.4.

(a)   Clearly, $\mathcal{J}[f] = \langle x, y^3 \rangle_\mathcal{E}$ contains m^3 due to (4.9). Then the cosets represented by $[y]$ and $[y^2]$ form a basis of $m/\mathcal{J}[f]$. This follows from (1.27) and from (10) for $k = 1$ and $l = 2$.

(b)   Since $[x^3 - y^k] \sim -[x^3 + y^k]$, it suffices to consider $f(x,y) := x^3 + y^k$. Clearly, $\mathcal{J}[f] = \langle x^2, y^{k-1} \rangle_\mathcal{E}$ contains m^k by (4.9). Then the cosets represented by $[x^a y^b]$ for $a = 0, 1$ and $b = 0, 1, \ldots, k-2$ with $a + b \geq 1$ form a basis of $m/\mathcal{J}[f]$. This is a consequence of (1.27) and (10). In particular, $\mathrm{cod}[f] = 2k - 3$.

(c)   Since $\mathcal{J}[f] = \langle x^4, y^4 \rangle_\mathcal{E}$ contains m^7 by (4.9), the cosets represented by

$$[x^a y^b] \quad \text{for} \quad a, b = 0, 1, 2, 3 \quad \text{and} \quad a + b \geq 1$$

form a basis of $m/\mathcal{J}[f]$. This follows from (1.27) and (10).

(d)   This example is a bit more involved than the previous ones. It will be shown that the cosets represented by the germs $[y^b]$ for $b = 1, \ldots, k-1$ and $[x^a y^b]$ for $a = 1, 2$ and $b = 0, \ldots, k-2$ form a basis of $m/\mathcal{J}[f]$. In particular, this implies that the codimension of $[f]$ is $3(k - 1)$.

   The Jacobi ideal $\mathcal{J}[f]$ is generated by $[3x^2 \pm y^k]$ and $[xy^{k-1}]$. Hence it contains the germ of

$$x^3 = (3x^2 \pm y^k)\tfrac{1}{3}x \mp (xy^{k-1})\tfrac{1}{3}y.$$

To find a basis of $m/\mathcal{J}[f]$, use the following diagram. Arrange the monomials $x^a y^b$ in Pascal's triangle:
The monomials lying in the angles with apex x^3 and xy^{k-1}, respectively are contained in $\mathcal{J}[f]$. Moreover, the germs of y^{k+b}, $b \geq 0$, are identified with $[x^2 y^b]$, because $[3x^2 \pm y^k]$ is in $\mathcal{J}[f]$. Since no other relations hold, the cosets represented by the germs of the remaining monomials form a basis of $m/\mathcal{J}[f]$.

   To complete the proof of the assertion, note that the $2k - 1$ germs $[y], \ldots, [y^{k-1}], [x], [xy], \ldots, [xy^{k-2}], [x^2]$ are linearly independent not only modulo $\mathcal{J}[f]$ but even modulo $\mathcal{J}[f] + m^{k+1}$. Since $x^2 y^b = (3x^2 \pm y^k)\tfrac{1}{3}y^b \mp y^{k+b}$ the germ $[x^2 y^b]$ represents the zero element in $m/(\mathcal{J}[f] + m^{k+1})$ if $b \geq 1$.

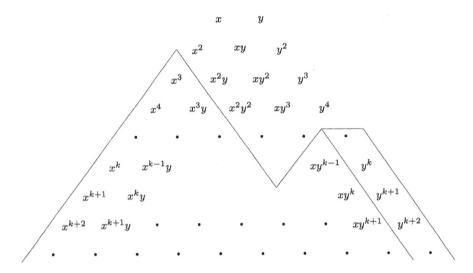

5.5. According to (7), there exists an integer $k \geq 0$ satisfying $m^k \subset \mathcal{A}$. Let k be the smallest nonnegative integer with this property. Due to $0 < \dim \mathcal{E}/\mathcal{A}$, the ideal $\mathcal{A}$ is different from $\mathcal{E}$. Hence $k \geq 1$ and $m^{k-1} \not\subset \mathcal{A}$ hold. It also follows from $\mathcal{A} \neq \mathcal{E}$ that $\mathcal{A}$ cannot contain any germ $[f]$ with $f(0) \neq 0$. Hence, $\mathcal{A} \subset m$ is valid. Assume there is no largest integer $l \geq 1$ satisfying $\mathcal{A} \subset m^l$. Then

$$\mathcal{A} \subset \bigcap_{l=0}^{\infty} m^l = m^{\infty},$$

and consequently $\dim \mathcal{E}/\mathcal{A} \geq \dim \mathcal{E}/m^{\infty}$. According to the remark right after the proof of (4.5), the latter is equal to $\dim \mathbf{R}[[x_1, x_2, \ldots, x_n]]$, which is infinite. This contradicts $\dim \mathcal{E}/\mathcal{A} < \infty$, and the desired integer l exists.

5.6. Clearly, $\dim \mathcal{E}_1/m_1^l = l$ by (4.8). Now let $\mathcal{A}$ be an ideal in $\mathcal{E}_1$ satisfying $\dim \mathcal{E}_1/\mathcal{A} < \infty$, and assume $\mathcal{A} \neq \mathcal{E}_1$. According to Exercise 5.5, there is a largest positive integer with $\mathcal{A} \subset m_1^l$. Then there is a germ $[f] \in \mathcal{A}$ not in m_1^{l+1}. Without restriction, $[f] = [x^l][1 + g]$ with $[g] \in m_1$, by Taylor's formula (1.27). Obviously, $[f]\mathcal{E}_1 = m_1^l$ after (4.9). Therefore $\mathcal{A} \supset m_1^l$, and thus $\mathcal{A} = m_1^l$ holds.

5.7. According to Exercise 5.5, there is a largest positive integer l such that $\mathcal{A} \subset m^l$. Therefore $\dim \mathcal{E}/\mathcal{A} \geq \dim \mathcal{E}/m^l = \binom{l+1}{l-1} = \frac{1}{2}l(l+1)$ by (4.8). The condition $\dim \mathcal{E}/\mathcal{A} = 3$ implies $l = 1$ or $l = 2$.

  (i) Let $l = 2$, i.e., $\mathcal{A} \subset m^2$. Then $\dim \mathcal{E}/m^2 = 3$ yields $\mathcal{A} = m^2$.

  (ii) Let $l = 1$, i.e., $\mathcal{A} \subset m$ but $\mathcal{A} \not\subset m^2$. Choose $[f]$ in $\mathcal{A}$ but not in m^2. Then $[f] = [\alpha x + \beta y] + [g]$ holds after (1.27) where $[g] \in m^2$ and $(\alpha, \beta) \neq (0, 0)$. According to (1.35), there is a $[\varphi] \in \mathcal{G}$ with $[f][\varphi] = [x]$. Hence it may be assumed without restriction that $[x]$ is in $\mathcal{A}$.

Denote all germs in $\mathcal{E}$ depending only on the second variable y by $\mathcal{E}_1$. For $[g] \in \mathcal{E}$, set $g_1(y) := g(0, y)$. Then $[g_1] \in \mathcal{E}_1$. Furthermore, $\mathcal{A}_1 := \mathcal{A} \cap \mathcal{E}_1$ is an ideal in $\mathcal{E}_1$. Now it will be shown that the map

(*) $$\mathcal{E}/\mathcal{A} \to \mathcal{E}_1/\mathcal{A}_1, \quad [g] + \mathcal{A} \mapsto [g_1] + \mathcal{A}_1$$

is linear and bijective.

From (1.28) it follows for $[g] \in \mathcal{E}$ that $[g] \in [g_1] + [x]\mathcal{E} \subset [g_1] + \mathcal{A}$. In particular, $[g] \in \mathcal{A}$ if and only if $[g_1] \in \mathcal{E}_1 \cap \mathcal{A} = \mathcal{A}_1$. This shows that the map (*) is well defined, linear, and injective. Trivially, (*) is also surjective.

Since $3 = \dim \mathcal{E}/\mathcal{A} = \dim \mathcal{E}_1/\mathcal{A}_1$ it follows from Exercise 5.6 that $\mathcal{A}_1 = \langle y^3 \rangle_{\mathcal{E}_1}$. Therefore, $\mathcal{A} \supset \langle x, y^3 \rangle_{\mathcal{E}}$. However, Exercise 5.4(a) shows that $\dim \mathcal{E}/\langle x, y^3 \rangle_{\mathcal{E}} = 3$, proving $\mathcal{A} = \langle x, y^3 \rangle_{\mathcal{E}}$.

5.8. The expansion of f up to terms of third order according to Taylor's formula (1.27) yields $f = p + R$ with $p(x, y, z, w) := y(x - z) + x(z^2 + w^2)$ and $[R] \in \mathfrak{m}_4^4$. Now p will be analyzed using the Reduction Lemma (3.2). The Hessian matrix

$$D^2 p(0) = \begin{pmatrix} 0 & 1 & 0 & 0 \\ 1 & 0 & -1 & 0 \\ 0 & -1 & 0 & 0 \\ 0 & 0 & 0 & 0 \end{pmatrix}$$

of p at the origin is diagonalized, yielding

$$\frac{1}{2} T^t D^2 p(0) T = \begin{pmatrix} -1 & 0 & 0 & 0 \\ 0 & 1 & 0 & 0 \\ 0 & 0 & 0 & 0 \\ 0 & 0 & 0 & 0 \end{pmatrix}$$

for

$$T := \begin{pmatrix} \sqrt{2} & \sqrt{2} & 1 & 0 \\ -1/\sqrt{2} & 1/\sqrt{2} & 0 & 0 \\ 0 & 0 & 1 & 0 \\ 0 & 0 & 0 & 1 \end{pmatrix}.$$

The matrix T defines a linear coordinate transformation τ. Instead of p, it is convenient to consider the equivalent polynomial $q := p \circ \tau$, which is

$$q(x, y, z, w) = -x^2 + y^2 + (\sqrt{2}x + \sqrt{2}y + z)(z^2 + w^2).$$

The Hessian matrix of q at the origin is the diagonal matrix $T^t D^2 p(0) T$; see (1.33). Using the notations in the proof of (3.2), one has

$$A = \begin{pmatrix} -1 & 0 \\ 0 & 1 \end{pmatrix} \quad \text{and} \quad F = \left(\frac{\partial}{\partial x} q, \frac{\partial}{\partial y} q \right);$$

that is,

$$F(x, y, z, w) = (-2x + \sqrt{2}(z^2 + w^2), 2y + \sqrt{2}(z^2 + w^2)).$$

Then

$$h(z, w) := \frac{1}{\sqrt{2}}(z^2 + w^2)(1, -1)$$

satisfies $h(0) = 0$ and $F(h(z, w), z, w) = 0$. According to (3.2), it follows that q and, hence, p are equivalent to $\tilde{p}(x, y, z, w) := -x^2 + y^2 + q(h(z, w), z, w) = -x^2 + y^2 + z(z^2 + w^2)$. Exercise 4.11 shows that the germ of $z(z^2 + w^2)$ is 3-determined, so that $[\tilde{p}]$ is also 3-determined due to Exercise 4.20. Therefore, $[p]$ is 3-determined, and $[f]$ is equivalent to $[\tilde{p}]$. It follows from (5.18) that the codimension of $[f]$ is equal to that of $[z(z^2 + w^2)]$, which is 3 by Exercise 5.1.

5.9. Since $[f]$ is finitely determined, there is a minimal positive integer k satisfying $m^k \subset \mathcal{J}[f]$ by (4.40). Then

$$\text{cod}[f] \le \dim m/m^k = \binom{n + k - 1}{n} - 1$$

by (4.8) and (5.1). On the other hand, $m^{k-1} \not\subset \langle m\mathcal{J}[f] \rangle$, because even $m^{k-1} \not\subset \mathcal{J}[f]$. Therefore (4.38) implies $d + 1 > k - 1$, and hence

$$\binom{n + k - 1}{n} \le \binom{n + d}{n}$$

follows by (4.8) since $m^{d+1} \subset m^k$.

5.10. Let $\mathcal{A}$ be the ideal in $\mathcal{E}$ generated by $[g_1], \ldots, [g_r]$. Then

$$\sum_{j=1}^{r} [g_j][h_j] = \sum_{j=1}^{r} [g_j][h_j - h_j(0)] + \sum_{j=1}^{r} h_j(0)[g_j]$$

for germs $[h_j]$ in $\mathcal{E}$, proving

$$\mathcal{A} = \langle m\mathcal{A} \rangle + \langle g_1, \ldots, g_r \rangle.$$

Applying this to $\langle m\mathcal{A} \rangle$ yields

$$\langle m\mathcal{A} \rangle = \langle m^2\mathcal{A} \rangle + \langle \{ [x_i g_j] : 1 \le i \le n, 1 \le j \le r \} \rangle.$$

By induction $\mathcal{A} = \langle m^{l+1}\mathcal{A} \rangle + L$ follows, where L is the linear space spanned by $[x^\nu g_j]$ for $0 \le |\nu| \le l$ and $1 \le j \le r$. Using (4.6), $\dim L \le r \binom{n+l}{l}$ follows.

When $[g_j] := [D_j f]$, $1 \le j \le n$, this implies $\mathcal{J}[f] + m^{k+l} = L + m^{k+l}$ because of $\langle m^{l+1}\mathcal{J}[f] \rangle \subset m^{k+l}$. Now the first assertion easily follows, since

$$\text{cod}[f] \ge \dim m/(\mathcal{J}[f] + m^{k+l})$$
$$= \dim m/m^{k+l} - \dim(L + m^{k+l})/m^{k+l}$$
$$\ge \dim m/m^{k+l} - \dim L = \binom{n + k + l - 1}{k + l - 1} - 1 - n\binom{n+l}{l}$$

by (4.8).

An immediate consequence of Taylor's formula (1.27) is that the cosets $[x^\nu] + \langle x_1^{k-1}, \ldots, x_n^{k-1} \rangle_\varepsilon$ form a basis of $m/\mathcal{J}[x_1^k + \ldots + x_n^k]$ for $|\nu| > 0$ and $0 \le \nu_1, \ldots, \nu_n \le k - 2$. This proves $\text{cod}[x_1^k + \ldots + x_n^k] = (k-1)^n - 1$.

In the table the bounds for the codimension cannot be improved because

$$C(n, k, l) := \binom{n + k + l - 1}{k + l - 1} - n \binom{n + l}{l} - 1 \quad \text{and} \quad (k-1)^n - 1$$

coincide in the following three cases: $n = 1, l = 1$; $n = 2, l = k - 2$; $n = k = 3, l = 1$.

Finally, note that

$$C(n, k, 1) = \binom{n + k}{k} - n(n + 1) - 1$$
$$= (k + 1) \ldots (k + n)/n! - n(n + 1) - 1$$

is strictly increasing in k and that

$$C(n, 3, 1) = \frac{n}{6}(n^2 + 5)$$

is strictly increasing in n. From this, $\text{cod}[f] \ge 7$ for $n \ge 3$ follows, since $C(3, 3, 1) = 7$. Moreover, (5.18) implies

$$\text{cod}[f] \ge C(s, 3, 1) = \frac{s}{6}(s^2 + 5)$$

for $[f] \in m_n^2$, if $s := \text{cor}[f]$.

5.11. Let $\mathcal{A}$ denote the Jacobi ideal of the germ of $p(x, y, z) := x^3 + y^3 + z^3 + 3cxyz$. It is generated by $[x^2 + cyz]$, $[y^2 + cxz]$, and $[z^2 + cxy]$. Suppose that $c = -1$. Then all points on the diagonal $x = y = z$ are critical, and the origin is a nonisolated critical point. Hence, determinacy and codimension are infinite by (4.41) and (8).

Now let $c \ne -1$. Due to

$$(x^2 + cyz)y - (y^2 + cxz)cz + (z^2 + cxy)c^2x = (1 + c^3)x^2y,$$

the germ of x^2y lies in $\langle m\mathcal{A} \rangle$. Then $[x^2z]$, $[xy^2]$, $[xz^2]$, $[y^2z]$, and $[yz^2]$ also lie in $\langle m\mathcal{A} \rangle$, because of the invariance of p under permutations of the variables.

It will be shown that $[x^4]$ and therefore, by symmetry, also $[y^4]$ and $[z^4]$, lie in $\langle m^2\mathcal{A} \rangle$. Indeed, after some guessing, one finds

$$(1 + c^3)x^4 = (x^2 + cyz)(x^2 - cyz + c^3x^2)$$
$$- (y^2 + cxz)c^3xy + (z^2 + cxy)c^2y^2.$$

In particular, $m^4 \subset \langle m^2\mathcal{A} \rangle$ follows from (4.9). Since a cubic form is obviously not 2-determined, the criterion (4.24) yields $\det[p] = 3$. The above considerations also imply that the cosets represented by $[x]$, $[y]$, $[z]$, $[xy]$, $[yz]$, $[xz]$, and $[xyz]$ span $m/\mathcal{A}$. On the other hand, by comparing Taylor coefficients it is not hard to show that these cosets are linearly independent, proving $\text{cod}[p] = 7$.

Chapter 6

The Classification Theorem for Germs of Codimension at Most 4

This chapter is devoted to the proof of René Thom's theorem classifying the degenerate critical points of functions whose codimension is at most 4. Roughly speaking, this theorem states that there are only seven different types of degenerate critical points for such functions; these are often referred to as the **7** elementary catastrophes. Using similar arguments, this theorem can be generalized to cover the critical points of functions with codimension at most 6; see Exercises 6.2 and 6.3. In this case, the classification gives 14 different standard types of such critical points. Furthermore, for functions of just two variables whose codimension is at most 7, a classification can also be achieved resulting in 17 different possibilities; see Exercise 6.4. A finite classification is no longer possible, however, for functions of two variables whose codimension exceeds 7 or for functions of more than two variables with codimension at least 7, as shown in Exercises 6.6 and 6.7.

> **Classification of Degenerate Critical Points** (1). A germ $[f]$ in m_n^2 satisfies $1 \leq \text{cod}[f] \leq 4$ if and only if it is equivalent to exactly one of the germs in Table 1 of polynomials in 1 or 2 variables up to the addition of a nondegenerate normal quadratic form in the remaining $n-1$ or $n-2$ variables, respectively. In the last three columns of the table the corank, codimension, and determinacy of the corresponding germs are listed.

Actually, there are 10 and not **7** elementary catastrophes, because three of the germs in Table 1 occur with both signs; the negative germs are called **duals**. All of these 10 germs are mutually inequivalent and, by Theorem (4.48), stay inequivalent after the addition of germs of nondegenerate normal quadratic forms in new variables. Before beginning the proof, recall that a germ $[f]$ in m_n^2 has a degenerate critical point at the origin exactly when its codimension is not zero;

TABLE 1

germ	cor	cod	det
$[x^3]$		1	3
$[x^4], [-x^4]$	1	2	4
$[x^5]$		3	5
$[x^6], [-x^6]$		4	6
$[x^3 - xy^2]$		3	3
$[x^3 + y^3]$	2	3	3
$[x^2y + y^4], [-x^2y - y^4]$		4	4

see (5.6). When the codimension is between 1 and 4, then the corank is either 1 or 2 by (5.19). Consequently, it is the Reduction Lemma (3.2) and the uniqueness of the residual singularity up to equivalence (4.48) that enables a reduction from n variables to just one or two variables. Notice that corank 2 implies that the codimension must be either 3 or 4 by (5.19). The proof of the classification theorem (1) treats the three cases, namely, corank 1, corank 2 with codimension 3, and corank 2 with codimension 4, separately in the following theorems (2), (8), and (9). For corank 2, the classification of cubic forms on $\mathbf{R}^2$ is needed, and that will also be proved in this chapter.

Theorem (2).   A germ $[f]$ in m_n^2 has corank 1 and satisfies $1 \leq \text{cod}[f] \leq 4$ if and only if it is equivalent to exactly one of the following mutually inequivalent germs of polynomials in one variable up to the addition of a nondegenerate normal quadratic form in the remaining $n - 1$ variables:

(3) $[x^3], [x^4], [-x^4], [x^5], [x^6], [-x^6].$

Proof. Due to (5.2) and (5.18), just the necessity need be shown. By (5.18) and (4.48) it suffices to consider a germ $[f]$ of a function of *one* variable in m_1^3. Due to (5.8), the germ $[f]$ is finitely determined. Let $k := \det[f]$. Then k is the smallest positive integer with $f^{(k)}(0) \neq 0$ after (4.13). It follows that $k \geq 3$ because $[f]$ is in m_1^3. According to (1.34), it may be assumed that $f(x) = (\pm 1)^{k-1} x^k$ holds in a neighborhood of the origin, implying $\text{cod}[f] = k - 2$ by (5.2). Now $1 \leq \text{cod}[f] \leq 4$ yields $k \in \{3, 4, 5, 6\}$. The mutual inequivalence of the germs $[x^3], \ldots, [-x^6]$ results from (1.34).

As already mentioned, the case corank 2 requires the classification of cubic forms on $\mathbf{R}^2$.

Definition.   A homogeneous polynomial $p(x, y) = ax^3 + bx^2y + cxy^2 + dy^3$ in $\mathbf{R}[x, y]$ of third degree is called a **cubic form on $\mathbf{R}^2$**.

It is easily checked that cubic forms on $\mathbf{R}^2$ are transformed into cubic forms by a linear coordinate transformation of $\mathbf{R}^2$.

Classification of Cubic Forms on $\mathbf{R}^2$ (4). Let p be a cubic form on $\mathbf{R}^2$. There is a linear coordinate transformation τ of $\mathbf{R}^2$ such that $p \circ \tau$ is one of the following five cubic forms on $\mathbf{R}^2$:

(5) $$0, \quad x^3, \quad x^2 y, \quad x^3 - xy^2, \quad x^3 + y^3,$$

and this form is uniquely determined by p.

Proof.

(i) First the uniqueness will be shown, i.e., no two forms in (5) are **linearly equivalent**, meaning they cannot be transformed into each other by a linear coordinate transformation $\tau: \mathbf{R}^2 \to \mathbf{R}^2$, $\tau(x,y) := (\alpha x + \beta y, \gamma x + \delta y)$ with $\alpha\delta - \beta\gamma \neq 0$.

Since the null form is clearly invariant under τ, only the other four forms have to be treated. Consider the zero set and the set of critical points of these forms:

form	zero set	set of critical points
x^3	$\{x = 0\}$	$\{x = 0\}$
$x^2 y$	$\{x = 0\} \cup \{y = 0\}$	$\{x = 0\}$
$x^3 - xy^2$	$\{x = 0\} \cup \{x = y\} \cup \{x = -y\}$	$\{0\}$
$x^3 + y^3$	$\{x = -y\}$	$\{0\}$

These sets are either a point, a line, a pair of lines, or three lines. Only x^3 and $x^3 + y^3$ have zero sets that can be transformed into each other by a linear coordinate transformation. However, the sets of their critical points cannot be mapped into each other. Hence, the mutual inequivalence of the four forms follows.

(ii) Now it will be shown that every cubic form p on $\mathbf{R}^2$ is linearly equivalent to one of the forms in (5).

The coefficients a', b', c', and d' of the transformed form $p'(x,y) := p(\tau(x,y))$ are

$$a' = a\alpha^3 + b\alpha^2\gamma + c\alpha\gamma^2 + d\gamma^3,$$
$$b' = 3a\alpha^2\beta + b(2\alpha\beta\gamma + \alpha^2\delta) + c(2\alpha\gamma\delta + \beta\gamma^2) + 3d\gamma^3\delta,$$
$$c' = 3a\alpha\beta^2 + b(2\alpha\beta\delta + \beta^2\gamma) + c(2\beta\gamma\delta + \alpha\delta^2) + 3d\gamma\delta^2,$$
$$d' = a\beta^3 + b\beta^2\delta + c\beta\delta^2 + d\delta^3.$$

Choose τ such that d' vanishes. If a vanishes, this is accomplished by taking $\alpha := \delta := 0$ and $\beta := \gamma := 1$. If $a \neq 0$, then set $\alpha := \delta := 1$, $\gamma := 0$, and let β be a real zero of the cubic equation $a\beta^3 + b\beta^2 + c\beta + d = 0$. Consequently, one may start with a cubic form $p(x,y) = ax^3 + bx^2 y + cxy^2$.

Now τ can be chosen to make both b' and d' vanish. If $c \neq 0$, then such a τ is found by setting $\alpha := \delta := 1$, $\beta := 0$, and $\gamma := -b/2c$. When c vanishes but not b, take $\alpha := 0$, $\beta := \gamma := 1$, $\delta := -a/b$, and for $c = b = 0$ set $\beta := 0$. Thus, it is no restriction to assume that both b and d vanish, i.e., $p(x, y) = ax^3 + cxy^2$.

By scaling x and y it is necessary to consider only one of the five forms 0, x^3, xy^2, $x^3 - xy^2$, and $x^3 + xy^2$, depending on whether a or c or both are zero.

Obviously, xy^2 is linearly equivalent to x^2y. Moreover, $x^3 + xy^2$ is linearly equivalent to $x^3 + y^3$, since the latter is equal to

$$(4^{-1/3}(x + y))^3 + (4^{-1/3}(x + y))(3^{1/2}4^{-1/3}(x - y))^2.$$

The form in (5) to which a cubic form p on $\mathbf{R}^2$ is uniquely linearly equivalent is called its **normal cubic form**. Notice that a cubic form p has the same normal form as $-p$, as seen by considering the transformation $(x, y) \mapsto (-x, -y)$.

Lemma (6). Let p be a cubic form on $\mathbf{R}^2$. Then $\det[p] = \text{cod}[p] = \infty$ holds if the normal form of p is either 0, x^3, or x^2y, and otherwise $\det[p] = \text{cod}[p] = 3$ is valid.

Proof. Since equivalent germs have the same determinacy and the same codimension by (4.12) and (5.5), it suffices to consider the normal form of p. The origin is a critical point of 0, x^3, and x^2y, which is not isolated, implying the first assertion after (4.41) and (5.8). The second assertion results from (4.26), (5.11), and (5.12).

The normal cubic forms (5) are not only linearly inequivalent, as has been proved in (4).

Lemma (7). The germs of the normal cubic forms on $\mathbf{R}^2$ are mutually inequivalent.

Proof. Let p and q be any two cubic forms on $\mathbf{R}^2$. For any $[\varphi] \in \mathcal{G}_2$ the homogeneous part of degree 3 of $[p][\varphi]$ is given by $[p][\tau]$, where τ is the linear part of φ. Suppose that $[p]$ and $[q]$ are equivalent, i.e., $[p][\varphi] = [q]$ for a $[\varphi] \in \mathcal{G}_2$. Then $p \circ \tau = q$, proving that p and q actually are linearly equivalent. The assertion follows immediately from this observation.

This result is generalized in Exercise 6.5.

Theorem (8). A germ in m_n^2 has corank 2 and codimension 3 if and only if it is equivalent to exactly one of the germs

$$[x^3 - xy^2], \quad [x^3 + y^3]$$

up to the addition of a nondegenerate normal quadratic form in the remaining $n - 2$ variables.

Proof. Due to (6) and (5.18), just the necessity need be proved. Lemma (5.18) and Theorem (4.48) imply that it is sufficient to consider a germ $[f]$ in m_2^3 with $\text{cod}[f] = 3$. First it will be shown that $\det[f] = 3$ holds. From $\mathcal{J}[f] \subset m_2^2$ it is clear that $\mathcal{A} := \langle m_2 \mathcal{J}[f] \rangle \subset m_2^3$ follows. Note that $\dim \mathcal{E}_2/\mathcal{A} = \dim \mathcal{E}_2/m_2^3 + \dim m_2^3/\mathcal{A}$ is true. By (4.8), $\dim \mathcal{E}_2/m_2^3 = 6$ holds, and (5.17) yields $\dim \mathcal{E}_2/\mathcal{A} = 3 + \text{cod}[f] = 6$. Therefore, $\dim m_2^3/\mathcal{A} = 0$ is valid, resulting in $\mathcal{A} = m_2^3$. From the corollary (4.25) of the sufficient criterion for determinacy (4.24) and the fact that $[f]$ is in m_2^3, one obtains $\det[f] = 3$.

Thus, $[f]$ is equivalent to the germ of the third Taylor polynomial T_f^3 of f at the origin, and then by (4.12), the determinacy of $[T_f^3]$ is 3. Now T_f^3 is a cubic form on $\mathbf{R}^2$ because of $[f] \in m_2^3$, and the assertion results from (6) and (7).

The most difficult case to prove is the following one.

Theorem (9).   A germ in m_n^2 has corank 2 and codimension 4 if and only if it is equivalent to exactly one of the germs

$$[x^2 y + y^4], \quad [-x^2 y - y^4]$$

up to the addition of a nondegenerate normal quadratic form in the remaining $n - 2$ variables.

Proof. Since the codimension of a germ is independent of its sign, the sufficiency follows from (5.13) and (5.18). To prove the necessity, it is enough to consider a germ $[f]$ in m_2^3 with $\text{cod}[f] = 4$ by (5.18) and (4.48).

(i) As in the proof of (8), it will be shown first that $\det[f] = 4$ holds. The inclusion $\mathcal{J}[f] \subset m_2^2$ implies $\mathcal{A} := \langle m_2 \mathcal{J}[f] \rangle \subset m_2^3$. Therefore

$$\dim \left(\mathcal{E}_2/\mathcal{A} \right) = 6 + \dim \left(m_2^3/\mathcal{A} \right) = 3 + \text{cod}[f] = 7$$

follows, from which $\dim m_2^3/\mathcal{A} = 1$ results. This means

$$c_3 := \dim \left((\mathcal{A} + m_2^3) / \mathcal{A} \right) = 1,$$

yielding

$$c_4 := \dim \left((\mathcal{A} + m_2^4) / \mathcal{A} \right) = 0$$

by (5.7), i.e., $m_2^4 \subset \mathcal{A}$. According to (4.39), one obtains $3 \le \det[f] \le 4$. However, $\det[f] = 3$ is excluded, since it would yield $[f] \sim [T_f^3]$, from which the contradiction $\text{cod}[f] = 3$ would result by (4.12) and (6). Thus $\det[f] = 4$ holds.

Hence, $[f]$ is equivalent to $[T_f^4]$. Since $[f]$ is in m_2^3, it suffices to consider the case that f is a polynomial that can be written as the sum $p + h$ of a cubic form p on $\mathbf{R}^2$ and a homogeneous polynomial h of degree 4 on $\mathbf{R}^2$. Since the degree of a polynomial is invariant under linear transformations, by (4) it is no restriction to assume that p is a normal cubic form.

(ii) It will be proved here that p has to be the form x^2y.

The forms $x^3 - xy^2$ and $x^3 + y^3$, which have determinacy 3, are easily excluded, because they would entail $[f] \sim [p]$ due to $p = T_f^3$. Then the contradiction $\det[f] = 3$ would follow by (4.12).

Also $p = 0$ cannot be true, since it would imply $[f] \in m_2^4$ and therefore $\mathcal{J}[f] \subset m_2^3$. This would lead to the contradiction $\mathrm{cod}[f] = \dim(m_2/\mathcal{J}[f]) \geq \dim(m_2/m_2^3) = \dim(\mathcal{E}_2/m_2^3) - 1 = 5$ by (4.8).

To verify that p cannot be the form x^3, set $h(x,y) := ay^4 + bxy^3 + x^2q(x,y)$ with $a, b \in \mathbf{R}$ and q a quadratic form on $\mathbf{R}^2$. By (1.29),

$$(x,y) \mapsto (x - \tfrac{1}{3}q(x,y), y)$$

defines a local diffeomorphism at the origin transforming f to $x^3 + ay^4 + bxy^3$ plus terms of higher order. Since $[f]$ is 4-determined, it follows from (4.12) that $[f]$ is equivalent to $[g] := [x^3 + ay^4 + bxy^3]$. It will be proved now that $\mathrm{cod}[g] \geq 5$ holds, yielding the desired contradiction due to (5.5).

If b is zero, then $\mathcal{J}[g] = \langle x^2, ay^3 \rangle \mathcal{E}$, and it is easily verified that the cosets represented by $[x]$, $[y]$, $[xy]$, $[y^2]$, and $[xy^2]$ are linearly independent in $m/\mathcal{J}[g]$; cf. Exercise 5.4(b). If b is nonzero, consider the cosets represented by $[x]$, $[y]$, $[xy]$, $[y^2]$, and $[y^3]$. In order to prove their linear independence in $m/\mathcal{J}[g]$, study the equation

$$\alpha x + \beta y + \gamma xy + \delta y^2 + \varepsilon y^3 + (3x^2 + by^3)h(x,y)$$
$$+ (4ay^3 + 3bxy^2)k(x,y) = 0$$

for small x and y with real coefficients $\alpha, \beta, \gamma, \delta, \varepsilon$ and germs $[h]$, $[k]$ in $\mathcal{E}$. Examining the Taylor coefficients for x, y, xy, y^2, x^2, and xy^2, one finds immediately that $\alpha, \beta, \gamma, \delta, h(0)$, and $k(0)$ are zero. It remains to show that ε also vanishes. This follows from looking at the Taylor coefficient for y^3.

(iii) Let h be as in (ii). A smooth coordinate transformation will be found to show that $[f] \sim [x^2y + ay^4]$ holds for a real.

For arbitrary quadratic forms q_1 and q_2 on $\mathbf{R}^2$, the map given by $x \mapsto x + q_1(x,y), y \mapsto y + q_2(x,y)$ is a local diffeomorphism at the origin after (1.29). It transforms f to $x^2y + x^2q_2(x,y) + 2xy\, q_1(x,y) + h(x,y)$ plus terms of higher order. Specify $q_1(x,y) := -(b/2)y^2$ and $q_2 := -q$, and the assertion follows as in part (ii).

(iv) A linear coordinate transformation $\tau(u,v) = (x,y)$ of $\mathbf{R}^2$ will be defined transforming $x^2y + ay^4$ into either $u^2v + v^4$ or $-u^2v - v^4$.

From (4.42) it follows that a is not zero. For $a > 0$, set $\tau(u,v) := (a^{1/8}u, a^{-1/4}v)$, and for $a < 0$, set $\tau(u,v) := (|a|^{1/8}u, -|a|^{-1/4}v)$.

(v) Finally, by (4.44) the germs $[x^2y + y^4]$ and $[-x^2y - y^4]$ in $\mathcal{E}_2$ are not equivalent.

Theorems (2), (8), and (9) prove the classification theorem (1). As already mentioned, the classification of critical points of germs of codimension 5 and 6 is given in Exercises 6.2 and 6.3. A finite classification is not possible for germs of codimension exceeding 6. The reason is the existence of so-called **unimodal germs**. A unimodal germ is a family of mutually inequivalent germs depending on one real parameter. Unimodal germs occur first for codimension 7 and corank 3 and then for codimension 8 and corank 2; see Exercises 6.6 and 6.7. It is interesting to note that a finite classification of germs of codimension 7 and corank 2 is still possible; see Exercise 6.4.

In summary, the classification of germs of codimension at most 7 and corank at most 2 gives 17 elementary catastrophes. Counting the 6 duals, one obtains a list of 23. A further finite classification is impossible. Using the notation of V.I. Arnol'd [A], the 23 elementary catastrophes are the first nine members $A_2, A_{\pm 3}, A_4, A_{\pm 5}, A_6, A_{\pm 7}$, of the infinite series $A_{\pm n} := [\pm x^{n+1}], n \geq 2$, the first ten members $D_{\pm 4}, D_{\pm 5}, D_{\pm 6}, D_{\pm 7}, D_{\pm 8}$ of the infinite series $D_{\pm n} := [x^2 y \pm y^{n-1}]$, and the four exceptional germs $E_{\pm 6} := [x^3 \pm y^4], E_7 := [x^3 + xy^3], E_8 := [x^3 + y^5]$. In order to compare this list with that given in Theorem (1) and Exercises 6.2, 6.3, and 6.4, notice that $D_{-4} \sim [x^3 - xy^2], D_4 \sim [x^3 + y^3], D_{-n} \sim [-x^2 y - y^{n-1}]$ for odd n, and $E_{-6} \sim [-x^3 - y^4]$; the equivalence of D_4 with $[x^3 + y^3]$ is shown at the end of the proof of (4). The series $A_{\pm n}$ and $D_{\pm n}$ are treated in (1.34), (4.13), (5.2), and Exercises 4.11 and 5.1. In Exercises 4.12(b), 4.13, and 5.4(b) and (d), the series $([x^3 \pm y^n])$ and $([x^3 \pm xy^{n-1}])$ for $n \geq 3$ are studied; these incorporate the exceptional germs $E_{\pm 6}, E_8$, and E_7.

Exercises

6.1. Let k be an integer greater than 3. Show that every finitely determined germ in

(a) $[x^3] + m_2^k$ is equivalent to a germ in exactly one of the sets

$$\{\eta^{l-1}[x^3 + y^l]\} \quad \text{and} \quad [x^3 + \eta^l xy^{l-1}] + m_2^{l+1}$$

with $\eta \in \{1, -1\}$ and $l \geq k$, and

(b) $[x^2 y] + m_2^k$ is equivalent to exactly one of the germs $[x^2 y + \eta y^l]$ with $\eta \in \{1, -1\}$ and $l \geq k$.

If a germ in $[x^2 y] + m_2^k$ is not finitely determined, then for every $l \geq k$ it is equivalent to a germ in $[x^2 y] + m_2^l$.

Hint: Write a germ in $[x^3] + m_2^k$ as

$$[x^3] + [ay^k + bxy^{k-1} + x^2 p(x, y)]$$

plus terms in m_2^{k+1}, and find a local diffeomorphism at the origin transforming it to $[x^3], \pm[x^3 + y^k]$, or $[x^3 \pm xy^{k-1}]$ up to terms in m_2^{k+1}. Treat case (b) analogously.

6.2. (Classification of Critical Points of Germs of Codimension 5). Prove that a germ in m_n^2 has codimension 5 if and only if it is equivalent to exactly one of the following germs of polynomials in 1 or 2 variables up to the addition of a nondegenerate normal quadratic form in the remaining $n - 1$ or $n - 2$ variables, respectively.

germ	cor	cod	det
$[x^7]$	1	5	7
$[x^3 + y^4], [-x^3 - y^4]$	2	5	4
$[x^2y + y^5]$	2	5	5
$[x^2y - y^5]$	2	5	5

Hint: Adapt the proof of Theorem (9). In the case of corank 2, conclude first that the determinacy is at most 5. In determining the cubic form, which is part of the germ, use Exercise 5.10 and Lemma (6). For the remaining part of the germ, apply Exercises 6.1, 5.4(b) and (d), 5.1, and Example (4.28).

6.3. (Classification of Critical Points of Germs of Codimension 6). Extend the table in Exercise 6.2 to codimension 6:

germ	cor	cod	det
$[x^8], [-x^8]$	1	6	8
$[x^3 + xy^3]$	2	6	4
$[x^2y + y^6], [-x^2y - y^6]$	2	6	6

Hint: Use the result of Exercise 5.10.

6.4. (Classification of Critical Points of Germs of Codimension 7 and Corank 2). Supplement the table in Exercise 6.3 by the following complete list of pairwise inequivalent germs in m_2^3 of codimension 7:

germ	cor	cod	det
$[x^3 + y^5]$	2	7	5
$[x^2y + y^7]$	2	7	7
$[x^2y - y^7]$	2	7	7

Hint: As in Exercise 6.3, show that a germ in $[x^3 \pm xy^4] + m^6$ is equivalent to a germ in $[x^3 \pm xy^4 + ay^6 + by^7] + m^8$ for real a and b.

6.5. Let p and q be two homogeneous polynomials in n variables of positive degrees k and l, respectively. Suppose that p and q are linearly inequivalent, meaning they cannot be transformed into each other by a linear coordinate transformation of $\mathbf{R}^n$. Show that no germ in $[p] + m^{k+1}$ is equivalent to a germ in $[q] + m^{l+1}$.

6.6. Consider the one-parameter family of germs

$$[x^3 + y^3 + z^3 + 3cxyz]$$

in m_3^3 studied in Exercise 5.11 with a real parameter c. Show that different values of c give inequivalent germs. The parameter c is called a **modulus**. This family is an example of a so-called **unimodal germ**. Note that the codimension is 7 for $c \neq -1$ by Exercise 5.11. Unimodal germs do not exist for codimension less than 7 according to Theorem (1) and Exercises 6.2 and 6.3. Moreover, unimodal germs of corank 2 do not exist even for codimension less than 8; see Exercise 6.4.

Hint: Show that the matrix

$$T := \begin{pmatrix} \alpha_1 & \beta_1 & \gamma_1 \\ \alpha_2 & \beta_2 & \gamma_2 \\ \alpha_3 & \beta_3 & \gamma_3 \end{pmatrix}$$

of a linear transformation between two germs of the family has to satisfy $\tilde{T}T = E_3$, where the coefficients of the transpose of $\tilde{T}$ are given by $\tilde{\alpha}_1 = \alpha_1^2 + c\alpha_2\alpha_3, \tilde{\alpha}_2 = \alpha_2^2 + c\alpha_1\alpha_3, \ldots$, and $\tilde{\gamma}_3 = \gamma_3^2 + c\gamma_1\gamma_2$. Forming suitable linear combinations of the 9 equations obtained from the matrix relation $T\tilde{T} = E_3$, the reduction to the case $c = 0$ is achieved. Then apply Cramer's rule for computing the inverse matrices to T and $\tilde{T}$. This results in 18 equations. First exclude the case that all coefficients of $\tilde{T}$ are nonzero.

6.7. Consider the one-parameter family of germs

$$[x^4 + 2cx^2y^2 + y^4]$$

in m_2^4 with a real parameter c. Recall that the codimension of such germs is infinite if $|c| = 1$ and is equal to 8 if $|c| \neq 1$; see Exercises 5.2 and 5.3. For $-\infty < c \leq 1$, show that the germs are mutually inequivalent. Germs corresponding to different parameter values c and c' are equivalent if and only if both c and c' are greater than -1 and $c' = (3 - c)/(1 + c)$ holds.

This example shows that unimodal germs of corank 2 exist for codimension 8. Compare this with the result of Exercise 6.6.

Solutions

6.1.

(a) A germ $[f]$ in $[x^3] + m_2^k$ can be written as $[x^3] + [h] + [j]$, where $[j]$ lies in m_2^{k+1} and h is a homogeneous polynomial of degree k. Furthermore, write $h(x,y) =: ay^k + bxy^{k-1} + x^2p(x,y)$ with a homogeneous polynomial p of degree $k-2$, and real coefficients a and b. Now replace x by $x - \frac{1}{3}p(x,y)$ but keep y unaltered. Obviously, this defines a local diffeomorphism at the origin transforming $[f]$ to $[x^3 + ay^k + bxy^{k-1}]$ plus a germ in m_2^{k+1}.

Which of these germs are equivalent? For given pairs of coefficients a, b and a', b' the question is whether

$$[\varphi_1^3 + a\varphi_2^k + b\varphi_1\varphi_2^{k-1}] \in [x^3 + a'y^k + b'xy^{k-1}] + m_2^{k+1}$$

holds for some local diffeomorphism $\varphi = (\varphi_1, \varphi_2)$ at the origin. It is clear that the terms of φ_1 with orders exceeding $k-2$ as well as the nonlinear terms of φ_2 can be omitted. Moreover, comparing the terms of third order, it follows immediately that the linear part of φ_1 has to be x. Thus the above relation reads

(*)
$$[(x + q)^3 + a(\gamma x + \delta y)^k + bx(\gamma x + \delta y)^{k-1}]$$
$$\in [x^3 + a'y^k + b'xy^{k-1}] + m_2^{k+1},$$

where γ, δ are real and q is a polynomial without linear terms of degree less than $k-1$. Moreover, δ is not zero, because φ is a local diffeomorphism.

Since (*) implies that $(x + q)^3$ does not contain terms of order greater than 3 and less than k, q is homogeneous of order $k-2$. A look at the y^k terms shows that the relation $a\delta^k = a'$ holds. Therefore, a necessary condition for equivalence is that a' is zero if and only if a vanishes and that a' has the same sign as a if k is even.

Consider first the case $a \neq 0, a' \neq 0$, and a' has the same sign as a if k is even. Then set $\delta := (a'/a)^{1/k}$, $\gamma := (\delta^{1-k}b' - b)/ka$, and

$$q(x,y) := -\frac{1}{3x^2}[a(\gamma x + \delta y)^k - a\delta^k y^k - ka\gamma\delta^{k-1}xy^{k-1}$$
$$+ bx(\gamma x + \delta y)^{k-1} - b\delta^{k-1}xy^{k-1}].$$

Check that q is a homogeneous polynomial of degree $k-2$. Further, note that $[(x + q)^3] \in [x^3 + 3x^2q] + m_2^{k+1}$ since $1 + (k-2)^2$ and $(k-2)^3$ exceed k. A short calculation shows that (*) is fulfilled. Thus, equivalence holds. Since $\pm[x^3+y^k]$ is k-determined by Exercise 4.12(b), it follows that $[f]$ is equivalent to $\eta^{k-1}[x^3 + y^k]$, where η denotes the sign of a.

Now the case $a = a' = 0$ will be treated. Because $[(x + q)^3] \in [x^3 + 3x^2q] + m_2^{k+1}$, the relation $b' = \delta^{k-1}b$ follows from (*). It is clear that this necessary condition for equivalence is also sufficient. Thus if b is nonzero, $[f]$ is equivalent to a germ in $[x^3 + \eta^k xy^{k-1}] + m_2^{k+1}$, where η denotes the sign of b.

Finally the last case, $a = b = 0$, will be considered. $[f]$ is equivalent to a germ in $[x^3] + m_2^{k+1}$, and the foregoing steps can be repeated with

k replaced by $k + 1$. Since $[f]$ is finitely determined, this procedure will stop after finitely many repetitions, because otherwise $[f]$ would be equivalent to $[x^3]$, contradicting (4.45). This proves the assertion.

(b)   Write $[f] = [x^2y] + [h] + [j]$ as in the proof of (a). Then

$$\varphi(x, y) := \left(x - \frac{b}{2} y^{k-2}, y - p(x, y) \right)$$

defines a local diffeomorphism at the origin that transforms $[f]$ to $[x^2y + ay^k]$ plus a germ in m_2^{k+1}. Moreover, by rescaling x and y, a is transformed to $\eta \in \{0, 1, -1\}$. If η is nonzero, then $\det[x^2y + \eta x^k] = k$ by Exercise 4.11. Hence, $[f]$ is equivalent to $[x^2y + \eta x^k]$. If η is zero, the foregoing procedure will be repeated with k replaced by $k + 1$. If $[f]$ is finitely determined, this iteration ceases after finitely many steps, because otherwise $[f]$ would be equivalent to $[x^2y]$, contradicting (4.42).

6.2.   According to (5.19), the corank of a germ of codimension 5 is either 1 or 2. For corank 1, the assertion follows immediately from the proof of Theorem (2), which also applies to the case of codimension 5. Thus, due to (5.18) and (4.48), it suffices to consider a germ $[f]$ in m_2^3 with $\text{cod}[f] = 5$.

(i) The inclusion $\mathcal{J}[f] \subset m_2^2$ implies $\mathcal{A} := \langle m\mathcal{J}[f] \rangle \subset m_2^3$. Therefore, $\dim \mathcal{E}_2/\mathcal{A} = \dim \mathcal{E}_2/m_2^3 + \dim m_2^3/\mathcal{A} = 6 + \dim m_2^3/\mathcal{A}$ by (4.8), and $\dim m_2^3/\mathcal{A} = 2$ follows from (5.17). According to (5.7), $c_3 = 2$ implies $m_2^5 \subset \mathcal{A}$. Therefore, $[f]$ is 5-determined by (4.25).

It is no restriction to assume that $[f] = [p] + [h]$, where p is a normal cubic form and $[h]$ is a germ in m_2^4. The forms $0, x^3 - xy^2$, and $x^3 + y^3$ do not occur, since the null form entails $\text{cod}[f] \geq 8$ by Exercise 5.10 and the other two forms yield $\det[f] = 3$, after (6). Thus only the forms x^3 and x^2y are left.

(ii) Consider first $p(x, y) = x^3$. According to Exercise 6.1(a), and since $[f]$ is 5-determined, $[f]$ is equivalent to exactly one of the following germs: $\pm[x^3 + y^4], [x^3 + y^5], [x^3 + xy^3] + [j]$, and $[x^3 \pm xy^4]$ with $[j]$ in m_2^5. By Exercises 5.4(b) and 4.12(b), the germs $\pm[x^3 + y^4]$ have codimension 5 and determinacy 4. The other germs are ruled out, because they have the wrong codimension. This follows immediately from Exercise 5.4(b) and (d) for $[x^3 + y^5]$ and $[x^3 \pm xy^4]$. Since $[x^3 + xy^3]$ is 4-determined after Example (4.28), this also follows for the remaining germ by Exercise 5.4(d).

(iii) Now assume $p(x, y) = x^2y$. According to Exercise 6.1(b), and because $[f]$ is 5-determined, $[f]$ is equivalent to exactly one of the following germs: $[x^2y \pm y^4]$ and $[x^2y \pm y^5]$. However, only the last two germs have the right codimension, according to Exercise 5.1. By Exercise 4.11, their determinacy is 5.

6.3.   Exercise 5.10 yields cor$[f] \leq 2$. The assertion for corank 1 follows as in Theorem (2). Hence, it is no restriction to assume that $[f]$ is in m_2^3. Then $\mathcal{A} := \langle m\mathcal{J}[f] \rangle \subset m_2^3$ and dim $\mathcal{E}/\mathcal{A} = 6 + \dim m_2^3/\mathcal{A} = 6 + 3$ by (4.8) and (5.17). Therefore dim $m_2^3/\mathcal{A} = 3$, so that m^6 is contained in $\mathcal{A}$ by (5.7). It follows from (4.25) that $[f]$ is 6-determined.

It is no restriction to assume $[f] = [p] + [h]$ with a normal cubic form p and a germ $[h]$ in m_2^4. The forms $p = 0, x^3 - xy^2, x^3 + y^3$ are excluded by Exercise 5.10 and Lemma (6). It remains to consider the forms x^3 and x^2y. From Exercise 6.1 it follows that $[f]$ is equivalent to exactly one of the following germs: $[x^3 + y^5], \pm[x^3 + y^6], [x^3 + xy^3] + [k], [x^3 \pm xy^4] + [j], [x^3 + xy^5]$, and $[x^2y \pm y^6]$ with $[k]$ in m^5 and $[j]$ in m^6. The germs of codimension less than 6 have been omitted; see Theorem (1) and Exercise 6.2. Germs $[h]$ in m^7 need not be considered, since $[f]$ is 6-determined. Exercise 5.4(b) and (d) exclude all the above germs except $[x^3 + xy^3] + [k]$ and $[x^2y \pm y^6]$, because of their codimension. According to (4.28), det$[x^3 + xy^3] = 4$, so that $[k]$ can be omitted. By Exercises 5.4(d) and 5.1 the germs $[x^3 + xy^3], [x^2y + y^6]$, and $[x^2y - y^6]$ have codimension 6. The last germ is obviously equivalent to $-[x^2y + y^6]$. According to Exercise 4.11, det$[x^2y + y^6] = 6$ holds. This completes the proof.

6.4.   As in Exercise 6.3 it follows that $[f]$ is 7-determined and that without restriction $[f] = [p] + [h]$, with $[h]$ in m_2^4 and p equal to x^3 or x^2y.

According to Exercise 6.1, $[f]$ has to be equivalent to exactly one of the following germs: $\pm[x^3 + y^4], [x^3 + y^5], \pm[x^3 + y^6], [x^3 + y^7], [x^3 + xy^3] + [j_5], [x^3 \pm xy^4] + [j_6], [x^3 + xy^5] + [j_7], [x^3 \pm xy^6], [x^2y \pm y^4], [x^2y \pm y^5], [x^2y \pm y^6]$, and $[x^2y \pm y^7]$ with $[j_k]$ in m^k. By (1) as well as Exercises 6.2 and 6.3, the germs $\pm[x^3 + y^4], [x^3 + xy^3] + [j_5], [x^2y \pm y^4], [x^2y \pm y^5]$, and $[x^2y \pm y^6]$ have codimension less than 7. Furthermore, Exercise 5.4 implies that cod$[x^3 + y^6] = 9$, cod$[x^3 + y^7] = 11$, cod$[x^3 + xy^5] + [j_7] \geq 9$, and cod$[x^3 \pm xy^6] = 15$. The germ $[x^3 + y^5]$ has codimension 7 by Exercise 5.4(b) and determinacy 5 by Exercise 4.12(b). The germs $[x^2y \pm y^7]$ have codimension and determinacy equal to 7; see Exercises 5.1 and 4.11.

Thus, it remains to rule out the germ $[x^3 + xy^4] + [j_6]$. Analogously, the germ $[x^3 - xy^4] + [j_6]$ is discarded. Write $[j_6]$ as $[ay^6 + a'xy^5 + x^2p] + [j_7]$ with a and a' real, p a homogeneous polynomial of degree 4, and $[j_7]$ a germ in m^7. First perform the transformation $x \mapsto x - \frac{1}{3}p, y \mapsto y$, and subsequently the transformation $x \mapsto x, y \mapsto y - (a'/4)y^2$. By the resulting local diffeomorphism, the original germ is transformed to $[x^3 + xy^4 + ay^6] + [j_7']$ for some $[j_7']$ in m^7. Put now $[j_7'] = [by^7 + b'xy^6 + x^2p'] + [j_8]$ with b and b' real, p' a homogeneous polynomial of degree 5, and $[j_8]$ in m^8. After performing the transformations $x \mapsto x - \frac{1}{3}p', y \mapsto y$, and $x \to x, y \mapsto y - (b'/4)y^3$, one finally obtains a germ in $[x^3 + xy^4 + ay^6 + by^7] + m^8$ equivalent to the original one.

As already mentioned, codimension 7 entails determinacy at most 7. Hence, it suffices to consider the germ $[x^3 + xy^4 + ay^6 + by^7]$ and to prove that its codimension is greater than 7. It will be shown that the eight germs $[x], [y], [x^2], [xy], [y^2], [x^2y], [xy^2]$, and $[y^3]$ in m are linearly independent modulo the

Jacobi ideal

$$\langle 3x^2 + y^4, 4xy^3 + 6ay^5 + 7by^6 \rangle \varepsilon$$

of $[x^3 + xy^4 + ay^6 + by^7]$. To verify this, it obviously suffices to consider the equation

$$\alpha x^2 + \beta x^2 y = (3x^2 + y^4)g(x,y) + (4xy^3 + 6ay^5 + 7by^6)h(x,y)$$

for small x and y with coefficients α and β and germs $[g]$ and $[h]$ in $\mathcal{E}$. Because of the y^4 term, $g(0)$ vanishes. Thus $\alpha = 3g(0) = 0$ holds. Now set $g(x,y) = \gamma x + \delta y + k(x,y)$ with $[k]$ in m^2. Then the equation

$$\beta x^2 y = 3\gamma x^3 + \gamma x y^4 + 3\delta x^2 y + \delta y^5 + (3x^2 + y^4)k(x,y)$$
$$+ y^3(4x + 6ay^2 + 7by^3)h(x,y)$$

follows. Clearly, it entails $\beta = 3\delta, \gamma = 0$, and, as a consequence, $k(x,y) = y^3 j(x,y)$ for some $[j]$ in $\mathcal{E}$. Therefore the reduced equation

$$-\frac{\beta}{3}y^2 = (3x^2 + y^4)j(x,y) + (4x + 6ay^2 + 7by^3)h(x,y)$$

is valid. Because of the $4x$ term, $h(0)$ vanishes. Hence $\beta = 0$, since $-\beta/3 = 6ah(0)$.

6.5. Consider the germs $[p] + [g]$ and $[q] + [h]$ with $[g]$ in m^{k+1} and $[h]$ in m^{l+1}. Assume that they are equivalent, i.e., $([p] + [g])[\varphi] = [q] + [h]$ for some $[\varphi] \in \mathcal{G}_n$. Then $([p] + [g])[\varphi]$ is in m^k. Furthermore, its homogeneous part of degree k is given by $[p][\tau]$, where τ is the linear part of φ. This implies $p \circ \tau = q$, which proves the assertion.

6.6. According to Exercise 6.5, it suffices to consider linear equivalence. Now $c = -1$ is the unique value for which the germ is not finitely determined (see Exercise 5.11) and hence is inequivalent to all other germs of the family.

Suppose that $c \neq -1$. Let T be the real invertible matrix given in the hint. The claim is that the equation

$$x^3 + y^3 + z^3 + 3c'xyz$$
$$= (\alpha_1 x + \beta_1 y + \gamma_1 z)^3 + (\alpha_2 x + \beta_2 y + \gamma_2 z)^3 + (\alpha_3 x + \beta_3 y + \gamma_3 z)^3$$
$$+ 3c(\alpha_1 x + \beta_1 y + \gamma_1 z)(\alpha_2 x + \beta_2 y + \gamma_2 z)(\alpha_3 x + \beta_3 y + \gamma_3 z)$$

for x, y, z in $\mathbf{R}$ can be satisfied only if c' is equal to c. After comparing Taylor coefficients, the relation $\tilde{T}T = E_3$ is implied by the foregoing equation, where

$$\tilde{T} := \begin{pmatrix} \tilde{\alpha}_1 & \tilde{\alpha}_2 & \tilde{\alpha}_3 \\ \tilde{\beta}_1 & \tilde{\beta}_2 & \tilde{\beta}_3 \\ \tilde{\gamma}_1 & \tilde{\gamma}_2 & \tilde{\gamma}_3 \end{pmatrix}$$

with $\tilde{\alpha}_1 := \alpha_1^2 + c\alpha_2\alpha_3$, $\tilde{\alpha}_2 := \alpha_2^2 + c\alpha_1\alpha_3$, $\tilde{\alpha}_3 := \alpha_3^2 + c\alpha_1\alpha_2$, and with $\tilde{\beta}_i, \tilde{\gamma}_i$ defined analogously. In the following it will be shown that this relation yields that T is a permutation matrix, proving the assertion.

Since $\tilde{T}$ is the inverse of T, $T\tilde{T} = E_3$ holds. Explicitly, this means that

$$\alpha_i^3 + c\alpha_1\alpha_2\alpha_3 + \beta_i^3 + c\beta_1\beta_2\beta_3 + \gamma_i^3 + c\gamma_1\gamma_2\gamma_3 = 1, \quad i = 1, 2, 3,$$

and

$$\alpha_1\alpha_2^2 + c\alpha_1^2\alpha_3 + \beta_1\beta_2^2 + c\beta_1^2\beta_3 + \gamma_1\gamma_2^2 + c\gamma_1^2\gamma_3 = 0,$$
$$\alpha_1^2\alpha_3 + c\alpha_2\alpha_3^2 + \beta_1^2\beta_3 + c\beta_2\beta_3^2 + \gamma_1^2\gamma_3 + c\gamma_2\gamma_3^2 = 0,$$
$$\alpha_2\alpha_3^2 + c\alpha_1\alpha_2^2 + \beta_2\beta_3^2 + c\beta_1\beta_2^2 + \gamma_2\gamma_3^2 + c\gamma_1\gamma_2^2 = 0,$$

as well as three further equations of this kind are valid. The first three equations immediately imply

$$\alpha_1^3 + \beta_1^3 + \gamma_1^3 = \alpha_2^3 + \beta_2^3 + \gamma_2^3 = \alpha_3^3 + \beta_3^3 + \gamma_3^3.$$

From the next three equations one deduces $\alpha_1\alpha_2^2 + \beta_1\beta_2^2 + \gamma_1\gamma_2^2 = 0$ by forming the linear combination of the left sides with coefficients 1, $-c$, and c^2, respectively, and dividing by $1 + c^3$. More generally, one finds

$$\alpha_i\alpha_j^2 + \beta_i\beta_j^2 + \gamma_i\gamma_j^2 = 0 \text{ for } i \neq j.$$

In matrix notation, these equations read $TS = \lambda E_3$ with $\lambda := \alpha_i^2 + \beta_i^2 + \gamma_i^2$ and where the entries of the transpose of S are the squares of those of T. Since T is invertible, λ cannot be zero. After dividing the coefficients α_i, β_i, and γ_i by the third root of λ, it is no restriction to assume $\lambda = 1$. Thus, the reduction to the case $c = 0$ is achieved.

Expressing the entries of the inverse matrix by Cramer's rule, the equations

$$\alpha_1^2 = D^{-1}(\beta_2\gamma_3 - \beta_3\gamma_2), \quad \alpha_1 = D(\beta_2^2\gamma_3^2 - \beta_3^2\gamma_2^2)$$

follow, where

$$D := \alpha_1\beta_2\gamma_3 + \alpha_2\beta_3\gamma_1 + \alpha_3\beta_1\gamma_2 - \alpha_1\beta_3\gamma_2 - \alpha_2\beta_1\gamma_3 - \alpha_3\beta_2\gamma_1$$

denotes the determinant of T.

Analogous equations are valid for α_2, α_3, and $\beta_i, \gamma_i, i = 1, 2, 3$. If α_1 is nonzero, then combining both equations yields the relation

$$D^{-2} = \alpha_1(\beta_2\gamma_3 + \beta_3\gamma_2).$$

At least one entry of T must be zero, since otherwise from this and from analogous relations one would obtain $\alpha_1\beta_2\gamma_3 = \alpha_2\beta_3\gamma_1 = \alpha_3\beta_1\gamma_2$ and $\alpha_1\beta_3\gamma_2 = \alpha_3\beta_2\gamma_1 = \alpha_2\beta_1\gamma_3$. This would imply $D = 3\alpha_1\beta_2\gamma_3 - 3\alpha_1\beta_3\gamma_2 = 3\alpha_1 D\alpha_1^2$, i.e., $\alpha_1 = 3^{-1/3}$. Then, by symmetry, all entries of T would be equal to $3^{-1/3}$, contradicting $D \neq 0$. Consequently, it is no restriction to assume that $\alpha_1 = 0$ and—since neither $(\alpha_1, \beta_1, \gamma_1)$ nor $(\alpha_1, \alpha_2, \alpha_3)$ is zero—that $\beta_1 \neq 0$ and $\alpha_2 \neq 0$.

First it will be shown that $\gamma_2 = 0$ follows. Assume the contrary. Then $D^{-2} = \alpha_3\beta_1\gamma_2$ and $D^{-2} = \beta_1(\alpha_2\gamma_3 + \alpha_3\gamma_2)$ are valid, implying $\alpha_2\beta_1\gamma_3 = 0$, i.e., $\gamma_3 = 0$. This contradicts $\gamma_3^2 = -D^{-1}\alpha_2\beta_1$. Analogously, $\beta_3 = 0$ follows. This implies $\alpha_3 = 0$ due to $0 = \gamma_2^2 = D^{-1}\alpha_3\beta_1$, and also $\gamma_1 = 0$, due to $0 = \beta_2^2 = D^{-1}\alpha_2\gamma_1$. Hence β_2 also vanishes, since $\beta_2^2 = D^{-1}\alpha_3\gamma_1$ holds. Finally, because of $1 = \alpha_i^3 + \beta_i^3 + \gamma_i^3, i = 1, 2, 3$, one obtains $\alpha_2 = \beta_1 = \gamma_3 = 1$.

6.7. According to Exercise 6.5, it suffices to consider linear transformations $(x, y) \mapsto (\alpha x + \beta y, \gamma x + \delta y)$ with $\alpha\delta - \beta\gamma \neq 0$. Comparing coefficients, it follows that two quartic forms with parameter values c and c' are equivalent if and only if the following five equations hold:

(i) $1 = \alpha^4 + 2c\alpha^2\gamma^2 + \gamma^4,$

(ii) $0 = \alpha^3\beta + c\alpha\gamma(\alpha\delta + \beta\gamma) + \gamma^3\delta,$

(iii) $0 = \alpha\beta^3 + c\beta\delta(\alpha\delta + \beta\gamma) + \gamma\delta^3,$

(iv) $1 = \beta^4 + 2c\beta^2\delta^2 + \delta^4,$

(v) $c' = 3\alpha^2\beta^2 + c(\beta^2\gamma^2 + 4\alpha\beta\gamma\delta + \alpha^2\delta^2) + 3\gamma^2\delta^2.$

Multiply (ii) by $\beta\delta$ and (iii) by $\alpha\gamma$, then subtract them from each other. Dividing the result by $\alpha\delta - \beta\gamma$ gives $\alpha^2\beta^2 = \gamma^2\delta^2$. Next, multiply (i) by δ^4 and replace $\gamma^2\delta^2$ by $\alpha^2\beta^2$. Using (iv) multiplied by α^4, one obtains $\delta^4 = \alpha^4$, and therefore

(1) $$\alpha^2 = \delta^2 \quad \text{and} \quad \beta^2 = \gamma^2.$$

Now the sum of (ii) and (iii) yields $0 = \alpha\beta(\alpha^2 + \beta^2) + c\alpha\beta(\gamma^2 + \delta^2) + c\gamma\delta(\alpha^2 + \beta^2) + \gamma\delta(\gamma^2 + \delta^2) = (\alpha^2 + \beta^2)(1 + c)(\alpha\beta + \gamma\delta)$, due to $\alpha^2 + \beta^2 = \gamma^2 + \delta^2$ by (1). Clearly $\alpha^2 + \beta^2 > 0$ holds, so that

(2) $$0 = (1 + c)(\alpha\beta + \gamma\delta)$$

follows. Similarly, the difference of (ii) and (iii) is

(3) $$0 = (\alpha^2 - \beta^2)(1 - c)(\alpha\beta - \gamma\delta).$$

Consider first $c = -1$. Then (i) reads $1 = (\alpha^2 - \gamma^2)^2 = (\alpha^2 - \beta^2)^2$ by (1), and (3) implies $\alpha\beta = \gamma\delta$. Using this equality and (1), Equation (v) yields $c' = 6\alpha^2\beta^2 - (\beta^4 + 4\alpha^2\beta^2 + \alpha^4) = -(\alpha^2 - \beta^2)^2$, i.e., $c' = -1$.

Next let $c = 1$. Then $\alpha\beta = -\gamma\delta$ holds by (2), and (i) yields $1 = \alpha^2 + \beta^2$, implying $c' = 1$ by (v).

Now assume $|c| \neq 1$. Then $\alpha\beta = -\gamma\delta$ and $0 = (\alpha^2 - \beta^2)(\alpha\beta - \gamma\delta)$ hold by (2) and (3). If $\alpha\beta - \gamma\delta = 0$, then $\alpha\beta = \gamma\delta = 0$ follows, and hence $\alpha^4 = 1$ or $\beta^4 = 1$ by (1) and (i). Therefore (v) implies $c = c'$.

Finally consider the last case, $\alpha^2 - \beta^2 = 0$. By (1), $\alpha^2 = \beta^2 = \gamma^2 = \delta^2$ holds. Then (i) yields $1 = 2\alpha^4(1 + c)$, implying $c > -1$ and $2\alpha^4 = (1 + c)^{-1}$. Since $\alpha\beta = -\gamma\delta$ is valid, $c' = 6\alpha^4 - 2c\alpha^4 = (3 - c)/(1 + c)$ follows from (v).

The function $]-1, \infty[\to]-1, \infty[$, $x \mapsto (3 - x)/(1 + x)$, is bijective and equal to its own inverse. Moreover, it maps the interval $]-1, 1[$ bijectively onto $]1, \infty[$.

For $c > -1$ verify that the linear transformation with coefficients $\alpha := (2 + 2c)^{-1/4}, \beta := \gamma := \alpha, \delta := -\alpha$ is an equivalence for the germs with parameter values c and $c' := (3 - c)/(1 + c)$.

Chapter 7

Unfoldings

The seven elementary catastrophes in (6.1) represent normal forms for degenerate critical points of arbitrary smooth functions whose codimension is at most 4. Although this classification certainly is an achievement, the main intention of Catastrophe Theory has not been addressed yet. The following observation is of fundamental importance in pointing out what Catastrophe Theory is really all about.

Every degenerate critical point of a smooth function is unstable in the following sense. When the function is slightly perturbed by adding a small term, it exhibits a different qualitative character—new critical points appear in the neighborhood of the original critical point, thereby reducing its degeneracy. In the case that the function represents a potential, for example the potential energy of a system, then its critical points correspond to equilibrium states, and a perturbation of the function corresponds to a change of the external parameters influencing the system. The fact that a minimum can disappear and become a maximum after perturbing the function corresponds to the transition from a stable to an unstable equilibrium. The relevance of this concerning applications was discussed in detail by means of examples in Chapter 2.

The present chapter is primarily dedicated to presenting an introduction to the goal of Catastrophe Theory, which is to classify systems according to their behavior under perturbation. The basic concepts for this classification are given here, but the proof will require the material developed in the following three chapters. The central notion, which was introduced by René Thom, is the mathematical expression for a perturbation, i.e., an unfolding. Let r be a nonnegative integer, and let $\mathbf{R}^{n+r}$ be identified with $\mathbf{R}^n \times \mathbf{R}^r$; the elements of $\mathbf{R}^{n+r}$ are usually denoted by (x, u).

Definition. Let $[f]$ be in $\mathcal{E}_n$. Then a germ $[F]$ in $\mathcal{E}_{n+r}$ is an r-**parameter unfolding** of $[f]$, if $F(x, 0) = f(x)$ holds for all x in a neighborhood of the origin of $\mathbf{R}^n$. The case $r = 0$ means $[F] = [f]$.

The germ $[f]$ is called the **center of organization** of the unfolding $[F]$. In

145

the standard literature, the term **deformation** is also commonly used for an unfolding. The variables $x_1, \ldots, x_n$ of an unfolding are often referred to as the **state variables**, because in applications their values often determine the state of the system. The parameters $u_1, \ldots, u_r$ are called **external** or **control parameters**, since they describe an external influence on the system. A special class of r-parameter unfoldings of $[f]$ is given by adding a linear combination of germs $[g_j]$ in $\mathcal{E}_n$ to $[f]$,

$$(1) \qquad\qquad [f] + u_1[g_1] + \cdots + u_r[g_r],$$

and considering the real coefficients $u_1, \ldots, u_r$ as parameters. A trivial case is the **constant r-parameter unfolding** $[F] \in \mathcal{E}_{n+r}$ of $[f]$ defined by $F(x, u) := f(x)$ for u in $\mathbf{R}^r$ and x in a neighborhood of the origin.

An unfolding $[F]$ of a germ $[f]$ in $\mathcal{E}_n$ gives rise to a family of germs $([F_u])_u$ in $\mathcal{E}_n$ by taking the partial functions $F_u(x) := F(x, u)$. These functions describe the perturbed system. The perturbation $F_u - f$, as well as its derivatives, vanish uniformly on compact sets as the parameters u approach zero.

A visualization of a one-parameter unfolding $[F]$ of a germ $[f] \in \mathcal{E}_1$ is a smooth surface in $\mathbf{R}^2$:

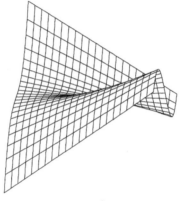

$$F_u(x) := x^3 + ux.$$

The curves on the surface correspond to fixed parameter values u and are the graphs of the partial functions F_u. They illustrate how the function f changes when the parameter varies. In applications, the parameters $u = (u_1, \ldots, u_r)$ may have various meanings, such as position coordinates or other physical entities, e.g., volume or pressure. In sociological and psychological applications more abstract meanings are assigned to the parameters; see [C1, C2, Sei, Sin, U, Z1].

Two examples of unfoldings, which were already discussed in Chapter 2, will be considered again, namely the fold and the cusp, which are special unfoldings of $[x^3]$ and $[x^4]$. These will later be proved to be universal, meaning, roughly speaking, that they describe all possible perturbations of $[x^3]$ and $[x^4]$, respectively. First let $f(x) := x^3$ and

$$(2) \qquad\qquad F \colon \mathbf{R} \times \mathbf{R} \to \mathbf{R}, \quad F(x, u) := x^3 + ux.$$

For every parameter value u, the function $F_u : \mathbf{R} \to \mathbf{R}$, $F_u(x) := x^3 + ux$, represents a perturbation of $f = F_0$. Whereas f has exactly one degenerate critical point, this is no longer true for F_u when u is not zero. For negative u, F_u has exactly two nondegenerate critical points—a maximum at $-|u/3|^{1/2}$ and a minimum at $|u/3|^{1/2}$. There is no critical point at all when u is positive.

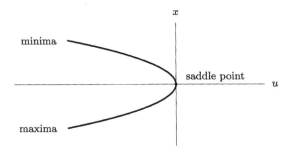

A less simple example is the two-parameter family

(3) $F : \mathbf{R} \times \mathbf{R}^2 \to \mathbf{R}, \quad F(x, u, v) := x^4 - ux^2 + vx,$

which is an unfolding of $f(x) := x^4$. Different parameter values result in partial functions $F_{uv} : \mathbf{R} \to \mathbf{R}$, $F_{uv}(x) := x^4 - ux^2 + vx$, whose critical points are qualitatively different. To see this, recall the discussion of the cusp in Chapter 2. The number and kind of critical points of F_{uv} are determined by the sign of the discriminant $\Delta := -8u^3 + 27v^2$ of the cubic equation

$$\frac{d}{dx} F_{uv}(x) = 4x^3 - 2ux + v = 0.$$

Explicitly, one has (see Figs. 6, 7, 8 in Chapter 2):

Δ	number and kind of critical points
< 0	three nondegenerate
> 0	one nondegenerate
$= 0, uv \neq 0$	one nondegenerate and one degenerate
$= 0, u = v = 0$	one degenerate

As seen in these examples, it is natural to consider the set of all critical points of the partial functions of an unfolding as well as the subset of just the degenerate critical points.

Definition. Let $[F] \in \mathcal{E}_{n+r}$ be an unfolding with a representative $F : W \to \mathbf{R}$ defined on an open neighborhood W of the origin in $\mathbf{R}^{n+r}$. The set

$$M_F := \{(x, u) \in W : DF_u(x) = 0\}$$

is called the **catastrophe surface** of F. Its subset

$$C_F := \{(x, u) \in M_F : D^2 F_u(x) \text{ is not invertible}\}$$

is called the **catastrophe set** of F, and the projection of C_F onto the parameter space $\mathbf{R}^r$

$$B_F := \{u \in \mathbf{R}^r : \text{ there is an } x \in \mathbf{R}^n \text{ with } (x, u) \in C_F\}$$

is referred to as the **bifurcation set** of F.

These notions are already familiar from Chapter 2, where they played a crucial role in the discussion of the fold and cusp. Note that the choice of a representative for the unfolding is not essential, because only the behavior of the unfolding around the origin is of interest. It is important to emphasize that the catastrophe surface M_F of F is composed of exactly the equilibrium states of a system for the various values of the external parameters. In general this surface is not a manifold. For example, the catastrophe surface of the unfolding $[x^3 + ux^2]$ of $[x^3]$ consists of two intersecting lines. Around points (x_0, u_0) not lying in the catastrophe set C_F, the catastrophe surface M_F can be uniquely represented as the graph of a smooth map defined on the control parameters. This follows immediately from the Implicit Function Theorem (see Appendix A.1) applied to $(x, u) \mapsto DF_u(x)$. However, for points (x_0, u_0) in the catastrophe set C_F, such a one-to-one assignment of states to parameter values is not possible in general. It may happen that several states correspond to the same control parameter, implying that a **multimodality** takes place at the catastrophe set. The examples discussed in Chapter 2 show how the study of the catastrophe surface and its catastrophe set gives an insight into the behavior of a system subjected to a **quasistatic change** of its external parameters. Since every germ obviously has a multitude of unfoldings, it is natural to try to obtain an overview. One way of achieving this is to classify all possible unfoldings of a germ. The first step in this direction is to introduce a notion for comparing unfoldings, called inducing.

Recall that for $s, r \in \mathbf{N}$ the set of germs $[\psi]$ of smooth maps $\psi \colon V \to \mathbf{R}^r$ at the origin of $\mathbf{R}^s$ is denoted by $\mathcal{E}_{s,r}$ (see Chapter 4), and set

$$m_{s,r} := \{[\psi] \in \mathcal{E}_{s,r} \colon \psi(0) = 0\}.$$

Definition. Let $[F] \in \mathcal{E}_{n+r}$ be an unfolding of $[f] \in \mathcal{E}_n$. Then an unfolding $[G] \in \mathcal{E}_{n+s}$ is said to be **induced** by $[F]$ if there are germs $[\varphi] \in m_{n+s,n}, [\psi] \in m_{s,r}$, and $[\gamma] \in m_s$ such that

(i) $\varphi(y, 0) = y$,

(ii) $G(y, v) = F(\varphi(y, v), \psi(v)) + \gamma(v)$

hold for (y, v) in an open neighborhood of the origin in $\mathbf{R}^{n+s}$; $[\varphi]$ is called the germ of the **variable transformation** and $[\psi]$ the germ of the **parameter transformation**.

Of course, $[G]$ is also an unfolding of $[f]$. Indeed, (ii) implies $G(y, 0) = F(\varphi(y, 0), \psi(0)) + \gamma(0) = f(y)$ for small y. The idea behind this definition is that an induced unfolding does not contain any new information about the organization center, and it may contain even less information. To see this, consider first an unfolding $[G]$ induced by $[F]$ resulting from only a parameter transformation, i.e., $\varphi(y, v) = y, \gamma = 0$. Since $G_v = F_{\psi(v)}$ holds, $([G_v])_v$ is only a subfamily of $([F_u])_u$. The general case is a bit more complicated. Clearly, $G_v = F_{\psi(v)} \circ \varphi_v + \gamma(v)$ holds. Since γ is independent of the y variables, the last summand is irrelevant when considering the critical points of the partial functions of F. It will be shown in the next lemma that φ_v is a local diffeomorphism at the origin for small v. Assuming this, it then follows that the partial functions of G form a subfamily of the partial functions of F up to local diffeomorphisms and translations. Consequently, no new information is obtained; see (1.31)–(1.33). The following lemma also provides an interesting comparison between the catastrophe surfaces of F and G.

Lemma (4). Let $[G] \in \mathcal{E}_{n+s}$ be an unfolding induced by an unfolding $[F] \in \mathcal{E}_{n+r}$ of a germ in $\mathcal{E}_n$ with respect to the germs $[\varphi] \in m_{n+s,n}, [\psi] \in m_{s,r}$ of the variable and parameter transformations. Then for small v the partial function φ_v is a local diffeomorphism on an open neighborhood of the origin of $\mathbf{R}^n$, which is independent of v. Moreover,

$$\Phi(y, v) := (\varphi(y, v), \psi(v))$$

defines a smooth map on an open neighborhood Z of the origin of $\mathbf{R}^{n+s}$ into $\mathbf{R}^{n+r}$ for which $M_G \cap Z = \Phi^{-1}(M_F)$ and $C_G \cap Z = \Phi^{-1}(C_F)$ hold.

Proof. Put $x := \varphi(y, v)$ and $u := \psi(v)$. Property (ii) of the last definition implies $DG_v(y) = DF_u(x)D\varphi_v(y)$, and when $DF_u(x) = 0$ is true it also implies that $D^2G_v(y) = [D\varphi_v(y)]^t D^2 F_u(x)D\varphi_v(y)$ holds. Thus, it suffices to prove the assertion concerning φ_v. Note that

$$D\Phi(y, v) = \begin{pmatrix} D\varphi_v(y) & * \\ 0 & D\psi(v) \end{pmatrix},$$

where the submatrix denoted by $*$ is not specified. In particular, if $r = s$,

(5) $$\det D\Phi(y, v) = \det D\varphi_v(y) \cdot \det D\varphi(v)$$

follows. Define $\tilde{\Phi}(y, v) := (\varphi(y, v), v)$. Then (5) yields $\det D\tilde{\Phi}(y, v) = \det D\varphi_v(y)$, implying $\det D\tilde{\Phi}(0, 0) = 1$. Therefore, by (1.29), $\tilde{\Phi}$ is a local diffeomorphism at the origin, proving the claim about φ_v.

Notice that Φ is not injective, and $\Phi(Z)$ is not open in general. The correspondence between the catastrophe surfaces as well as between the catastrophe sets of F and G is bijective if the following more restrictive condition is fulfilled.

Definition. Two unfoldings $[G] \in \mathcal{E}_{n+s}$ and $[F] \in \mathcal{E}_{n+r}$ of a germ in $\mathcal{E}_n$ are said to be **equivalent** if $r = s$ and if $[G]$ is induced by $[F]$ so that the germ $[\psi]$ of the parameter transformation is a local diffeomorphism at the origin, i.e., $[\psi] \in \mathcal{G}_r$.

The term isomorphic is frequently found in the literature, meaning equivalent. In this book the term isomorphic will be used in a more general sense, see after (9).

The relationship (5) implies that $[\psi]$ is in $\mathcal{G}_r$ exactly when $[\Phi]$ is in $\mathcal{G}_{n+r}$ for Φ defined as in (4). Consequently, the catastrophe surfaces of equivalent unfoldings, as well as their catastrophe sets, are mapped onto each other by the local diffeomorphism Φ. Thus, Φ preserves the phenomenology, since the essentials, such as modality, hysteresis, and sudden jumps, are unaltered. Indeed, if $[\Phi]$ is in $\mathcal{G}_{n+r}$, then a local inverse X of Φ has the form $X(x,u) = (\chi(x,u), \eta(u))$ for small x and u, where $[\eta] = [\psi]^{-1}$ and $\chi(x,0) = x$. This follows from

$$(x,u) = \Phi(X(x,u)) = (\varphi(X(x,u)), \psi(X_2(x,u))),$$

since $u = \psi(X_2(x,u))$ implies $X_2(x,u) = \eta(u)$, and therefore, $x = \varphi(X(x,u))$ yields $x = \varphi(X(x,0)) = \varphi(\chi(x,0),0) = \chi(x,0)$. Consequently, $F \circ \Phi(y,v) = G(y,v) - \gamma(v)$ gives

$$F(x,u) = G(X(x,u)) - \gamma(\eta(u)),$$

which means that $[F]$ is induced by $[G]$. This, together with the following theorem, proves that the equivalence of unfoldings defines an equivalence relation on the set of all unfoldings of a given germ.

Transitivity of Inducing Unfoldings (6). Let $[F] \in \mathcal{E}_{n+r}$, $[G] \in \mathcal{E}_{n+s}$, and $[H] \in \mathcal{E}_{n+t}$ be unfoldings of a germ in $\mathcal{E}_n$. Suppose that $[H]$ is induced by $[G]$ and $[G]$ is induced by $[F]$. Then $[H]$ is induced by $[F]$.

Proof. Let y, v, z, and w be sufficiently small. By definition, $G(y,v) = F(\Phi(y,v)) + \gamma(v)$ and $H(z,w) = G(\Psi(z,w)) + \beta(w)$ with $[\Phi] \in m_{n+s,n+r}$, $[\gamma] \in m_s$, $[\Psi] \in m_{n+t,n+s}$, and $[\beta] \in m_t$. Then

$$H(z,w) = F(\Phi(\Psi(z,w))) + \gamma(\Psi_2(w)) + \beta(w)$$

follows if Ψ_2 is given by $\Psi(z,w) = (\Psi_1(z,w), \Psi_2(w))$. Hence, $H(z,w) = F(X(z,w)) + \alpha(w)$ with $\alpha(w) := \gamma(\Psi_2(w)) + \beta(w)$ and $X = \Phi \circ \Psi$. Obviously, $[\alpha]$ is in m_t and $[X]$ is in $m_{n+t,n+r}$. Finally, using $\Phi(y,v) = (\Phi_1(y,v), \Phi_2(v))$, one obtains $X(z,w) = (\Phi_1(\Psi(z,w)), \Phi_2(\Psi_2(w)))$ and $\Phi_1(\Psi(z,0)) = \Phi_1(z,0) = z$.

Note that the parameter transformation $X_2 := \Phi_2 \circ \psi_2$ for the induction of $[H]$ by $[F]$ is obviously a local diffeomorphism if Φ_2 and ψ_2 are local diffeomorphisms.

Given two unfoldings of a germ, a simple construction of a third unfolding inducing the first two is to build their sum. This construction will be frequently used in subsequent chapters.

Sum of Unfoldings (7). Let $[F] \in \mathcal{E}_{n+r}$ and $[G] \in \mathcal{E}_{n+s}$ be two unfoldings of a germ $[f] \in \mathcal{E}_n$. Then

$$H(x; u, v) := F(x,u) + G(x,v) - f(x),$$

for $(x; u, v)$ in an open neighborhood of the origin in $\mathbf{R}^{n+(r+s)}$, defines an $(r + s)$-parameter unfolding $[H] \in \mathcal{E}_{n+(r+s)}$ of $[f]$ called the **sum of** $[F]$ **and** $[G]$. The unfoldings $[F]$ and $[G]$ are induced by their sum $[H]$.

Proof. Obviously, $[H]$ lies in $\mathcal{E}_{n+(r+s)}$, and $H(x; 0, 0) = F(x, 0) + G(x, 0) - f(x) = 2f(x) - f(x) = f(x)$ holds. Hence, $[H]$ is an $(r + s)$-parameter unfolding of $[f]$. Let $[\Phi] \in m_{n+r,n+(r+s)}$ be given by $\Phi(x, u) := (x; u, 0)$ and let $[\gamma] := 0 \in m_r$. Then $H(\Phi(x, u)) + \gamma(u) = H(x; u, 0) = F(x, u) + G(x, 0) - f(x) = F(x, u)$ follows, and $[F]$ is induced by $[H]$. Similarly, $[G]$ is induced by $[H]$.

The question now is whether a germ has an unfolding by which all others can be induced. The fundamental importance of such an unfolding is evident, since it determines a classification of all unfoldings of the given germ.

Definition. An unfolding of a germ in $\mathcal{E}_n$ is said to be **versal** when it induces all unfoldings of the germ. If the number of parameters of a versal unfolding is minimal, then it is called a **universal unfolding**.

Of course, a universal unfolding of a germ exists if and only if there is a versal one. In Chapter 12 the notion of a **stable unfolding** is introduced denoting a certain kind of persistence under the influence of small perturbations. It is proven that versality and stability imply each other. Therefore in the literature the term stable is often used to mean versal.

A brief summary of the main results proved in the following three chapters is given here. First of all, the **Fundamental Theorem on Universal Unfoldings** states that a universal unfolding of a germ $[f]$ in m^2 exists if and only if the codimension or, equivalently, the determinacy, of $[f]$ is finite. Moreover, a universal unfolding can be calculated explicitly as follows. If a linear basis of the quotient space of m with respect to the Jacobi ideal of $[f]$ has representatives $[g_1], \ldots, [g_r]$, then

(8) $$F(x, u) := f(x) + u_1 g_1(x) + \cdots + u_r g_r(x)$$

defines a universal unfolding of $[f]$; the number of parameters is equal to the codimension of $[f]$. Universal unfoldings are unique in the sense that any two universal unfoldings of a germ are equivalent. Indeed, even a more general result holds; namely, two versal unfoldings of a germ are equivalent if and only if they have the same number of parameters. In this way, a classification of the unfoldings of germs with finite codimension will be achieved.

Using these results, the list in (6.1) of the **7** elementary catastrophes is easily supplemented by the list of their universal unfoldings given in Table 1. These unfoldings are constructed by means of the computations in (5.2), (5.11), (5.12), and (5.13). For the cusp, butterfly, and parabolic umbilic the negative unfolding

TABLE 1: CLASSIFICATION OF UNFOLDINGS

germ	cod	universal unfolding	
$[x^3]$	1	$[x^3 + ux]$	fold
$\pm[x^4]$	2	$\pm[x^4 - ux^2 + vx]$	cusp
$[x^5]$	3	$[x^5 + ux^3 + vx^2 + wx]$	swallowtail
$\pm[x^6]$	4	$\pm[x^6 + tx^4 + ux^3 + vx^2 + wx]$	butterfly
$[x^3 - xy^2]$	3	$[x^3 - xy^2 + w(x^2 + y^2) - ux - vy]$	elliptic umbilic
$[x^3 + y^3]$	3	$[x^3 + y^3 + wxy - ux - vy]$	hyperbolic umbilic
$\pm[x^2y + y^4]$	4	$\pm[x^2y + y^4 + wx^2 + ty^2 - ux - vy]$	parabolic umbilic

is referred to as the **dual**. Computer-generated graphics of several elementary catastrophes follow.

Catastrophe Surface of the Cusp

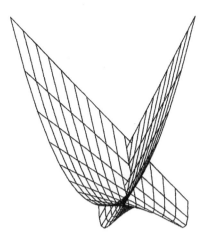

Bifurcation Set of the Swallowtail

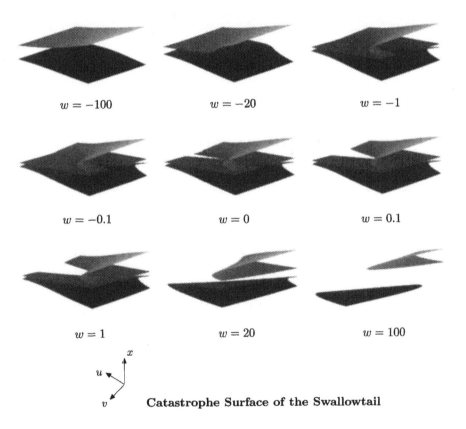

$w = -100$ $w = -20$ $w = -1$

$w = -0.1$ $w = 0$ $w = 0.1$

$w = 1$ $w = 20$ $w = 100$

Catastrophe Surface of the Swallowtail

Notice that if $[F]$ is a versal, respectively universal, unfolding of a germ $[f]$ and if q is a nondegenerate quadratic form in other variables, then the sum $[q] + [F]$ is a versal, respectively universal, unfolding of $[q] + [f]$.

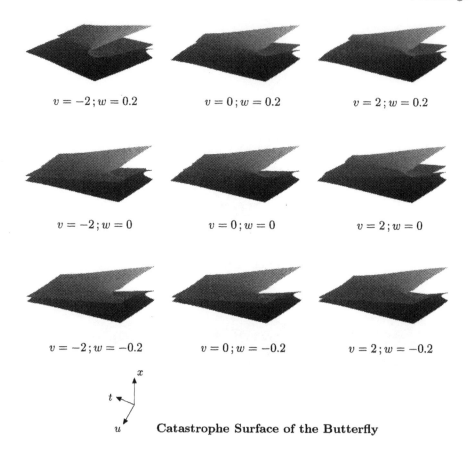

$$v = -2 \,;\, w = 0.2 \qquad\qquad v = 0 \,;\, w = 0.2 \qquad\qquad v = 2 \,;\, w = 0.2$$

$$v = -2 \,;\, w = 0 \qquad\qquad v = 0 \,;\, w = 0 \qquad\qquad v = 2 \,;\, w = 0$$

$$v = -2 \,;\, w = -0.2 \qquad\qquad v = 0 \,;\, w = -0.2 \qquad\qquad v = 2 \,;\, w = -0.2$$

Catastrophe Surface of the Butterfly

Unfoldings of equivalent germs are closely related, as will be explained now. Let $[f]$ and $[g]$ be in $\mathcal{E}_n$ and $[\chi]$ in $\mathcal{G}_n$ such that $[g] = [f][\chi]$. Then for an r-parameter unfolding $[F]$ of $[f]$ consider the germ $[G]$ in $\mathcal{E}_{n+r}$ given by $G(x, u) := F(\chi(x), u)$ for small x and u. Since $G(x, 0) = F(\chi(x), 0) = f(\chi(x)) = g(x)$, $[G]$ is an r-parameter unfolding of $[g]$, which will be denoted by $\chi^*[F]$. Hence χ^* is a map from the set of unfoldings of $[f]$ into the set of unfoldings of $[g]$ preserving the number of parameters. Obviously, χ^* is a bijection and its inverse is given by $[G] \mapsto \xi^*[G]$ if ξ is a local inverse of χ at the origin. It will be shown now that χ^* preserves induction.

Suppose that $[H] \in \mathcal{E}_{n+s}$ is an unfolding of $[f]$ induced by $[F]$. Then $H(y, v) = F(\zeta(y, v), \psi(v)) + \gamma(v)$ holds for some $[\zeta] \in m_{n+s,n}$ satisfying $\zeta(y, 0) = y$, $[\psi] \in m_{s,r}$, and $[\gamma] \in m_s$. Therefore

$$H(\chi(y), v) = F(\zeta(\chi(y), v), \psi(v)) + \gamma(v) = G(\varphi(y, v), \psi(v)) + \gamma(v)$$

is valid with $\varphi(y, v) := \eta(\zeta(\chi(y), v))$ for small y and v and $[\eta] := [\chi]^{-1}$. Clearly, $[\varphi]$ lies in $m_{n+s,n}$ and satisfies $\varphi(y, 0) = y$, showing that $\chi^*[H]$ is induced by $\chi^*[F]$, as asserted. As a consequence, versal and universal unfoldings are mapped bijectively onto versal and universal ones, respectively.

These considerations are important for applications, because unfoldings usually do not appear in a normal form. If $[G] \in \mathcal{E}_{n+s}$ is an unfolding of a germ $[g]$, which is equivalent to a germ $[f]$ whose universal unfolding $[F]$ is listed in Table 1, what is the relationship between $[G]$ and $[F]$? To answer this, consider a $[\chi] \in \mathcal{G}_n$ with $[g] = [f][\chi]$. As shown above, $\chi^*[F]$ is a universal unfolding of $[g]$, so that $[G]$ is induced by $\chi^*[F]$. Therefore, $G(y, v) = F(\chi(\zeta(\zeta(y, v)), \psi(v)) + \gamma(v)$ holds for some $[\psi] \in m_{s,r}, [\gamma] \in m_s$, and $[\zeta] \in m_{n+s,n}$ satisfying $\zeta(y, 0) = y$. This is equivalent to

$$(9) \qquad\qquad G(y, v) = F(\omega(y, v), \psi(v)) + \gamma(v)$$

for small x and v, where $[\psi] \in m_{s,r}$, $[\gamma] \in m_s$, and $[\omega] \in m_{n+s,n}$ is such that $x \mapsto \omega_0(x) := \omega(x, 0)$ is a local diffeomorphism at the origin satisfying $[g] = [f][\omega_0]$. If, more generally, $[g]$ is equivalent to $[f]$ plus the germ of a nondegenerate quadratic form q in other variables, then the unfolding $[F]$ is replaced in (9) by $[q] + [F]$ as remarked after Table 1.

Notice that the relationship (9) between $[F]$ and $[G]$ generalizes the concept of induction to unfoldings of equivalent germs. If $r = s$ and ψ is in $\mathcal{G}_r$, then $[F]$ and $[G]$ are said to be **isomorphic** and $\Phi(y, v) := (\omega(y, v), \psi(v))$ defines a local diffeomorphism at the origin that maps M_G onto M_F and C_G onto C_F.

In conclusion, it should be noted that in applications an unfolding $[G]$ frequently arises as a germ at some point in $\mathbf{R}^n \times \mathbf{R}^s$ different from the origin. This necessitates a translation in $\mathbf{R}^n \times \mathbf{R}^s$ and an additional obvious modification of (9). Furthermore, in applications $[g]$ is frequently equivalent to $[f]$ only up to a nonzero constant, but this can easily be taken into account by replacing γ with $\gamma + g(0)$ in (9).

These considerations on unfoldings of equivalent germs at different points will become particularly important when treating the stability of unfoldings in Chapter 12.

Exercises

7.1. Let $[f]$ be a germ in m^2. If 0 is a nondegenerate critical point of f, show that the origin of every unfolding $[F]$ of $[f]$ is not a cluster point of the catastrophe set C_F. If 0 is degenerate, prove the existence of an unfolding $[F]$ of $[f]$ such that the origin is a cluster point of $M_F \backslash C_F$.

7.2. Consider the two polynomial unfoldings $[P]$ and $[Q]$ of the germ $[x^r]$ in $\mathcal{E}_1$ for $r \geq 3$ given by

$$P(x, u) := x^r + u_{r-2}x^{r-2} + \cdots + u_1 x,$$
$$Q(y, v) := y^r + v_{r-1}y^{r-1} + \cdots + v_1 y + v_0.$$

Show that $[P]$ and $[Q]$ induce each other but are not equivalent. Furthermore, show that the sum of $[P]$ and the constant two-parameter unfolding of $[x^r]$ is equivalent to $[Q]$.

Notice that according to the Fundamental Theorem on Universal Unfoldings, $[P]$ is universal due to (5.2), and hence $[Q]$ is versal.

Hint: Use the variable transformation

$$\varphi \colon \mathbf{R} \times \mathbf{R}^r \to \mathbf{R}, \quad \varphi(y, v) := y + \frac{v_{r-1}}{r}.$$

7.3.   Show that the germ of $F(x, u) := x^3 + u$ is not a versal unfolding of $[x^3]$.

7.4.   Let $[F] \in \mathcal{E}_{n+r}$ be an unfolding of a germ $[f]$ in m_n^2, and denote by $\mathcal{I}[F]$ the ideal in $\mathcal{E}_{n+r}$ generated by the germs of the partial derivatives $[\partial F / \partial x_i]$. As usual, let $\mathcal{E}_n$ and $\mathcal{E}_r$ be embedded in $\mathcal{E}_{n+r}$. Show that the origin in $\mathbf{R}^n$ is a nondegenerate critical point of f if and only if

$$\mathcal{E}_{n+r} = \mathcal{I}[F] + \mathcal{E}_r$$

holds. Moreover, when this decomposition of $\mathcal{E}_{n+r}$ is valid, then the sum is direct, i.e.,

$$\mathcal{I}[F] \cap \mathcal{E}_r = \{0\}.$$

Hint: Show that $\mathcal{E}_{n+r} = \mathcal{I}[F] + \mathcal{E}_r$ implies $m_n = \mathcal{J}[f]$, and therefore 0 is nondegenerate by (5.6). To prove the converse of this, use the map $\varphi(x, u) :=$

$$\left(\frac{\partial F}{\partial x_1}(x, u), \ldots, \frac{\partial F}{\partial x_n}(x, u), u_1, \ldots, u_r \right),$$

and apply its pullback $\varphi^* \colon \mathcal{E}_{n+r} \to \mathcal{E}_{n+r}, [G] \mapsto \varphi^*[G] := [G \circ \varphi]$ to the direct sum

$$\mathcal{E}_{n+r} = \langle m_n \rangle_{\mathcal{E}_{n+r}} \oplus \mathcal{E}_r.$$

7.5.   The following notations refer to Lemma (4). Find simple examples for which

(a) Φ is not injective and $\Phi(Z)$ is open,

(b) Φ is injective and $\Phi(Z)$ is not open, and

(c) Φ is injective and $\Phi(Z)$ is open, but Φ is not a local diffeomorphism at the origin.

7.6.   Extend Table 1 of this chapter to include the universal unfoldings of the germs with codimensions 5 and 6 treated in Exercises 6.2 and 6.3.

Hint: Apply the Fundamental Theorem on Universal Unfoldings as cited in (8).

7.7.   Find the universal unfoldings of the codimension 7 germs in m_2^3 given in Exercise 6.4.

Hint: Again apply (8).

7.8. Let $[F]$ be an unfolding of $[f] \in m^2$, and denote by $[G]$ an unfolding induced by $[F]$. Prove the following:

(a) If $[G]$ is versal, then $[F]$ is versal.

(b) If $[F]$ is versal, then there is a $[G]$ that is universal.

7.9.

(a) Consider germs $[F] \in \mathcal{E}_{n+1}$ and $[\varphi] \in m_{n+1,n}$ satisfying

$$F(x,u) = F(\varphi(x,u),0), \quad \varphi(x,0) = x$$

for small $x \in \mathbf{R}^n$ and $u \in \mathbf{R}$. Show that there are germs $[h_i] \in \mathcal{E}_{n+1}, i = 1,\ldots,n$, such that $X[F] = 0$, where $X: \mathcal{E}_{n+1} \to \mathcal{E}_{n+1}$ denotes the differential operator given by

$$[G] \mapsto \left[\frac{\partial}{\partial u}G\right] + \sum_{i=1}^{n}[h_i]\left[\frac{\partial}{\partial x_i}G\right].$$

(b) Conversely, show that if X is a differential operator as in (a) and $[F]$ is a germ in $\mathcal{E}_{n+1}$ with $X[F] = 0$, then there is a germ $[\varphi]$ in $m_{n+1,n}$ satisfying

$$\varphi(x,0) = x, \quad F(x,u) = F(\varphi(x,u),0)$$

for small $x \in \mathbf{R}^n$ and $u \in \mathbf{R}$.

Note that if $[h_j] = 0$ for one j, then the derivative with respect to x_j does not appear in X, and x_j can be considered as a parameter on which F, the h_i's, and X itself smoothly depend. Then the solution φ also depends smoothly on that parameter by (b).

Hint: For (b), consider the system of differential equations

$$\frac{d}{du}\xi_i = h_i(\xi,u), \quad 1 \le i \le n,$$

and verify that

$$\frac{\partial}{\partial u}F(\xi_1(u),\ldots,\xi_n(u),u) = 0$$

for every solution $\xi_1,\ldots,\xi_n$. Apply Appendix A.3. This is essentially the method of characteristic curves used to solve certain types of partial differential equations.

7.10. Let $[f]$ be in m_n^2. Prove that the origin is a nondegenerate critical point of $[f]$ if and only if $[f]$ is a versal, and hence a universal, unfolding of itself.
Hint: This is an immediate consequence of (8) and (5.6). It also results trivially from (10.6), as mentioned in the comment following (10.6). Here a proof is requested that is based on (5.6) and on Exercises 7.4 and 7.9(b).

The sufficiency of the condition is easy to verify using (5.6). For the proof of the necessity, apply Exercise 7.4 to $\partial F/\partial u_r$, and show that after an obvious modification of $[F]$ one obtains an unfolding $[F_1]$ of $[f]$ satisfying

$$\frac{\partial F_1}{\partial u_r} = \sum_{i=1}^{n} h_i \frac{\partial F_1}{\partial x_i}.$$

Now apply Exercise 7.9(b). Repeat this procedure replacing $\partial F/\partial u_r$ by $\partial F_1/\partial u_{r-1}$, etc.

7.11. Let the origin of $[f] \in m_n^2$ be a nondegenerate critical point. Suppose that $[F] \in \mathcal{E}_{n+1}$ is a one-parameter unfolding of $[f]$. By Exercise 7.10, $[F]$ is induced by $[f]$, i.e., there exists a germ $[\varphi] \in m_{n+1,n}$ satisfying $\varphi(x,0) = x$ for small x and a germ $[\gamma] \in m_1$ such that

$$F(x,u) = f(\varphi(x,u)) + \gamma(u)$$

holds for small x and u.

    (a)   Show that $[\gamma]$ is uniquely determined.

    (b)   For $n = 1$ and $f(x) := x^2$ consider the unfoldings $[F] \in \mathcal{E}_{1+1}$ of $[f]$ given by

$$F(x,u) := x^2 + ux,$$
$$F(x,u) := x^2 + u(x + x^2),$$
$$F(x,u) := x^2 + u(x + x^3).$$

For each unfolding $[F]$ compute the corresponding $[\gamma]$ and then find $[\varphi]$.

Hint: Apply Exercise 7.9(a) to $\tilde{F} := F - \gamma$ and solve $D_1\tilde{F}(x,u) = 0$ for x by the Implicit Function Theorem (Appendix A.1).

7.12. Let $[f]$ be in $\mathcal{E}$ such that the origin is not a critical point for $[f]$. Show that $\mathcal{J}[f] = \mathcal{E}$ and that $[f]$ is a universal unfolding of itself.

Solutions

7.1. Let 0 be a nondegenerate critical point of f, and suppose that $[F]$ is an unfolding of $[f]$. Since $D^2f(0)$ is invertible and $D^2F_u(x) \to D^2f(0)$ as $(x,u) \to (0,0)$, then $D^2F_u(x)$ also is invertible for small x and u.

Now assume that 0 is degenerate. By the remark preceding (1.37) it is no restriction to assume $D^2 f(0) = 2 \begin{pmatrix} A & 0 \\ 0 & 0 \end{pmatrix}$, where A is an invertible $(r \times r)$ matrix and r is the rank of f at the origin. Consider the unfolding of $[f]$ given by

$$F(x, u) := f(x) + u(x_{r+1}^2 + \cdots + x_n^2)$$

for a real parameter u. It satisfies

$$D^2 F(0, u) = 2 \begin{pmatrix} A & 0 \\ 0 & uE_{n-r} \end{pmatrix},$$

which is an invertible matrix for nonzero u. Consequently, $(0, u)$ is in $M_F \backslash C_F$ and converges to the origin when u approaches zero.

7.2. Obviously, $[P]$ and $[Q]$ are not equivalent, since $[P]$ is an $(r-2)$-parameter unfolding and $[Q]$ has r parameters.

To see that $[Q]$ induces $[P]$, choose $\varphi(x, u) := x$, $\psi_j(u) := u_j$ for $1 \le j \le r - 2$, $\psi_0(u) := \psi_{r-1}(u) := 0$, and $\gamma := 0$.

In order to show that $[P]$ induces $[Q]$, take the variable transformation φ proposed in the hint. For the following computations it is more convenient to insert $y = x - v_{r-1}/r$ in $Q(y, v)$ instead of $x = y + v_{r-1}/r$ in $P(x, u)$. Setting $v_r := 1$, one obtains

$$Q\left(x - \frac{v_{r-1}}{r}, v\right) = \sum_{j=0}^{r} v_j \left(x - \frac{v_{r-1}}{r}\right)^j = \sum_{j=0}^{r} v_j \sum_{\iota=0}^{j} \binom{j}{\iota} \left(-\frac{v_{r-1}}{r}\right)^{j-\iota} x^\iota$$

$$= \sum_{\iota=0}^{r} x^\iota \left[\sum_{j=\iota}^{r} \binom{j}{\iota} v_j \left(-\frac{v_{r-1}}{r}\right)^{j-\iota} \right]$$

$$= x^r + \sum_{\iota=1}^{r-2} x^\iota \psi_\iota(v) + \gamma(v),$$

where

$$\psi_\iota(v) := \sum_{j=0}^{r} \binom{j}{\iota} v_j \left(-\frac{v_{r-1}}{r}\right)^{j-\iota}, \quad 1 \le \iota \le r - 2,$$

$$\gamma(v) := \sum_{j=0}^{r} v_j \left(-\frac{v_{r-1}}{r}\right)^j.$$

Clearly, $[\psi] \in m_{r,r-2}$, $[\gamma] \in m_r$, and

$$P(\varphi(y, v), \psi(v)) + \gamma(v) = Q(y, v).$$

Now the last assertion will be proved. The germ of $R(x, u_0, u, u_{r-1}) := P(x, u)$ is the sum of $[P]$ and the constant two-parameter unfolding of $[x^r]$. The parameters of the constant unfolding are denoted by u_0 and u_{r-1}. For ψ as defined above, the germ of $\chi(v) := (v_0, \psi(v), v_{r-1})$ is in $m_{r,r}$ and satisfies

$$R(\varphi(y, v), \chi(v)) + \gamma(v) = P(\varphi(y, v), \psi(v)) + \gamma(v) = Q(y, v).$$

Moreover, $D\chi(0) = E_r$ holds, and χ is a local diffeomorphism at the origin proving the equivalence of $[R]$ and $[Q]$.

7.3. Assume the contrary. Then germs $[\psi] \in m$, $[\gamma] \in m$, and $[\varphi] \in m_{2,1}$ exist with $\varphi(x, 0) = x$ for small x, satisfying $F(\varphi(x, u), \psi(u)) + \gamma(u) = x^3 + ux$, i.e.,

$$\varphi(x, u)^3 + \psi(u) + \gamma(u) = x^3 + xu$$

for small x and u. Differentiate this equation with respect to x. Evaluating at $x = 0$ yields $3\varphi(0, u)^2 D_1\varphi(0, u) = u$. Because of $\varphi(x, 0) = x$, the partial derivative $D_1\varphi(0, u)$ is equal to 1 at $u = 0$ and is hence positive for small u. Of course $\varphi(0, u)^2$ is nonnegative, and consequently for small negative u a contradiction is obtained.

7.4.

(i) Suppose that $\mathcal{E}_{n+r} = \mathcal{I}[F] + \mathcal{E}_r$ is valid. Let $[g]$ be a germ in m_n. Considering $[g]$ as an element in $\mathcal{E}_{n+r}$, the decomposition of $\mathcal{E}_{n+r}$ means there are germs $[G_i]$ in $\mathcal{E}_{n+r}, i = 1, \ldots, n$, and $[h]$ in $\mathcal{E}_r$ satisfying

$$g(x) = \sum_{i=1}^{n} G_i(x, u)\partial_i F(x, u) + h(u)$$

for small x and u. Evaluating at $u = 0$ yields $m_n = \mathcal{J}[f]$, since $\partial_i F(x, u) = D_i f(x)$ and $[D_i f] \in m_n$ for $i = 1, \ldots, n$. Hence $\mathrm{cod}[f] = 0$, and 0 is a nondegenerate critical point of $[f]$ by (5.6).

(ii) Assume now that 0 is a nondegenerate critical point of $[f]$. Then $\det D^2 f(0) \neq 0$ holds. First it will be shown that the map φ given in the hint is a local diffeomorphism at the origin with $\varphi(0) = 0$. Indeed, φ is smooth, and $\varphi(0) = 0$ holds since $\partial_i F(0, 0) = D_i f(0) = 0$ for $i = 1, \ldots, n$. Moreover, $D\varphi(0)$ has the form

$$\begin{pmatrix} D_1^2 F(0, 0) & * \\ 0 & E_r \end{pmatrix},$$

where D_1 denotes the derivation with respect to the x variables. Thus, $\det D\varphi(0) = \det D_1^2 F(0, 0) = \det D^2 f(0) \neq 0$.

For $[G]$ in $\mathcal{E}_{n+r}$ consider the decomposition $G = \tilde{G} + G_0$ with $\tilde{G}(x, u) := G(x, u) - G(0, u)$ and $G_0(u) := G(0, u)$ for small x and u. Then

$[G_0]$ lies in $\mathcal{E}_r$. Due to $\tilde{G}(0, u) = 0$, (1.28) can be applied, yielding $[\tilde{G}] \in \langle m_n \rangle_{\mathcal{E}_{n+r}}$. This proves

$$\mathcal{E}_{n+r} = \langle m_n \rangle_{\mathcal{E}_{n+r}} + \mathcal{E}_r.$$

Now apply φ^* to this sum, which obviously is direct; φ^* is an algebra isomorphism (see Exercise 4.6). Therefore, $\varphi^*(\mathcal{E}_{n+r}) = \mathcal{E}_{n+r}$ and

$$\varphi^*(\langle m_n \rangle_{\mathcal{E}_{n+r}}) = \langle \{\varphi^*[x_1], \ldots, \varphi^*[x_n]\} \varphi^*(\mathcal{E}_{n+r}) \rangle$$
$$= \langle \{[\partial_1 F], \ldots, [\partial_n F]\} \mathcal{E}_{n+r} \rangle = \mathcal{I}[F]$$

hold. Furthermore $\varphi^*(\mathcal{E}_r) = \mathcal{E}_r$ is valid, because $\varphi^*[h] = [h]$ holds for $[h]$ in $\mathcal{E}_r$ by definition of φ. Thus $\mathcal{E}_{n+r} = \mathcal{I}[F] + \mathcal{E}_r$ is true. It also follows that this sum is direct, and the final assertion follows using (i).

7.5. In the following three examples n and s are equal to 1. Moreover let $\varphi(y, v) := y$ and put $Z := \{|y| < 1, |v| < 1\}$.

(a) For $r = 2$ and $\psi(v) := (v, 0)$, it follows immediately that Φ is injective and $\Phi(Z) = Z \times \{0\}$ is not open in $\mathbf{R}^3$.

(b) Setting $r = 1$ and $\psi(v) := \sin \pi v$, one obtains $\Phi(Z) = Z$, which is open in $\mathbf{R}^2$. Obviously, Φ is not injective.

(c) For $r = 1$ and $\psi(v) := v^3$, Φ is injective and $\Phi(Z) = Z$ is open in $\mathbf{R}^2$. But Φ is not a local diffeomorphism at the origin, since ψ is not.

7.6.

germ	cod	universal unfolding	
$[x^7]$	5	$[x^7 + sx^5 + tx^4 + ux^3 + vx^2 + wx]$	wigwam
$\pm[x^3 + y^4]$	5	$\pm[x^3 + y^4 + sxy^2 + txy + uy^2 + vy + wx]$	symbolic umbilic
$[x^2y + y^5]$	5	$[x^2y + y^5 + sy^4 + ty^3 + uy^2 + vy + wx]$	2nd hyperbolic umbilic
$[x^2y - y^5]$	5	$[x^2y - y^5 + sy^4 + ty^3 + uy^2 + vy + wx]$	2nd elliptic umbilic
$\pm[x^8]$	6	$\pm[x^8 + rx^6 + sx^5 + tx^4 + ux^3 + vx^2 + wx]$	star
$[x^3 + xy^3]$	6	$[x^3 + xy^3 + rx^2y + sxy + ty^2 + ux^2 + vy + wx]$	
$\pm[x^2y + y^6]$	6	$\pm[x^2y + y^6 + ry^5 + sy^4 + ty^3 + uy^2 + vy + wx]$	

According to (8), a universal unfolding of a germ $[f]$ is given by adding to $[f]$ a general linear combination of representatives of a basis of $m/\mathcal{J}[f]$. The coefficients of the linear combination are the parameters of the unfolding. For the germs listed in the tables of Exercises 6.2 and 6.3, a basis of $m/\mathcal{J}[f]$ is computed in (5.2) and Exercises 5.4(b), 5.1, and 5.4(d).

7.7.

germ	cod	universal unfolding
$[x^3 + y^5]$	7	$[x^3 + y^5 + qxy^3 + rxy^2 + sy^3 + txy + uy^2 + vy + wx]$
$[x^2y + y^7]$	7	$[x^2y + y^7 + qy^6 + ry^5 + sy^4 + ty^3 + uy^2 + vy + wx]$
$[x^2y - y^7]$	7	$[x^2y - y^7 + qy^6 + ry^5 + sy^4 + ty^3 + uy^2 + vy + wx]$

A basis of $m/\mathcal{J}[f]$ for the germs $[f]$ listed in the table of Exercise 6.4 is computed in Exercises 5.4(b) and 5.1.

7.8.

(a) Since $[G]$ is versal, every unfolding $[H]$ of $[f]$ is induced by $[G]$. Because $[G]$ is induced by $[F]$, $[H]$ is also induced by $[F]$ due to (6).

(b) By hypothesis, the set of versal unfoldings of $[f]$ is not empty. Let $[G]$ be an element of this set with a minimal number of parameters; then $[G]$ is universal. Since $[F]$ is versal, $[G]$ is induced by $[F]$.

7.9.

(a) Set $f(x) := F(x,0)$ for small x. Then $F(x,u) = f(\varphi(x,u))$ holds and also

$$\frac{\partial}{\partial u}F(x,u) = \sum_{j=1}^{n} D_j f(\varphi(x,u))\frac{\partial}{\partial u}\varphi_j(x,u),$$

$$\frac{\partial}{\partial x_i}F(x,u) = \sum_{j=1}^{n} D_j f(\varphi(x,u))\frac{\partial}{\partial x_i}\varphi_j(x,u),$$

for $i = 1, \ldots, n$. Denote by $\Phi(x,u)$ the $n \times n$ matrix

$$\left(\frac{\partial}{\partial x_i}\varphi_j(x,u)\right).$$

It is invertible for small x and u, because $\Phi(x,0) = E_n$ follows from $\varphi(x,0) = x$. Solving the second of the above equations for $Df(\varphi(x,u))$, one obtains $Df(\varphi(x,u)) = \Phi(x,u)^{-1}D_1F(x,u)$, where D_1 denotes the gradient with respect to x. Inserting this result into the first equation yields the assertion $X[F] = 0$ for

$$h_i(x,u) := -\sum_{j=1}^{n}(\Phi(x,u)^{-1})_{ji}\frac{\partial}{\partial u}\varphi_j(x,u),$$

where $i = 1, \ldots, n$, and x and u are small.

(b)   Let $0 \in U \subset \mathbf{R}^n$ and $0 \in J \subset \mathbf{R}$ be open such that F and $h_i, i = 1, \ldots, n$, are defined on $U \times J$. For

$$H : U \times J \to \mathbf{R}^n, \quad H(x, u) := (h_1(x, u), \ldots, h_n(x, u)),$$

consider the ordinary differential equation

$$\frac{d}{du} \xi = H(\xi, u).$$

According to Appendix A.3, there are open sets V in U and I in J containing the origin with the following property: For every (x_0, u_0) in $V \times I$ there exists a unique solution $I \to U, u \mapsto \xi(x_0, u_0; u)$, of the above differential equation satisfying $\xi(x_0, u_0; x_0) = x_0$; the map $V \times I \times I \to \mathbf{R}^n, (x_0, u_0; u) \mapsto \xi(x_0, u_0; u)$, is smooth.

For any solution ξ in U on I of the above differential equation, the function $u \mapsto F(\xi(u), u)$ is constant, since

$$\frac{d}{du} F(\xi(u), u) = \frac{\partial}{\partial u} F(\xi(u), u)$$

$$+ \sum_{i=1}^{n} \frac{\partial}{\partial x_i} F(\xi(u), u) h_i(\xi(u), u) = 0$$

due to $X[F] = 0$. Therefore,

$$F(x_0, u_0) = F(\xi(x_0, u_0; u_0), u_0)$$
$$= F(\xi(x_0, u_0; 0), 0) \text{ for } (x_0, u_0) \in V \times I.$$

Set $\varphi(x, u) := \xi(x, u; 0)$ for $(x, u) \in V \times I$. Then $F(\varphi(x, u), 0) = F(x, u)$ and $\varphi(x, 0) = \xi(x, 0; 0) = x$.

7.10.   First the sufficiency will be shown. Consider the n-parameter unfolding of $[f]$ given by

$$H(x, u) := f(x) + \sum_{j=1}^{n} u_j x_j.$$

Since $[f]$ is versal, there exist $[\varphi] \in m_{n+n,n}$ and $[\gamma] \in m_n$ with $\varphi(x, 0) = x$ and $H(x, u) = f(\varphi(x, u)) + \gamma(u)$ for small x and u. Then

$$\frac{\partial}{\partial u_j} H(x, 0) = x_j = \sum_{i=1}^{n} D_i f(x) \frac{\partial}{\partial u_j} \varphi_i(x, 0) + \frac{\partial}{\partial u_j} \gamma(0)$$

follows. Since $D_i f(0) = 0$, this yields $(\partial \gamma / \partial u_j)(0) = 0$ and thus $m \subset \mathcal{J}[f]$. Therefore, $\text{cod}[f] = 0$, proving the assertion by (5.6).

For the proof of the necessity, let $[F] \in \mathcal{E}_{n+r}$ be an unfolding of $[f]$. Applying Exercise 7.4 to $\partial F / \partial u_r$, one finds a $[g_1] \in \mathcal{E}_r$ and $[h_{1i}] \in \mathcal{E}_{n+r}$ such that

$$\frac{\partial}{\partial u_r} F(x, u) = g_1(u) + \sum_{i=1}^{n} h_{1i}(x, u) \frac{\partial}{\partial x_i} F(x, u)$$

for small x and u. Define

$$\gamma_1(u) := \int_0^{u_r} g_1(u_1, \dots, u_{r-1}, \lambda) d\lambda \text{ and}$$

$$F_1(x, u) := F(x, u) - \gamma_1(u).$$

Then $[F_1]$ is still an unfolding of $[f]$ due to $\gamma_1(0) = 0$. Moreover,

$$\frac{\partial}{\partial u_r} F_1(x, u) = \sum_{i=1}^{n} h_{1i}(x, u) \frac{\partial}{\partial x_i} F_1(x, u)$$

holds. Therefore, by Exercise 7.9(b), there is a $[\varphi_1] \in m_{n+1,n}$ satisfying

$$\varphi_1(x, 0) = x \text{ and } F_1(x, u) = F_1(\varphi_1(x, u); u_1, \dots, u_{r-1}, 0).$$

Notice that the smoothness of φ_1 with respect to the parameters $u_1, \dots, u_{r-1}$ is also a consequence of Exercise 7.9(b), as the remark at the end of that exercise shows.

Now replace $[F]$ by $[F_1]$. Similarly, one obtains from $[F_1]$ an unfolding $[F_2]$ of $[f]$: According to Exercise 7.4,

$$\frac{\partial}{\partial u_{r-1}} F_1(x, u) = g_2(u) + \sum_{i=1}^{n} h_{2i}(x, u) \frac{\partial}{\partial x_i} F_1(x, u)$$

holds. Define $F_2(x, u) := F_1(x, u) - \gamma_2(u)$ for

$$\gamma_2(u) := \int_0^{u_{r-1}} g_2(u_1, \dots, u_{r-2}, \lambda, u_r) d\lambda.$$

By Exercise 7.9(b), it follows that

$$F_2(x, u) = F_2(\varphi_2(x, u); u_1, \dots, u_{r-2}, 0, u_r)$$

for some $[\varphi_2] \in m_{n+1,n}$ with $\varphi_2(x, 0) = x$. $[F_2]$ can be expressed by $[F]$ as follows:

$$\begin{aligned}
F_2(x, u) &= F_2(\varphi_2(x, u); u_1, \dots, u_{r-2}, 0, u_r) \\
&= F_1(\varphi_2(x, u); u_1, \dots, u_{r-2}, 0, u_r) \\
&\quad - \gamma_2(u_1, \dots, u_{r-2}, 0, u_r) \\
&= F_1(\varphi_1(\varphi_2(x, u); u_1, \dots, u_{r-2}, 0, u_r); u_1, \dots, u_{r-2}, 0, 0) \\
&\quad - \gamma_2(u_1, \dots, u_{r-2}, 0, u_r) \\
&= F(\tilde{\varphi}_2(x, u); u_1, \dots, u_{r-2}, 0, 0) - \gamma_1(u_1, \dots, u_{r-2}, 0, 0) \\
&\quad - \gamma_2(u_1, \dots, u_{r-2}, 0, u_r),
\end{aligned}$$

where $\tilde{\varphi}_2(x, u) := \varphi_1(\varphi_2(x, u); u_1, \ldots, u_{r-2}, 0, u_r)$. On the other hand, $F_2(x, u) = F_1(x, u) - \gamma_2(u) = F(x, u) - \gamma_1(u) - \gamma_2(u)$. Therefore,

$$F(x, u) = F(\tilde{\varphi}_2(x, u); u_1, \ldots, u_{r-2}, 0, 0) + \tilde{\gamma}_2(u)$$

holds, where

$$\tilde{\gamma}_2(u) := \gamma_1(u) + \gamma_2(u) - \gamma_1(u_1, \ldots, u_{r-2}, 0, 0)$$
$$- \gamma_2(u_1, \ldots, u_{r-2}, 0, u_r).$$

Moreover, $\tilde{\varphi}_2(x, 0) = \varphi_1(\varphi_2(x, 0); 0) = \varphi_2(x, 0) = x$ and $\tilde{\gamma}_2(0) = 0$ are valid.

Proceeding in this way, one finally obtains $F(x, u) = F(\tilde{\varphi}_r(x, u), 0) + \tilde{\gamma}_r(u) = f(\tilde{\varphi}_r(x, u)) + \tilde{\gamma}_r(u)$ for some $[\tilde{\varphi}_r] \in m_{n+1}, {}_n$ satisfying $\tilde{\varphi}_r(x, 0) = x$ and some $[\tilde{\gamma}_r] \in m_r$. Hence $[F]$ is induced by $[f]$.

7.11.

(a) Put $\tilde{F}(x, u) := F(x, u) - \gamma(u)$ for small x and u. Then $\tilde{F}(x, u) = f(\varphi(x, u))$ and $\tilde{F}(x, 0) = f(x)$. Hence $\tilde{F}(x, u) = \tilde{F}(\varphi(x, u), 0)$ holds, and by Exercise 7.9(a), there are germs $[h_i] \in \mathcal{E}_{n+1}, i = 1, \ldots, n$, satisfying

$$\frac{\partial \tilde{F}}{\partial u} = \sum_{i=1}^{n} h_i \frac{\partial}{\partial x_i} \tilde{F}$$

around the origin of $\mathbf{R}^n \times \mathbf{R}$.

Now consider $g(x, u) := D_1 \tilde{F}(x, u)$, where D_1 denotes the gradient with respect to x. Then $g(0, 0) = Df(0) = 0$ and $D_1 g(0, 0) = D^2 f(0)$ is invertible, since the origin is a nondegenerate critical point of f. The Implicit Function Theorem (Appendix A.1) can be applied to g, yielding a $[\xi] \in m_1, {}_n$ with $g(\xi(u), u) = 0$ for small u.

Combining this result with the above formula for $\partial \tilde{F}/\partial u$, one has

$$0 = \left(\frac{\partial}{\partial u} \tilde{F}\right)(\xi(u), u) = \left(\frac{\partial}{\partial u} F\right)(\xi(u), u) - \gamma'(u),$$

and hence

$$\gamma(u) = \int_0^u \left(\frac{\partial}{\partial u} F\right)(\xi(\lambda), \lambda) \, d\lambda.$$

This proves (a). With regard to the following applications of this formula, note that ξ just has to satisfy $\xi(0) = 0$ and $D_1 F(\xi(u), u) = 0$ for small u.

(b) For $F(x, u) := x^2 + ux$, one has $D_1 F(x, u) = 2x + u$ and hence $2\xi(u) + u = 0$, i.e., $\xi(u) = -u/2$. Clearly, ξ is smooth and vanishes at the origin. Furthermore, $(\partial/\partial u) F(x, u) = x$. Thus

$$\gamma(u) = \int_0^u (-\lambda/2) d\lambda = -u^2/4.$$

The equation $F(x, u) = f(\varphi(x, u)) + \gamma(u)$ means $x^2 + ux = (\varphi(x, u))^2 - u^2/4$, and thus $\varphi(x, u) = \varepsilon(x, u)(x + u/2)$ follows for some function ε with values in $\{1, -1\}$. Since φ is continuous, ε is constant on $\{(x, u) : x + u/2 > 0\}$ and on $\{(x, u) : x + u/2 < 0\}$. Denote by ε_+ and ε_- the values of ε on these regions. Because $\varepsilon(x, x) = 2\varphi(x, x)/3x$ has a limit as x tends to zero, $\varepsilon_+ = \varepsilon_-$ follows. Finally, the condition $\varphi(x, 0) = x$ singles out the positive sign, and $\varphi(x, u) = x + u/2$ is valid on all of $\mathbf{R} \times \mathbf{R}$.

For $F(x, u) := x^2 + u(x + x^2)$, one has $D_1 F(x, u) = 2x(1 + u) + u$, and hence $2\xi(u)(1 + u) + u = 0$, i.e., $\xi(u) = -u/2(1 + u)$ for $u > -1$. Consequently, ξ is smooth and vanishes at the origin. Furthermore, from $(\partial/\partial u)F(x, u) = x + x^2$, one obtains

$$\gamma(u) = \int_0^u \left(-\frac{\lambda}{2(1 + \lambda)} + \frac{\lambda^2}{4(1 + \lambda)^2} \right) d\lambda = \frac{-u^2}{4(1 + u)} \quad \text{for}$$

$u > -1$.

Now the condition on φ is

$$(\varphi(x, u))^2 = (1 + u)\left(x + \frac{1}{2}\frac{u}{1 + u} \right)^2.$$

Smoothness, together with the property $\varphi(x, 0) = x$, implies that

$$\varphi(x, u) = \sqrt{1 + u}\left(x + \frac{1}{2}\frac{u}{1 + u} \right)$$

is valid for $x \in \mathbf{R}$ and $u > -1$.

The previous two examples are so simple that one could guess $[\gamma]$ by completing the square. However, this is hardly possible for $F(x, u) := x^2 + u(x + x^3)$.

The zeros of $D_1 F(x, u) = 3ux^2 + 2x + u$ are

$$x = \frac{1}{3u}(-1 \pm \sqrt{1 - 3u^2}) \quad \text{for} \quad u \neq 0$$

and $x = 0$ for $u = 0$. Since the branch belonging to the negative root is not continuous at $u = 0$, one has

$$\xi(u) = \frac{1}{3u}(-1 + \sqrt{1 - 3u^2}) = -u(1 + \sqrt{1 - 3u^2})^{-1}.$$

The second expression shows that ξ is smooth for $|u| < 1/\sqrt{3}$ and that ξ vanishes at the origin. Then $(\partial/\partial u)F(x, u) = x + x^3$ and hence

$$\gamma(u) = \int_0^u (\xi(\lambda) + \xi(\lambda)^3)d\lambda.$$

This integral can be computed by substituting $\mu := 1 + \sqrt{1 - 3\lambda^2}$, and

$$\gamma(u) = \frac{2}{9}\frac{1 - 3u^2}{1 + \sqrt{1 - 3u^2}} - \frac{1}{9} \quad \text{for} \quad |u| < \frac{1}{\sqrt{3}}$$

follows.

The condition φ has to satisfy is $(\varphi(x,u))^2 = F(x,u) - \gamma(u)$. The problem is to find the appropriate square root of $F(x,u) - \gamma(u)$. To achieve this, its zeros are needed. First differentiate the equation with respect to x. Substituting x by $\xi(u)$ yields $\varphi(\xi(u),u)D_1\varphi(\xi(u),u) = 0$. Because of $D_1\varphi(x,0) = 1$ and $\xi(0) = 0$, the second factor does not vanish for small u. Therefore, $\varphi(\xi(u),u) = 0$ for small u. Consequently, $x - \xi(u)$ is a double root of $F(x,u) - \gamma(u)$. Knowing this, one easily finds

$$(\varphi(x,u))^2 = (x - \xi(u))^2 \left(ux + \frac{1}{3} + \frac{2}{3}\sqrt{1 - 3u^2} \right).$$

Since the second factor is greater than a positive constant for small ux, the square root is defined and is a smooth function of x and u. Moreover, the zeros of the product are those of the first factor. Because of $\xi(0) = 0$, the right root is determined by $\varphi(x,0) = x$. It follows that

$$\varphi(x,u) = \left(x + \frac{u}{1 + \sqrt{1 - 3u^2}} \right) \left(ux + \frac{1}{3} + \frac{2}{3}\sqrt{1 - 3u^2} \right)^{1/2}$$

for small x and u.

7.12. Without restriction assume that $f(0) = 0$, i.e., $[f] \in m_n$. Because of (1.35) and due to the relation (9) it suffices to consider the case $f(x) = x_1$. Then obviously $\mathcal{J}[f] = \mathcal{E}_n$ holds. Let now $[G] \in \mathcal{E}_{n+r}$ be an unfolding of $[f]$. Define $\varphi(x,u) := (G(x,u), x_2, \cdots, x_n)$ for small $(x,u) \in \mathbf{R}^{n+r}$. Then $[\varphi] \in m_{n+r,n}$ satisfies $\varphi(x,0) = x$ since $G(x,0) = f(x) = x_1$. Moreover, $G(x,u) = f(\varphi(x,u))$ holds trivially. This proves the assertion.

Chapter 8

Transversality

Transversality is the main tool to prove the existence of versal and universal unfoldings as recognized by Thom. For this purpose, a pragmatic definition of transversality suffices. The geometrical meaning of this concept will be discussed to some extent in Exercise 8.1. More about transversality will be needed only in Chapter 11 where the notion of a smooth map transverse to a submanifold is central. Recall that the variables in $\mathbf{R}^{n+r} = \mathbf{R}^n \times \mathbf{R}^r$ will usually be denoted by (x, u) as in Chapter 7.

> **Definition.** Suppose that $[F] \in \mathcal{E}_{n+r}$ is an r-parameter unfolding of a germ $[f]$ in m_n^2. Denote by $\mathcal{V}[F]$ the linear subspace of m_n spanned by the germs of the functions
>
> $$\frac{\partial}{\partial u_j} F(x, 0) - \frac{\partial}{\partial u_j} F(0, 0), \quad j = 1, \dots, r,$$
>
> defined for small $x \in \mathbf{R}^n$. Then $[F]$ is called a **k-transversal unfolding of** $[f]$ for a positive integer k if
>
> $$m_n = \mathcal{J}[f] + \mathcal{V}[F] + m_n^{k+1}$$
>
> holds.

Clearly, $\mathcal{J}[f] + \mathcal{V}[F] + m_n^{k+1} \subset m_n$ is true for every unfolding $[F]$ of $[f]$, and when $[F]$ is k-transversal, then it is l-transversal for $1 \leq l \leq k$.

Lemma (1). Let $[F]$ be an r-parameter unfolding of a germ $[f]$ in m_n^2.

(a) If $m_n = \mathcal{J}[f] + \mathcal{V}[F]$ is true, then $[F]$ is l-transversal for every $l \in \mathbf{N}$; furthermore, $\text{cod}[f] \leq r$ holds, and $[f]$ is finitely determined.

(b) If $[f]$ is k-determined and $[F]$ is k-transversal for some $k \in \mathbf{N}$, then $m_n = \mathcal{J}[f] + \mathcal{V}[F]$ holds.

Proof.   Assertion (a) follows from (5.8) and from $\dim m_n/\mathcal{J}[f] \leq \dim \mathcal{V}[F] \leq r$. Assertion (b) is a consequence of the inclusion $m_n^{k+1} \subset \langle m\mathcal{J}[f]\rangle \subset \mathcal{J}[f]$, which results from (4.38).

The existence of k-transversal unfoldings is easily seen:

Lemma (2).   Let k be a positive integer and $[f]$ be in m_n^2.   Then $m_n/(\mathcal{J}[f] + m_n^{k+1})$ is finite dimensional. Suppose that $[g_j], j = 1,\ldots,r$, are germs in m_n whose cosets span $m_n/(\mathcal{J}[f] + m_n^{k+1})$. Then

$$F(x, u) := f(x) + u_1 g_1(x) + \ldots + u_r g_r(x)$$

defines a k-transversal unfolding of $[f]$.

Proof.   The dimension of $m_n/(\mathcal{J}[f] + m_n^{k+1})$ is at most $\binom{n+k}{k} - 1$ by (4.8). Choose $[g_j], j = 1,\ldots,r$, as indicated. Then

$$\frac{\partial}{\partial u_j}F(x,0) = g_j(x) \quad \text{and} \quad [g_j] \in m_n$$

imply $\mathcal{V}[F] = \langle g_1,\ldots,g_r\rangle$ and thus the k-transversality of $[F]$.

The objective now is to study the relationship between versal and k-transversal unfoldings of a germ.

Lemma (3).   A versal unfolding of a germ in m_n^2 is k-transversal for every positive integer k.

Proof.   Let $[F] \in \mathcal{E}_{n+r}$ be a versal unfolding of $[f] \in m_n^2$, and let k be any positive integer. According to (2), there is a k-transversal unfolding $[G] \in \mathcal{E}_{n+s}$ of $[f]$. Since $[G]$ is induced by $[F]$, there exist germs $[\varphi] \in m_{n+s,n}, [\psi] \in m_{s,r}$, and $[\gamma] \in m_s$ satisfying $\varphi(y,0) = y$ and $G(y,v) = F(\varphi(y,v),\psi(v)) + \gamma(v)$ for small y and v. A straightforward computation yields

$$\frac{\partial}{\partial v_l}G(y,0) = \sum_{i=1}^{n} D_i f(y)\frac{\partial}{\partial v_l}\varphi_i(y,0) + \sum_{j=1}^{r} \frac{\partial}{\partial u_j}F(y,0)\frac{\partial}{\partial v_l}\psi_j(0)$$
$$+ \frac{\partial}{\partial v_l}\gamma(0),$$

which proves $\mathcal{V}[G] \subset \mathcal{J}[f] + \mathcal{V}[F]$. Because of the k-transversality of $[G]$, the equation $m_n = \mathcal{J}[f] + \mathcal{V}[G] + m_n^{k+1}$ is valid. The last inclusion then implies the analogous equation with $\mathcal{V}[G]$ replaced by $\mathcal{V}[F]$, proving the assertion.

An aim of the subsequent chapters is to prove the converse of (3) and to show that only finitely determined germs have versal unfoldings. The proof is based on the Main Lemma of Chapter 10, which requires the Malgrange–Mather Preparation Theorem, to which the next chapter is devoted. Using this result, (2) then gives a method for constructing versal—and, as it will turn out, even universal—unfoldings of finitely determined germs.

Exercises

8.1. Let $[F] \in \mathcal{E}_{n+r}$ be an r-parameter unfolding of a germ $[f] \in \mathcal{E}_n$, and let F be defined on an open neighborhood W of the origin in $\mathbf{R}^{n+r}$. For $(a, u) \in W$ consider the smooth function

$$x \mapsto F_u^a(x) := F(a + x, u) - F(a, u)$$

for small x. Let k be a nonnegative integer. Recall that J_n^k can be identified with $\mathbf{R}^N$ for $N = \dbinom{n+k}{k}$ by taking the Taylor coefficients at the origin; see (4.6). Then the map

$$(*) \qquad\qquad W \to J_n^k, \quad (a, u) \mapsto j^k[F_u^a]$$

is smooth. Its germ is denoted by $j_e^k[F]$ and is called the **k-extension of the unfolding** $[F]$.

(i) Show that the components of $(*)$ as a map into $\mathbf{R}^N$ are

$$\begin{array}{ll} 0 & \text{for} \quad \nu = 0, \\ \frac{1}{\nu!} D_1^\nu F(a, u) & \text{for} \quad 0 < |\nu| \le k, \end{array}$$

where

$$D_1^\nu := \left(\frac{\partial}{\partial x_1}\right)^{\nu_1} \cdots \left(\frac{\partial}{\partial x_n}\right)^{\nu_n} \quad \text{and} \quad \nu = (\nu_1, \dots, \nu_n) \in \mathbf{N}_0^n.$$

(ii) The derivative of $j_e^k[F]$ at the origin is a linear map from $\mathbf{R}^{n+r}$ into J_n^k. Taken as a map into $\mathbf{R}^N$, show that its matrix elements with respect to the standard bases of $\mathbf{R}^{n+r}$ and $\mathbf{R}^N$ are

$$\begin{array}{ll} 0 & \text{for} \quad \nu = 0, \\[2mm] \dfrac{1}{\nu!} \dfrac{\partial}{\partial x_i} D^\nu f(0) & \text{for} \quad 1 \le i \le n, \quad 0 < |\nu| \le k, \\[2mm] \dfrac{1}{\nu!} \dfrac{\partial}{\partial u_j} D_1^\nu F(0, 0) & \text{for} \quad 1 \le j \le r, \quad 0 < |\nu| \le k. \end{array}$$

(iii) Denote by Im $Dj_e^k[F](0)$ the image of $\mathbf{R}^{n+r}$ in J_n^k with respect to the derivative at the origin of $j_e^k[F]$. This is, of course, a linear subspace of $j^k(m_n)$. Suppose now that $[f]$ is in m_n^2 and put $z := j^k[f]$. Show that $[F]$ is k-transversal if and only if

$$j^k(m_n) = T_z(z\boldsymbol{\mathcal{G}}_n^k) + \text{Im } Dj_e^k[F](0).$$

To elucidate this equation, recall that two linear subspaces of a linear space are referred to as being **transversal** if they span the whole space. Using this geometrical notion, the result in (iii) means that the unfolding $[F]$ of $[f]$ in m_n^2 is k-transversal if and only if the following two subspaces of $j^k(m_n)$ are transversal, namely, the tangent space at $j^k[f]$ to the orbit of $\boldsymbol{\mathcal{G}}_n^k$ and the image space of the derivative at the origin of the k-extension of $[F]$. A later result (10.6) will show that $[F]$ is versal if and only if $\boldsymbol{\mathcal{J}}[f]$ and $\boldsymbol{\mathcal{V}}[F]$ are transversal in m_n.

Hint: Use Exercise 4.16.

8.2. Let $[F]$ be an unfolding of $[f] \in m^2$. If $[G]$ is an unfolding induced by $[F]$, prove that

$$\boldsymbol{\mathcal{V}}[G] \subset \boldsymbol{\mathcal{J}}[f] + \boldsymbol{\mathcal{V}}[F]$$

follows.

8.3. Suppose that k is a positive integer and $[f]$ is in m^2. Set $r := \dim m/(\boldsymbol{\mathcal{J}}[f] + m^{k+1})$. Clearly, r is less than $\binom{n+k}{k}$ by (4.8) and hence finite. Show that there exists a k-transversal unfolding of $[f]$ with r parameters and that every k-transversal unfolding of $[f]$ has at least r parameters.

8.4. Let k and r be positive integers, and let $[F]$ be a k-transversal r-parameter unfolding of a germ $[f]$ in m^2. If 0 is a flat point of f, show that $r \geq \binom{n+k}{k} - 1$ is true. If $[f]$ has infinite codimension, prove that there is no versal unfolding of $[f]$.

8.5. Determine all integers k for which $[F] \in \mathcal{E}_{2+7}$ given by

$$\begin{aligned}
F(x, y; u_1, u_2, u_3, v_1, v_2, v_3, w) &= \exp(x^2 + y^2 + u_1 x + u_2 x^2 + u_3 x^3) \\
&\quad + \exp(-x^2 - y^2 - v_1 y - v_2 y^2 - v_3 y^3) \\
&\quad - \exp(wxy) - 1
\end{aligned}$$

is a k-transversal unfolding of $[f]$ where $f(x, y) := F(x, y; 0)$.

8.6. Let k be a positive integer. Find a k-transversal unfolding of $[x^4 + x^2 y]$ in m_2^2 with a minimal number of parameters.

Solutions

8.1.

(i) As a map into $\mathbf{R}^N$, the components of (*) are just the Taylor coefficients of F_u^a at the origin, which depend on a and u. Clearly, $F_u^a(0) = 0$ so that the constant term vanishes. For the higher Taylor coefficients, $F(a, u)$ in the expression of F_u^a is irrelevant, since it does not depend on x.

(ii) In order to obtain the matrix elements of the linear map $Dj_e^k[F](0)$ one simply has to take the partial derivatives with respect to the variables a and u of the components determined in (i). Then the assertion follows from $F(x + a, 0) = f(x + a)$ and from the fact that the derivatives with respect to a and x are equal.

(iii) According to Exercise 4.16,

$$T_z(z\mathcal{G}_n^k) = j^k(\langle m_n \mathcal{J}[f]\rangle).$$

Moreover, it follows from (ii) by interchanging the order of differentiation that Im $Dj_e^k[F](0)$ is spanned by the k-jets of the germs $[D_1 f], \ldots, [D_n f]$ and of the germs in $\mathcal{V}[F]$. Hence, the assertion is that $[F]$ is k-transversal if and only if

$$j^k(m_n) = j^k(\langle m_n \mathcal{J}[f]\rangle) + j^k(\langle D_1 f, \ldots, D_n f\rangle)$$
$$+ j^k(\mathcal{V}[F])$$

holds. This equation is equivalent to

$$m_n = \langle m_n \mathcal{J}[f]\rangle + \mathbf{R}[D_1 f] + \cdots + \mathbf{R}[D_n f] + \mathcal{V}[F]$$
$$+ m_n^{k+1},$$

which indeed is equivalent to the k-transversality of $[F]$, due to (5.15).

8.2. Let r and s denote the number of parameters of $[F]$ and $[G]$, respectively. Since $[G]$ is induced by $[F]$, there are germs $[\varphi] \in m_{n+s,n}$, $[\psi] \in m_{s,r}$, and $[\gamma] \in m_s$ satisfying $\varphi(y, 0) = y$ and

$$G(y, v) = F(\varphi(y, v), \psi(v)) + \gamma(v)$$

for small y and v. Using $f(y) = F(\varphi(y, 0), \psi(0))$, the chain rule yields

$$\frac{\partial}{\partial v_l} G(y, 0) = \sum_{i=1}^{n} D_i f(y) \frac{\partial}{\partial v_l} \varphi_i(y, 0) + \sum_{j=1}^{r} \frac{\partial}{\partial u_j} F(y, 0) \frac{\partial}{\partial v_l} \psi_j(0)$$
$$+ \frac{\partial}{\partial v_l} \gamma(0).$$

In particular,

$$\frac{\partial}{\partial v_l}G(0,0) = \sum_{j=1}^{r}\frac{\partial}{\partial u_j}F(0,0)\frac{\partial}{\partial v_l}\psi_j(0) + \frac{\partial}{\partial v_l}\gamma(0),$$

since $D_i f(0) = 0$. Therefore,

$$\frac{\partial}{\partial v_l}G(y,0) - \frac{\partial}{\partial v_l}G(0,0) = \sum_{i=1}^{n}\frac{\partial}{\partial v_l}\varphi_i(y,0)D_i f(y)$$

$$+ \sum_{j=1}^{r}\frac{\partial}{\partial v_l}\psi_j(0)\left[\frac{\partial}{\partial u_j}F(y,0) - \frac{\partial}{\partial u_j}F(0,0)\right],$$

which proves the assertion.

8.3. The existence of a k-transversal unfolding with r parameters follows from (2). Now suppose that $[F]$ is an s-parameter unfolding of $[f]$ and k-transversal. By definition,

$$m = \mathcal{J}[f] + \mathcal{V}[F] + m^{k+1}$$

holds. Then $r \leq \dim \mathcal{V}[F]$ follows by taking the quotient space with respect to $\mathcal{J}[f] + m^{k+1}$. On the other hand, $\dim \mathcal{V}[F] \leq s$ is valid, since by definition $\mathcal{V}[F]$ is spanned by s elements.

8.4. If 0 is a flat point of $[f]$, then $\mathcal{J}[f] \subset m^\infty \subset m^{k+1}$, so that $r \geq \dim m/m^{k+1} = \binom{n+k}{k} - 1$ follows from Exercise 8.3 and from (4.8).

Suppose now that $\mathrm{cod}[f] = \infty$. For the nonexistence of a versal unfolding of $[f]$, it suffices to show that the increasing sequence

$$(\dim m/(\mathcal{J}[f] + m^{k+1}))_k$$

of integers is not bounded; see Exercise 8.3. Assume the contrary. Then $\mathcal{J}[f] + m^{k+1} = \mathcal{J}[f] + m^{k+2}$ holds for some k. By (4.18), one obtains $m^{k+1} \subset \mathcal{J}[f]$, from which the contradiction $\mathrm{cod}[f] = \dim m/\mathcal{J}[f] \leq \dim m/m^{k+1} < \infty$ follows.

8.5. $[F]$ is an unfolding of the germ of $f(x,y) = \exp(x^2+y^2) + \exp(-x^2-y^2) - 2$. The Jacobi ideal $\mathcal{J}[f]$ is

$$\langle(\exp(x^2+y^2) - \exp(-x^2-y^2))x, (\exp(x^2+y^2) - \exp(-x^2-y^2))y\rangle_\varepsilon$$
$$= \langle x(x^2+y^2), y(x^2+y^2)\rangle_\varepsilon,$$

since the germ of

$$h(x,y) := (\exp(x^2+y^2) - \exp(-x^2-y^2))/(x^2+y^2)$$

for $(x, y) \neq (0, 0)$ and $h(0, 0) := 2$ is invertible in $\mathcal{E}$.

The space $\mathcal{V}[F]$ is spanned by the germs of $x^l \exp(x^2 + y^2)$, $y^l \exp(-x^2 - y^2)$ for $l = 1, 2, 3$, and xy. Clearly,

$$[x^l \exp(x^2 + y^2)] \in [x^l] + \langle x(x^2 + y^2) \rangle_{\mathcal{E}}$$

and

$$[y^l \exp(x^2 + y^2)] \in [y^l] + \langle y(x^2 + y^2) \rangle_{\mathcal{E}}$$

for $l = 1, 2, 3$. Therefore

$$\mathcal{J}[F] + \mathcal{V}[F] = \langle x(x^2 + y^2), y(x^2 + y^2) \rangle_{\mathcal{E}} + \langle x, x^2, x^3, y, y^2, y^3, xy \rangle$$

holds. By (5.10) it follows immediately that $m = \mathcal{J}[F] + \mathcal{V}[F] + m^4$, so that $[F]$ is 3-transversal and hence also 2- and 1-transversal. But $[F]$ is not k-transversal for $k \geq 4$, because otherwise $[x^k]$ would be contained in $[x^2 + y^2]\mathcal{E}$, which is a contradiction, as shown in the solution of Exercise 4.8.

8.6. According to (2) and Exercise 8.3, such an unfolding is given by

$$x^4 + x^2 y + u_1 g_1(x, y) + \cdots + u_r g_r(x, y)$$

if $[g_j], j = 1, \ldots, r$, are germs in m whose cosets form a basis of

(*) $$m/(\mathcal{J}[x^4 + x^2 y] + m^{k+1}).$$

One easily checks that $\mathcal{J}[x^4 + x^2 y] = \langle xy, x^2 \rangle_{\mathcal{E}}$.

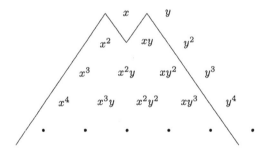

The diagram shows that only the germs of the monomials x and $y^l, l \in \mathbf{N}$, are not contained in $\mathcal{J}[x^4 + x^2 y]$. Moreover, m^{k+1} contains all germs $[x^l y^m]$ with $l + m \geq k + 1$, which are the germs of the monomials lying below the kth horizontal line of the diagram. Hence, by (5.10), the cosets of the germs $[x]$ and $[y^l], l = 1, \ldots, k$, form a basis of (*). Thus the germ $[F] \in \mathcal{E}_{2+(k+1)}$ given by

$$x^4 + x^2 y + u_1 x + u_2 y + \ldots + u_{k+1} y^k$$

is an unfolding, as desired.

Chapter 9

The Malgrange–Mather Preparation Theorem

The Malgrange–Mather Preparation Theorem is used to prove the Main Lemma of the next chapter. More specifically, a consequence of this theorem, which is similar in content to a corollary of Nakayama's Lemma, is what will be needed. In order to formulate it, let k and s be positive integers and, as usual, consider $\mathcal{E}_s$ as the subset of $\mathcal{E}_{k+s}$ whose elements do not depend on the first k variables. Furthermore, let $\mathcal{A}$ be an ideal of $\mathcal{E}_{k+s}$ and B a finite subset of $\mathcal{E}_{k+s}$. The objective of this chapter is to prove that

(1) $\mathcal{E}_{k+s} \subset \mathcal{A} + \langle \mathcal{E}_s B \rangle + \langle m_s \mathcal{E}_{k+s} \rangle$ implies $\mathcal{E}_{k+s} \subset \mathcal{A} + \langle \mathcal{E}_s B \rangle$.

This result will be needed in part (vii) of the proof of the Main Lemma (10.1). Actually, (1) is an immediate consequence of (4.16) if $\tilde{\mathcal{E}} := \mathcal{E}_{k+s}/\mathcal{A}$ is a finitely generated module over $\mathcal{E}_s$. To see this, observe that

$$\left(\langle m_s \mathcal{E}_{k+s} \rangle + \mathcal{A} \right) / \mathcal{A} = \langle m_s \tilde{\mathcal{E}} \rangle$$

holds, because every element of the left-hand side has the form

$$\sum_j [h_j][f_j] + \mathcal{A} = \sum_j [h_j]([f_j] + \mathcal{A})$$

with $[h_j] \in m_s, [f_j] \in \mathcal{E}_{k+s}$. Therefore, the hypothesis in (1) is equivalent to the following inclusion of modules over $\mathcal{E}_s$:

$$\tilde{\mathcal{E}} \subset \tilde{\mathcal{B}} + \langle m_s \tilde{\mathcal{E}} \rangle, \quad \text{for} \quad \tilde{\mathcal{B}} := \left(\langle \mathcal{E}_s B \rangle + \mathcal{A} \right) / \mathcal{A}.$$

Now if $\tilde{\mathcal{E}}$ is finitely generated over $\mathcal{E}_s$, then (4.16) yields $\tilde{\mathcal{E}} \subset \tilde{\mathcal{B}}$ or, equivalently, $\mathcal{E}_{k+s} \subset \mathcal{A} + \langle \mathcal{E}_s B \rangle$.

Observe that $\tilde{\mathcal{E}}$ is finitely generated over $\mathcal{E}_{k+s}$, since

$$\tilde{\mathcal{E}} = \langle [1] + \mathcal{A} \rangle_{\mathcal{E}_{k+s}}$$

holds, where $[1]$ is the identity of $\mathcal{E}_{k+s}$. Furthermore, $\tilde{\mathcal{E}}/\langle m_s \tilde{\mathcal{E}} \rangle$ is a finite dimensional vector space. Indeed, note that $\langle \mathcal{E}_s B \rangle \subset \langle B \rangle + \langle m_s \mathcal{E}_{k+s} \rangle$ holds because any element $\sum_j [h_j][b_j]$ in $\langle \mathcal{E}_s B \rangle$ for $[h_j] \in \mathcal{E}_s$ and $[b_j] \in B$ can be written as a sum

$$\sum_j h_j(0)[b_j] + \sum_j [h_j - h_j(0)][b_j].$$

From this and from

$$\langle m_s \tilde{\mathcal{E}} \rangle = \left(\langle m_s \mathcal{E}_{k+s} \rangle + \mathcal{A} \right) / \mathcal{A} \quad \text{and} \quad \tilde{\mathcal{E}} \subset \tilde{\mathcal{B}} + \langle m_s \tilde{\mathcal{E}} \rangle,$$

it follows that $\tilde{\mathcal{E}} \subset (\langle B \rangle + \mathcal{A})/\mathcal{A} + \langle m_s \tilde{\mathcal{E}} \rangle$ and hence that the cosets $([b] + \mathcal{A}) + \langle m_s \tilde{\mathcal{E}} \rangle$, $[b] \in B$, span the quotient space $\tilde{\mathcal{E}}/\langle m_s \tilde{\mathcal{E}} \rangle$. Now the Malgrange–Mather Preparation Theorem given below in (2) can be applied, yielding that $\tilde{\mathcal{E}}$ is even finitely generated over $\mathcal{E}_s$.

The reader who wants to continue along the main road of the development of the theory may skip the rest of this chapter and proceed to Chapter 10.

To formulate the Malgrange–Mather Preparation Theorem, let n and s be positive integers, let U be an open neighborhood of the origin in $\mathbf{R}^n$, and suppose that $\varphi\colon U \to \mathbf{R}^s$ is a smooth map preserving the origin. The **pullback** $\varphi^*\colon \mathcal{E}_s \to \mathcal{E}_n$ of φ, defined by $[g] \mapsto \varphi^*[g] := [g \circ \varphi]$, is a ring homomorphism preserving the identity and a linear map. By means of φ^*, any module M over $\mathcal{E}_n$ can be considered as a module over $\mathcal{E}_s$ in a natural way; the outer product $\mathcal{E}_s \times M \to M, ([g], a) \mapsto [g]a$, is given by $[g]a := (\varphi^*[g])a$. Notice that the vector space structure of M does not depend on whether M is regarded as a module over $\mathcal{E}_n$ or $\mathcal{E}_s$.

> **Malgrange–Mather Preparation Theorem** (2). Let M be a finitely generated module over $\mathcal{E}_n$. Then M is finitely generated over $\mathcal{E}_s$ (with respect to the module structure given by a map φ as above) if and only if the vector space $M/\langle m_s M \rangle$ is finite dimensional.

Before the proof of this important theorem is presented, notice that (2) really implies (1). As mentioned above, to show (1) it suffices to verify that $\tilde{\mathcal{E}}$ is finitely generated over $\mathcal{E}_s$. Set $n := k + s$, $M := \tilde{\mathcal{E}}$, and let $\varphi\colon \mathbf{R}^n \to \mathbf{R}^s$,

$$\varphi(x_1, \ldots, x_k, x_{k+1}, \ldots, x_{k+s}) := (x_{k+1}, \ldots, x_{k+s}).$$

Then φ^* is the usual embedding of $\mathcal{E}_s$ into $\mathcal{E}_{k+s}$. Thus, (2) can be applied to $\tilde{\mathcal{E}}$. In the same way the following generalization of (1) is verified.

> **Corollary** (3). Let $\mathcal{C}$ be a finitely generated module over $\mathcal{E}_n, \mathcal{A}$ a submodule of $\mathcal{C}$, and B a finite subset of $\mathcal{C}$. Then $\mathcal{C} \subset \mathcal{A} + \langle \mathcal{E}_s B \rangle + \langle m_s \mathcal{C} \rangle$ implies $\mathcal{C} \subset \mathcal{A} + \langle \mathcal{E}_s B \rangle$.

The proof of the Malgrange–Mather Preparation Theorem is rather lengthy and requires several auxiliary results that are interesting in their own right. These results are treated now. The first four results involve extensions of smooth functions. The following extension statement is an application of the generalized version of Borel's Lemma (4.5).

Lemma (4). Let V and W be subspaces of $\mathbf{R}^n$ with $V + W = \mathbf{R}^n$. Suppose that g and h are functions defined on an open neighborhood U of the origin in $\mathbf{R}^n$ such that $D^k g = D^k h$ on $U \cap V \cap W$ for every nonnegative integer k. Then there exists a smooth function f on an open neighborhood $\tilde{U}$ of the origin in $\mathbf{R}^n$ contained in U so that $D^k f = D^k g$ on $\tilde{U} \cap V$ and $D^k f = D^k h$ on $\tilde{U} \cap W$ for every nonnegative integer k.

Proof. It is no restriction to assume $h = 0$, because if f_1 is the required extension for $g_1 := g - h$ and 0, then $f := f_1 + h$ is the desired extension for g and h. By a linear transformation, new coordinates $y_1, \ldots, y_n$ of $\mathbf{R}^n$ can be found such that

$$V = \{y \in \mathbf{R}^n : y_1 = \cdots = y_j = 0\} \text{ and}$$
$$W = \{y \in \mathbf{R}^n : y_{j+1} = \cdots = y_k = 0\}$$

for some j and k with $1 \le j < k \le n$. Clearly, it suffices to prove the lemma for functions of these new coordinates.

Group together $\xi := (y_1, \ldots, y_j)$ and $\eta := (y_{j+1}, \ldots, y_n)$ and define

$$a_\nu(\eta) := D_1^\nu g(0, \eta)$$

for $\nu \in \mathbf{N}_0^j$ and sufficiently small η, where D_1^ν denotes partial derivation with respect to the first j variables ξ. Following the proof of (4.5), define

$$f(\xi, \eta) := \sum_\nu \frac{1}{\nu!} a_\nu(\eta) u(\xi/t_\nu) \xi^\nu.$$

On an open neighborhood $\tilde{U}$ of the origin in $\mathbf{R}^n$, f is smooth and its derivatives are given by the series of the corresponding derivatives of the summands. In addition,

$$D_1^\nu D_2^\mu f(0, \eta) = D^\mu a_\nu(\eta) = D_1^\nu D_2^\mu g(0, \eta),$$

where D_2^μ denotes partial derivation with respect to the last $n - j$ variables η. Hence, the derivatives of f and g coincide on $\tilde{U} \cap V$. Moreover, if $y := (\xi, \eta) \in \tilde{U} \cap W$, then $y_{j+1} = \cdots = y_k = 0$, and $D^\mu a_\nu(\eta) = D_1^\nu D_2^\mu g(0, \eta)$ vanishes, since $(0, \eta) \in \tilde{U} \cap V \cap W$. Thus, $D_1^\nu D_2^\mu f(\xi, \eta)$ vanishes, too, and f is the desired extension.

Another straightforward application of Borel's Lemma (4.5) and its proof is treated now for complex-valued functions. A complex-valued function is said to be smooth whenever its real and its imaginary parts are smooth. Consider a first-order differential operator $X := \sum_{i=1}^{n} h_i D_i$, where h_i are smooth complex-valued functions defined on an open neighborhood U of the origin in $\mathbf{R}^n$, i.e., X assigns to every smooth function $f: U \to \mathbf{C}$ the smooth function $U \to \mathbf{C}$ given by $Xf := \sum_{i=1}^{n} h_i D_i f$.

Lemma (5). Let $f: U \to \mathbf{C}$ be smooth and X a differential operator as defined above. Then there is a smooth complex-valued function F defined on an open neighborhood of the origin in $\mathbf{R} \times \mathbf{R}^n$ such that $F(0, x) = f(x)$ and all derivatives of $D_1 F$ and XF coincide at $(0, x)$ for small x in $\mathbf{R}^n$.

Proof. Let $a_0 := f$ and $a_k := X^k f$ for $k \in \mathbf{N}$ on a sufficiently small neighborhood of the origin in $\mathbf{R}^n$. As in (4.5), define

$$F(s, x) := \sum_{k=0}^{\infty} \frac{1}{k!} a_k(x) u(s/t_k) s^k.$$

Then F is smooth and $D_1^{k+1} D_2^\beta F(0, x) = D^\beta a_{k+1}(x)$ for $k \in \mathbf{N}_0$ and $\beta \in \mathbf{N}_0^n$. The latter term is equal to $D^\beta X a_k(x)$ by the definition of a_k and hence is $D_1^k D_2^\beta XF(0, x)$. Finally, $F(0, x) = a_0(x) = f(x)$ holds.

The last two lemmas are the preliminaries needed for the proof of the following crucial technical point. In the following, let $P_0 := 1$ and

(6) $$P_k(z, \lambda) := z^k + \sum_{j=0}^{k-1} \lambda_j z^j$$

for a positive integer k and complex numbers $z, \lambda_0, \ldots, \lambda_{k-1}$. If $z := x + iy$, the abbreviation

$$\frac{\partial}{\partial \bar{z}} := \frac{1}{2} \left(\frac{\partial}{\partial x} + i \frac{\partial}{\partial y} \right)$$

will be used. In particular, if $F: \mathbf{C} \to \mathbf{C}$ is smooth (meaning with respect to the real variables x and y), then

$$\frac{\partial}{\partial \bar{z}} F = \frac{\partial}{\partial \bar{z}} f + i \frac{\partial}{\partial \bar{z}} g = \frac{1}{2} \left(\frac{\partial f}{\partial x} - \frac{\partial g}{\partial y} \right) + \frac{i}{2} \left(\frac{\partial f}{\partial y} + \frac{\partial g}{\partial x} \right),$$

where $F = f + ig$, i.e., f is the real part of F and g the imaginary part. Note that **derivative** will continue to mean a derivative with respect to the real variables, and the 0th derivative is the function itself.

Nirenberg Extension Lemma (7). Suppose that G is a smooth complex-valued function defined on an open neighborhood of the origin in $\mathbf{R} \times \mathbf{R}^n$. Let k be a nonnegative integer. Then there exists a smooth complex-valued function $\tilde{G}$ defined on a neighborhood of the origin in $\mathbf{C} \times \mathbf{R}^n \times \mathbf{C}^k$ such that

(a) $\tilde{G}(z, x, \lambda) = G(z, x)$ for real z, and

(b) all derivatives of $\partial \tilde{G}/\partial \bar{z}$ vanish at (z, x, λ) if z is real or if $P_k(z, \lambda) = 0$.

Proof.

(i) Consider first the case $k = 0$. Since $P_0 = 1$, the claim is that there is a smooth complex-valued function $\tilde{G}$ defined for small $z := t + is$ in $\mathbf{C}$ and x in $\mathbf{R}^n$ such that $\tilde{G}(t, x) = G(t, x)$ and

$$D^\alpha \frac{\partial}{\partial \bar{z}} \tilde{G}(t, x) = 0 \quad \text{for} \quad \alpha \in \mathbf{N}^{n+2}.$$

By definition, the latter means that

$$D^\alpha \frac{\partial}{\partial s} \tilde{G}(t, x) = D^\alpha X \tilde{G}(t, x) \quad \text{for} \quad X := i \frac{\partial}{\partial t}.$$

The existence of such a function $\tilde{G}$ follows from (5).

(ii) Now let $k \geq 1$. Then the transformation given by

$$\varphi(z, x, \lambda_0, \ldots, \lambda_{k-1}) := (z, x, P_k(z, \lambda), \lambda_1, \ldots, \lambda_{k-1})$$

is a local diffeomorphism at the origin in $\mathbf{C} \times \mathbf{R}^n \times \mathbf{C}^k$ leaving the origin invariant. The reason is that the Jacobi matrix of φ at the origin is a lower triangular matrix with diagonal elements equal to 1. The only nonvanishing off-diagonal elements arise from

$$\frac{\partial}{\partial t} P_k(z, \lambda) \quad \text{and} \quad \frac{\partial}{\partial s} P_k(z, \lambda)$$

where $z = t + is$. Hence $(z, x, \mu) := \varphi(z, x, \lambda)$ are suitable new coordinates. In these coordinates, the zero set of $P_k(z, \lambda)$ is given by the simple equation $\mu_0 = 0$. Consider

$$P_k' := \frac{1}{2} \left(\frac{\partial}{\partial t} - i \frac{\partial}{\partial s} \right) P_k.$$

Explicitly, $P_k'(z, \lambda) = kz^{k-1} + \lambda_1 + 2\lambda_2 z + \cdots + (k-1)\lambda_{k-1}z^{k-2}$. Let $\overline{P_k'}$ be its complex conjugate. Since $\overline{P_k'}$ does not depend on λ_0, it can be regarded as a complex-valued function of the new variables $z, \mu_1, \ldots, \mu_{k-1}$. Thus

$$L := \frac{\partial}{\partial \bar{z}} + \overline{P_k'} \frac{\partial}{\partial \bar{\mu}_0}$$

is a differential operator acting on smooth functions of these variables, as well as of μ_0.

The assertion holds, if there is a smooth complex-valued function H defined on a neighborhood of the origin in $\mathbf{C} \times \mathbf{R}^n \times \mathbf{C}^k$ such that

(a') $H(z, x, \mu) = G(z, \mu)$ for real z, and

(b′) all derivatives of LH vanish at (z, x, μ) if z is real or if $\mu_0 = 0$.

To see this, set $\tilde{G} := H \circ \varphi$. Then (a) follows immediately from (a′). Next, verify

$$\frac{\partial}{\partial \bar{z}} \tilde{G}(z, x, \lambda) = (LH)(\varphi(z, x, \lambda));$$

see Exercise 4. Then (b) follows from (b′) by the chain rule.

(iii) In this step it is shown that the existence of H satisfying (a′) and (b′) follows from the existence of two smooth complex-valued functions E and F defined for small (z, x, μ) satisfying

(α) E is independent of μ_0,

(β) $E(z, x, \mu) = G(z, x)$ for real z,

(γ) all derivatives of $\partial E/\partial \bar{z}$ vanish at (z, x, μ) if z is real,

(δ) all derivatives of LF vanish at (z, x, μ) if $\mu_0 = 0$, and

(ε) all derivatives of E and F coincide at (z, x, μ) if z is real and $\mu_0 = 0$.

Let E and F be as indicated. Note that the two subspaces of $\mathbf{C} \times \mathbf{R}^n \times \mathbf{C}^k$ defined by the imaginary part of z vanishing on the one hand and by μ_0 vanishing on the other hand are complementary. Thus, they span the whole space. By (ε), Lemma (4) can be applied to the real and imaginary parts of E and F, yielding a smooth complex-valued function H such that around the origin all derivatives of H coincide with those of E if z is real, and they coincide with those of F if $\mu_0 = 0$. Hence (a′) follows from (β). Since $\partial E/\partial \bar{z} = LE$ follows from (α), (γ) implies that all derivatives of LE vanish for real z. Consequently, this is true for LH, too, showing the first part of (b′). The second part of (b′) follows from (δ).

(iv) The assertion can be proved now using induction on k. The case $k = 0$ was shown in (i). For $k \geq 1$ assume that the assertion holds for the case $k - 1$. Then there exists a smooth complex-valued function $\tilde{G}_{k-1}$ defined for small (z, x, λ) in $\mathbf{C} \times \mathbf{R}^n \times \mathbf{C}^{k-1}$, such that $\tilde{G}_{k-1}(z, x, \lambda) = G(z, x)$ for real z and such that all derivatives of $\partial \tilde{G}_{k-1}/\partial \bar{z}$ vanish at (z, x, λ) if z is real or if $P_{k-1}(z, \lambda) = 0$. Define E by

$$E(z, x, \mu_0, \ldots, \mu_{k-1}) := \tilde{G}_{k-1}\left(z, x, \frac{1}{k}\mu_1, \frac{2}{k}\mu_2, \ldots, \frac{k-1}{k}\mu_{k-1}\right).$$

This definition ensures that (α), (β), and (γ) are valid. In addition, it follows that all derivatives of $\partial E/\partial \bar{z}$ vanish at (z, x, μ) if $P_k'(z, \mu) = 0$, since

$$P_k'(z, \mu) = kP_{k-1}\left(z, \frac{1}{k}\mu_1, \ldots, \frac{k-1}{k}\mu_{k-1}\right).$$

To construct F, set $\mu' := (\mu_1, \ldots, \mu_{k-1})$ and define complex-valued functions a_j, $j \in \mathbf{N}_0$, for small $(z, x, \mu') \in \mathbf{C} \times \mathbf{R}^n \times \mathbf{C}^{k-1}$ by

$$a_j(z, x, \mu') := \left(-\left(\frac{1}{\overline{P_k'(z, \mu)}}\right)\frac{\partial}{\partial \bar{z}}\right)^j E(z, x, \mu).$$

if $P'_k(z, \mu) \neq 0$ and $a_j(z, x, \mu') := 0$ at the zeros of P'_k. Let X denote any differential operator with smooth coefficient functions of the real and imaginary parts of the variables (z, x, μ'), and let $m \in \mathbf{N}_0$. Since all derivatives of $X(\partial E/\partial \bar{z})$ vanish at the zeros of P'_k, it follows from the rule of De L'Hospital that

$$\frac{X \dfrac{\partial}{\partial \bar{z}} E}{(P'_k)^m}$$

extends continuously to the zeros of the denominator and vanishes there. This shows that the functions a_j are smooth.

Following (4.5), define

$$F(z, x, \mu) := \sum_{j=0}^{\infty} \frac{1}{j!} a_j(z, x, \mu') u(\bar{\mu}_0/t_j)(\bar{\mu}_0)^j.$$

Although a_j and $\bar{\mu}_0$ are complex-valued, the proof of (4.5) can be carried over quite analogously, showing that F is a smooth complex-valued function for small (z, x, μ) in $\mathbf{C} \times \mathbf{R}^n \times \mathbf{C}^k$ satisfying

$$X \left(\frac{\partial}{\partial \bar{\mu}_0} \right)^m F(z, x, 0, \mu') = X a_m(z, x, \mu').$$

It remains to check (δ) and (ε) for E and F. Concerning (δ), one computes

$$X \left(\frac{\partial}{\partial \bar{\mu}_0} \right)^m LF(z, x, 0, \mu') = X \frac{\partial}{\partial \bar{z}} \left(\frac{\partial}{\partial \bar{\mu}_0} \right)^m F(z, x, 0, \mu')$$

$$+ X \overline{P'_k} \left(\frac{\partial}{\partial \bar{\mu}_0} \right)^{m+1} F(z, x, 0, \mu')$$

$$= X \frac{\partial}{\partial \bar{z}} \left(-\left(\frac{1}{\overline{P'_k}} \right) \frac{\partial}{\partial \bar{z}} \right)^m E(z, x, \mu)$$

$$+ X \overline{P'_k} \left(-\left(\frac{1}{\overline{P'_k}} \right) \frac{\partial}{\partial \bar{z}} \right)^{m+1} E(z, x, \mu) = 0.$$

Property (ε) is easier to show. For $\mu_0 = 0$ and z real,

$$X \left(\frac{\partial}{\partial \bar{\mu}_0} \right)^m F(z, x, 0, \mu') = 0$$

holds due to (γ) and

$$X \left(\frac{\partial}{\partial \bar{\mu}_0} \right)^m E(z, x, 0, \mu') = 0$$

results from (α).

The goal now is to prove the so-called Mather Division Theorem. The heart of its proof lies in showing that a special case of it is true. This case will be proved using Lemma (7) and the following Inhomogeneous Cauchy Integral Formula, in which the notations $dz := dx + i\,dy$ and $dz\,d\bar{z} := -2i\,dx\,dy$ are used.

Inhomogeneous Cauchy Integral Formula (8). Let $F: \mathbf{C} \to \mathbf{C}$ be smooth and suppose that D is a relatively compact subset of $\mathbf{C}$ whose boundary ∂D is a simple closed curve that is positively oriented. Then for every w in the interior of D,

$$F(w) = \frac{1}{2\pi i} \int_{\partial D} \frac{F(z)}{z - w} \, dz + \frac{1}{2\pi i} \iint_{D} \frac{\partial F(z)/\partial \bar{z}}{z - w} \, dz \, d\bar{z}$$

holds, where the integral along ∂D is taken counterclockwise.

Proof. Let $\varepsilon > 0$ be sufficiently small so that the disc D_ε of radius ε about w lies in D. Apply Green's Theorem (see Appendix A.5) to the function $z \mapsto F(z)/(z - w)$ for $z \in D \backslash D_\varepsilon$. Since

$$\frac{\partial}{\partial \bar{z}} \left(\frac{F(z)}{z - w} \right) = \frac{\partial F(z)/\partial \bar{z}}{z - w},$$

it follows that

$$-\iint_{D \backslash D_\varepsilon} \frac{\partial F(z)/\partial \bar{z}}{z - w} \, dz \, d\bar{z} = \int_{\partial D} \frac{F(z)}{z - w} \, dz - \int_{\partial D_\varepsilon} \frac{F(z)}{z - w} \, dz.$$

In order to take the limit as ε approaches 0, it will be convenient to use polar coordinates centered at w. Then

$$\int_{\partial D_\varepsilon} \frac{F(z)}{z - w} \, dz = \int_0^{2\pi} F(w + \varepsilon e^{i\varphi}) i \, d\varphi$$

and this limit is $2\pi i F(w)$ as $\varepsilon \to 0$. Furthermore,

$$\iint_{D_\varepsilon} \left| \frac{\partial F/\partial \bar{z}}{z - w} \right| \, dx \, dy \le c \int_0^\varepsilon \int_0^{2\pi} d\varphi \, dr \quad \text{for} \quad c := \sup_{z \in D_\varepsilon} \left| \frac{\partial}{\partial \bar{z}} F(z) \right|,$$

and this vanishes for $\varepsilon \to 0$. Hence the assertion follows.

Note that the Inhomogeneous Cauchy Integral Formula reduces to the usual Cauchy integral formula if F is holomorphic, because $\partial F/\partial \bar{z}$ vanishes due to the Cauchy–Riemann differential equations.

Polynomial Division Theorem (9). Let G be a smooth complex-valued function defined on an open neighborhood of the origin in $\mathbf{R} \times \mathbf{R}^n$. Suppose that k is a nonnegative integer. Then there exist smooth complex-valued functions q and r defined on an open neighborhood of the origin in $\mathbf{R} \times \mathbf{R}^n \times \mathbf{R}^k$ as well as smooth complex-valued functions r_j, $j = 0, \ldots, k-1$, defined on an open neighborhood of the origin in $\mathbf{R}^n \times \mathbf{R}^k$ satisfying

(a) $G(t, x) = q(t, x, \lambda) P_k(t, \lambda) + r(t, x, \lambda),$

(b) $r(t, x, \lambda) = \sum_{j=0}^{k-1} r_j(x, \lambda) t^j,$

for small (t, x, λ) in $\mathbf{R} \times \mathbf{R}^n \times \mathbf{R}^k$. Moreover, if G is real-valued then q and r may be chosen to be real-valued, in which case all r_j are real-valued, too.

Proof. First note that for a real-valued G, Equation (a) is satisfied by the real parts of q and r, since P_k is real-valued. In the case that q and r are real-valued it follows easily from Equation (b) that all r_j are real-valued, too.

Let $\tilde{G}$ correspond to G and P_k as in the Nirenberg Extension Lemma (7). Consider a closed disc D around the origin in $\mathbf{C}$ small enough so that $\tilde{G}$ is defined on a neighborhood of $D \times \{0\} \times \{0\}$ in $\mathbf{C} \times \mathbf{R}^n \times \mathbf{C}^k$. According to the Inhomogeneous Cauchy Integral Formula (8), if w is in the interior of D,

$$\tilde{G}(w, x, \lambda) = \frac{1}{2\pi i} \int_{\partial D} \frac{\tilde{G}(z, x, \lambda)}{z - w} \, dz + \frac{1}{2\pi i} \int\int_D \frac{(\partial \tilde{G}/\partial \bar{z})(z, x, \lambda)}{z - w} \, dz \, d\bar{z}$$

holds for small (x, λ) in $\mathbf{R}^n \times \mathbf{C}^k$. By definition,

$$P_k(z, \lambda) - P_k(w, \lambda) = z^k - w^k + \sum_{j=0}^{k-1} \lambda_j (z^j - w^j).$$

Using the summation formula for the finite geometric series, one finds polynomials $s_j(z, \lambda)$ in z with coefficients depending linearly on λ that satisfy

$$P_k(z, \lambda) - P_k(w, \lambda) = (z - w) \sum_{j=0}^{k-1} s_j(z, \lambda) w^j.$$

When inserting

$$1 = \frac{P_k(z, \lambda)}{P_k(z, \lambda)} = \frac{P_k(w, \lambda)}{P_k(z, \lambda)} + (z - w) \sum_{j=0}^{k-1} \frac{s_j(z, \lambda)}{P_k(z, \lambda)} w^j$$

as a factor in the integrand of the above integrals, the desired decomposition of G results, where

$$q(w, x, \lambda) := \frac{1}{2\pi i} \int_{\partial D} \frac{\tilde{G}(z, x, \lambda)}{P_k(z, \lambda)(z - w)} \, dz$$

$$+ \frac{1}{2\pi i} \int\int_D \frac{(\partial \tilde{G}/\partial \bar{z})(z, x, \lambda)}{P_k(z, \lambda)(z - w)} \, dz \, d\bar{z}$$

$$r_j(x, \lambda) := \frac{1}{2\pi i} \int_{\partial D} \frac{\tilde{G}(z, x, \lambda) s_j(z, \lambda)}{P_k(z, \lambda)} \, dz$$

$$+ \frac{1}{2\pi i} \int\int_D \frac{(\partial \tilde{G}/\partial \bar{z})(z, x, \lambda) s_j(z, \lambda)}{P_k(z, \lambda)} \, dz \, d\bar{z}.$$

For real w it is of course necessary to check that the integrals are well defined and yield smooth complex-valued functions. Because of the compactness of ∂D and D, this follows from the smoothness of the integrands, since differentiation and integration are interchangeable. Concerning the integrals over ∂D, it suffices to know that P_k has no zeros when z lies on ∂D and λ is in a neighborhood of the origin in $\mathbf{C}^k$. This is true, because for small λ obviously

$$R^k > \sum_{j=1}^{k-1} |\lambda_j| R^j$$

holds, where R denotes the radius of the circle ∂D. For the integrals over D, note that all derivatives of $\partial \tilde{G}/\partial \bar{z}$ vanish on the zeros of P_k as well as whenever z is real. By the rule of l'Hospital it follows for real w that the integrands extend to smooth functions.

The foregoing result can be generalized to the Mather Division Theorem.

Mather Division Theorem (10). Let k be a nonnegative integer. Suppose that F is a smooth function on an open neighborhood of the origin in $\mathbf{R} \times \mathbf{R}^n$ such that $F(t,0) = g(t)t^k$, where g is a smooth function on an open neighborhood of the origin in $\mathbf{R}$ that does not vanish there. Given any smooth function G on an open neighborhood of the origin in $\mathbf{R} \times \mathbf{R}^n$, there exist smooth functions q and r also on an open neighborhood of the origin in $\mathbf{R} \times \mathbf{R}^n$ as well as smooth functions r_j, $j = 0, \ldots, k-1$, on an open neighborhood of the origin in $\mathbf{R}^n$ satisfying

(a) $G = qF + r$ on a neighborhood of the origin in $\mathbf{R} \times \mathbf{R}^n$,

(b) $r(t,x) = \sum_{j=0}^{k-1} r_j(x)t^j$ for small (t, x) in $\mathbf{R} \times \mathbf{R}^n$.

Proof. Apply the Polynomial Division Theorem (9) to F and G to obtain smooth functions q^F, r^F, q^G, and r^G such that

$$F = q^F P_k + r^F \quad \text{and} \quad G = q^G P_k + r^G.$$

For the remainder r^F, note that

$$r^F(t, x, \lambda) = \sum_{j=0}^{k-1} r_j^F(x, \lambda)t^j$$

holds. An analogous expression is also valid for r^G.

The first claim is that

(α) $$q^F(0) \neq 0$$

and

(β) $$r_j^F(0) = 0, \quad j = 0, \ldots, k-1.$$

This follows from the Taylor expansion of $t^k g(t)$ because

$$t^k g(t) = F(t,0) = q^F(t,0,0) P_k(t,0) + r^F(t,0,0)$$

$$= q^F(t,0,0) t^k + \sum_{j=0}^{k-1} r_j^F(0) t^j$$

and $g(0) \neq 0$. Next it will be shown that the matrix

(γ) \qquad\qquad\qquad $D_2 r_j^F(0)$ is invertible,

where D_2 denotes the derivation with respect to the second set of variables $\lambda \in \mathbf{R}^k$. Start again with $t^k g(t)$. At $x = 0$ and for small λ,

$$t^k g(t) = F(t,0) = q^F(t,0,\lambda) P_k(t,\lambda) + r^F(t,0,\lambda)$$

$$= q^F(t,0,\lambda) \left(t^k + \sum_{j=0}^{k-1} \lambda_j t^j \right) + \sum_{j=0}^{k-1} r_j^F(0,\lambda) t^j$$

holds. Differentiate this equation with respect to λ_l and evaluate the result at $\lambda = 0$. Then

$$0 = t^k \frac{\partial q^F}{\partial \lambda_l}(t,0,0) + t^l q^F(t,0,0) + \sum_{j=0}^{k-1} t^j \frac{\partial r_j^F}{\partial \lambda_l}(0)$$

follows. Comparing the coefficients of t^j, one finds for $j < l$ that $(\partial r_j^F / \partial \lambda_l)(0)$ vanishes and that $(\partial r_l^F / \partial \lambda_l)(0) = -q^F(0)$. Thus, the matrix $D_2 r_j^F(0)$ is lower triangular and has nonzero diagonal terms by (α).

Due to (β) and (γ), the Implicit Function Theorem (see Appendix A.1) ensures the existence of a germ $[\varphi] \in \mathcal{E}_{n,k}$ satisfying

(δ) \qquad\qquad\qquad $\varphi(0) = 0$

and

(ε) \qquad\qquad $r_j^F(x, \varphi(x)) = 0$ \quad for small x and $j = 0, \dots, k-1$.

Replacing λ by $\varphi(x)$ yields $F(t,x) = q^F(t,x,\varphi(x)) P_k(t,\varphi(x))$, since the remainder vanishes as a result of (ε). Hence $P_k(t,\varphi(x)) = F(t,x)/q^F(t,x,\varphi(x))$ for small t and x, because by (α) and (δ) the denominator does not vanish at the origin. Inserting this expression into the formula for G yields

$$G(t,x) = \frac{q^G(t,x,\varphi(x))}{q^F(t,x,\varphi(x))} F(t,x) + \sum_{j=0}^{k-1} r_j^G(x,\varphi(x)) t^j,$$

which is the desired representation for G.

In the proof of the Malgrange–Mather Preparation Theorem a homogeneous system of equations is considered with coefficients in $\mathcal{E}_n$ and unknowns in a module over $\mathcal{E}_n$. A generalization of Cramer's rule is needed to solve these equations. More generally, let $\mathcal{R}$ be a commutative ring with identity 1, and let M be a module over $\mathcal{R}$. For an $l \times l$ matrix $r := (r_{ij})$ with coefficients in $\mathcal{R}$, its determinant is given by

$$(11) \qquad \det r := \sum_{\sigma \in S_l} \operatorname{sgn}(\sigma) r_{\sigma(1)1} \cdots r_{\sigma(l)l}$$

where S_l is the group of all permutations of $\{1, \ldots, l\}$ and $\operatorname{sgn}(\sigma)$ denotes the sign of $\sigma \in S_l$. Of course, $\det r$ is an element of $\mathcal{R}$. The matrix $\tilde{r} := (\tilde{r}_{ij})$ **complementary** to r is defined as follows. For fixed i and $j, \tilde{r}_{ij}$ is given by the determinant of the matrix derived from r by replacing the coefficient r_{ji} by 1 and all other coefficients in the jth row or in the ith column by 0. Finally, let r^j be the jth column of the matrix r. Then $r = (r^1, \ldots, r^l)$. Columns of matrices with coefficients in $\mathcal{R}$ are added to each other and multiplied by elements in $\mathcal{R}$ componentwise.

Lemma (12).

(a) If $r^j = s^j + \alpha t^j$ for $s^j, t^j \in \mathcal{R}^l$ and $\alpha \in \mathcal{R}$, then

$$\det(r^1, \ldots, r^j, \ldots, r^l) = \det(r^1, \ldots, s^j, \ldots, r^l)$$
$$+ \alpha \det(r^1, \ldots, t^j, \ldots, r^l).$$

(b) If two columns of r are equal, then $\det r = 0$.

(c) The determinant of the identity matrix E_l is 1.

(d) The products $\tilde{r}r$ and $r\tilde{r}$ of the matrices $\tilde{r}$ and r are both equal to $(\det r)E_l$.

(e) Let $e_1, \ldots, e_l$ be elements in M such that

$$\sum_{j=1}^{l} r_{ij}e_j = 0 \quad \text{and} \quad i = 1, \ldots, l.$$

Then $(\det r)e_i = 0$ holds for $i = 1, \ldots, l$.

Proof.

(a) $\det(\ldots, r^i, \ldots) = \sum_\sigma \operatorname{sgn}(\sigma) r_{\sigma(1)1} \cdots (s_{\sigma(j)j} + \alpha t_{\sigma(j)j}) \cdots r_{\sigma(l)l}$

$= \sum_\sigma \operatorname{sgn}(\sigma) r_{\sigma(1)1} \cdots s_{\sigma(j)j} \cdots r_{\sigma(l)l}$

$\quad + \alpha \sum_\sigma \operatorname{sgn}(\sigma) r_{\sigma(1)1} \cdots t_{\sigma(j)j} \cdots r_{\sigma(l)l}$

$= \det(\ldots, s^j, \ldots) + \alpha \det(\ldots, t^j, \ldots).$

(b) Suppose the kth and mth columns to be equal. Let τ be the permutation in S_l that interchanges k and m and leaves all other elements in $\{1, \ldots, l\}$ fixed. Let A_l be the subgroup of S_l consisting of all permutations with positive sign. Then S_l is the disjoint union of A_l and $A_l\tau$, since $\operatorname{sgn}(\tau) = -1$ and $\tau^{-1} = \tau$. From (11) it follows that

$$\det r = \sum_{\sigma \in A_l} r_{\sigma(1)1} \cdots r_{\sigma(l)l} - \sum_{\sigma \in A_l} r_{\sigma(\tau(1))1} \cdots r_{\sigma(\tau(l))l}.$$

Now $r_{\sigma(\tau(1))1} \cdots r_{\sigma(\tau(l))l} = r_{\sigma(1)1} \cdots r_{\sigma(l)l}$, since τ interchanges only indices denoting equal columns. This proves $\det r = 0$.

(c) The assertion follows immediately from (11).

(d) Let s^j be the column in $\mathcal{R}^l$ with the jth coefficient equal to 1 and all other coefficients equal to 0. When suitable multiples of s^j are added to the other columns of the matrix $(r^1, \ldots, r^{i-1}, s^j, r^{i+1}, \ldots, r^l)$, zeros are generated everywhere in the jth row. Only the coefficient in the jith position remains equal to 1. Since by (a) and (b) these operations do not alter the determinant, it follows that

$$\tilde{r}_{ij} = \det(r^1, \ldots, r^{i-1}, s^j, r^{i+1}, \ldots, r^l).$$

Using this for the computation of the coefficients of the product $\tilde{r}r$ one finds

$$(\tilde{r}r)_{ij} = \sum_{k=1}^{l} \tilde{r}_{ik} r_{kj} = \det\left(r^1, \ldots, r^{i-1}, \sum_{k=1}^{l} r_{kj} s^j, r^{i+1}, \ldots, r^l\right)$$
$$= \det(r^1, \ldots, r^{i-1}, r^j, r^{i+1}, \ldots, r^l)$$

because of the linearity (a) of the determinant. The last expression is equal to $\det r$ if $i = j$ and to 0 if $i \neq j$ by (b). The assertion concerning the product $r\tilde{r}$ follows by an analogous computation.

(e) Turning to the last assertion compute

$$0 = \sum_{k=1}^{l} \tilde{r}_{ik} \left(\sum_{j=1}^{l} r_{kj} e_j\right) = \sum_{j=1}^{l} \left(\sum_{k=1}^{l} \tilde{r}_{ik} r_{kj}\right) e_j = \sum_{j=1}^{l} \delta_{ij}(\det r) e_j$$
$$= (\det r) e_i$$

by (d) using the Kronecker delta symbol.

Proof of the Malgrange–Mather Preparation Theorem (13). The general assumption is that M is finitely generated over $\mathcal{E}_n$.

(i) Suppose that M is finitely generated over $\mathcal{E}_s$. It will be shown that $\dim M/\langle m_s M\rangle < \infty$ holds. Let $M = \langle e_1, \ldots, e_l\rangle_{\mathcal{E}_s}$, i.e., any b in M has the form

$$b = \sum_{i=1}^{l} [g_i] e_i \quad \text{for} \quad [g_i] \in \mathcal{E}_s.$$

Then

$$b = \sum_{i=1}^{l} \alpha_i e_i + \sum_{i=1}^{l} [\tilde{g}_i] e_i \quad \text{with} \quad \alpha_i := g_i(0), \tilde{g}_i := g_i - \alpha_i.$$

Since $[\tilde{g}_i]$ is in m_s, this proves that $M/\langle m_s M \rangle$ is spanned by $e_1 + \langle m_s M \rangle, \ldots,$ $e_l + \langle m_s M \rangle$.

To prove the converse, suppose that $\dim M/\langle m_s M \rangle < \infty$ holds. The proof that M is finitely generated over $\mathcal{E}_s$ will be given first for three special cases of a germ $[\varphi] \in \mathcal{E}_{n,s}$ with $\varphi(0) = 0$ and then for a general φ.

(ii) Suppose that $s = n$ and that $[\varphi]$ is in $\mathcal{G}_n$. Let $M = \langle e_1, \ldots, e_l \rangle_{\mathcal{E}_n}$, i.e., any b in M can be written as

$$b = \sum_{i=1}^{l} [f_i] e_i \quad \text{for} \quad [f_i] \in \mathcal{E}_n.$$

If $[g_i] := [f_i][\varphi]^{-1}$, then $b = \Sigma_{i=1}^{l} [g_i] e_i$ holds with respect to the new outer product for M induced by φ^*. Hence, M is finitely generated with respect to the new structure, too.

(iii) Next let $s = n - 1$, and suppose that the rank of the matrix $D\varphi(0)$ is $n - 1$. There are $n - 1$ columns forming an invertible submatrix of $D\varphi(0)$. By continuity, the determinant of this submatrix does not vanish on a whole neighborhood of the origin. Hence, for x in this neighborhood, the rank of $D\varphi(x)$ remains maximal, i.e., equal to $n - 1$. The Rank Theorem (see Appendix A.4) can be applied to φ, resulting in a diffeomorphism u of an open neighborhood W of 0 in $\mathbf{R}^n$ onto the open cube $K := \{x \in \mathbf{R}^n : |x_i| < 1, i = 1, \ldots, n\}$ preserving the origin and in a diffeomorphism v of an open neighborhood V of 0 in $\mathbf{R}^{n-1}$ onto the open cube $L := \{x \in \mathbf{R}^{n-1} : |x_i| < 1, i = 1, \ldots, n - 1\}$ preserving the origin, such that

$$\varphi(x) = v^{-1}(\pi(u(x))) \quad \text{for} \quad x \in W,$$

where $\pi \colon \mathbf{R}^n \to \mathbf{R}^{n-1}$ denotes the projection $(x_1, \ldots, x_n) \mapsto (x_1, \ldots, x_{n-1})$.

By (ii), it suffices to prove the assertion for π instead of φ. The pullback π^* is the usual embedding of $\mathcal{E}_{n-1}$ in $\mathcal{E}_n$, where $\mathcal{E}_{n-1}$ consists of those elements in $\mathcal{E}_n$ that are independent of the last variable x_n.

Now choose $e_1, \ldots, e_l$ in M such that $e_1 + \langle m_{n-1} M \rangle, \ldots, e_l + \langle m_{n-1} M \rangle$ is a basis for $M/\langle m_{n-1} M \rangle$. Then

(α) $$M = \langle e_1, \ldots, e_l \rangle_{\mathcal{E}_n}$$

holds. To see this, note that $m_{n-1} \subset m_n$ implies

$$M = \langle e_1, \ldots, e_l \rangle + \langle m_{n-1} M \rangle \subset \langle e_1, \ldots, e_l \rangle_{\mathcal{E}_n} + \langle m_n M \rangle \subset M.$$

By assumption, M is finitely generated over $\mathcal{E}_m$; therefore, Nakayama's Lemma (4.16) applies, yielding (α). Furthermore,

$$(\beta) \qquad M = \langle e_1, \ldots, e_l \rangle + \langle \{e_1, \ldots, e_l\} m_{n-1} \mathcal{E}_n \rangle$$

is valid. This is verified by inserting (α) in $\langle m_{n-1} M \rangle$, which gives

$$\langle m_{n-1} M \rangle = \langle m_{n-1} \{e_1, \ldots, e_l\} \mathcal{E}_n \rangle,$$

and using $M = \langle e_1, \ldots, e_l \rangle + \langle m_{n-1} M \rangle$.

The generators for M over $\mathcal{E}_s$ can now be constructed. By (β), there are real numbers α_{ij} and germs $[f_{ij}]$ in $m_{n-1} \mathcal{E}_n$ such that

$$[x_n] e_i = \sum_{j=1}^{l} (\alpha_{ij} + [f_{ij}]) e_j$$

holds. Using the Kronecker delta symbol, this means

$$\sum_{j=1}^{l} ([x_n] \delta_{ij} - \alpha_{ij} - [f_{ij}]) e_j = 0$$

for $i = 1, \ldots, l$. Let $[F] \in \mathcal{E}_n$ be the determinant of the matrix

$$([x_n] \delta_{ij} - \alpha_{ij} - [f_{ij}]).$$

By Lemma (12) part (e), the equations $[F] e_i = 0$, $i = 1, \ldots, l$, follow. Note that $f_{ij}(0, \ldots, 0, x_n) = 0$ for small x_n, since $[f_{ij}]$ is in $m_{n-1} \mathcal{E}_n$. Then (11) yields $F(0, \ldots, 0, x_n) = \det(x_n \delta_{ij} - \alpha_{ij})$ for small x_n. This expression is a polynomial in x_n of degree l. Therefore $F(0, \ldots, 0, x_n)$ can be written in the form $x_n^k g(x_n)$ with $k \leq l$ and $g(0) \neq 0$.

For any element a in M, one has

$$a = \sum_{j=1}^{l} [f_j] e_j$$

according to (α). Apply the Mather Division Theorem (10) to F and f_j to obtain

$$f_j(x_1, \ldots, x_n) = q_j(x_1, \ldots, x_n) F(x_1, \ldots, x_n) + \sum_{i=0}^{k-1} r_{ji}(x_1, \ldots, x_{n-1}) x_n^i$$

for some $[r_{ji}] \in \mathcal{E}_{n-1}$. Then $[F] e_j = 0$ implies

$$[f_j] e_j = \sum_{i=0}^{k-1} [r_{ji}][x_n^i] e_j$$

and hence,

$$a = \sum_{j=1}^{l} \sum_{i=0}^{k-1} [r_{ji}][x_n^i]e_j.$$

Thus, M is generated as a module over $\mathcal{E}_{n-1}$ by the lk elements $[x_n^i]e_j$, $i = 0, \ldots, k-1, j = 1, \ldots, l$.

(iv) For the last special case of φ, let $n \leq s$ and suppose that the rank of $D\varphi(0)$ is n. An application of the Rank Theorem shows that it suffices to prove the assertion for

$$\eta \colon \mathbf{R}^n \to \mathbf{R}^s, \quad (x_1, \ldots, x_n) \mapsto (x_1, \ldots, x_n, 0, \ldots, 0)$$

instead of for φ. For η the assertion follows, because η^* is surjective. To verify the surjectivity of η^*, let $[f]$ be in $\mathcal{E}_n$ and let f be defined on the open neighborhood V of 0 in $\mathbf{R}^n$. Define $g \colon V \times \mathbf{R}^{s-n} \to \mathbf{R}$, $g(x_1, \ldots, x_s) := f(x_1, \ldots, x_n)$. Obviously, $[g]$ lies in $\mathcal{E}_s$ with $\eta^*[g] = [g \circ \eta] = [f]$.

(v) Finally, the general case can be treated. Define $\tilde{\varphi} \colon U \to \mathbf{R}^n \times \mathbf{R}^s$ by $\tilde{\varphi}(x) := (x, \varphi(x))$. Let $\pi_i \colon \mathbf{R}^i \times \mathbf{R}^s \to \mathbf{R}^{i-1} \times \mathbf{R}^s$ be given by $(x_1, \ldots, x_i, y) \mapsto (x_1, \ldots, x_{i-1}, y)$. Then $\varphi = \pi_1 \circ \pi_2 \circ \ldots \circ \pi_n \circ \tilde{\varphi}$ holds. Clearly, $\tilde{\varphi}$ fulfills the premises of (iv), so that M is finitely generated over $\mathcal{E}_{n+s}$. Since $m_s \subset m_{(n-1)+s}$ is true and $M/\langle m_s M \rangle$ is finite dimensional, then $M/\langle m_{(n-1)+s} M \rangle$ is also finite dimensional. Hence, (iii) can be applied to π_n, so that M is finitely generated over $\mathcal{E}_{(n-1)+s}$. An induction argument completes the proof.

Exercises

9.1. A germ $[f]$ in $\mathcal{E}_1$ is said to be **even** if $f(x) = f(-x)$ holds in a neighborhood of 0. If $[f]$ is even, show that there is an $[h] \in \mathcal{E}_1$ satisfying $f(x) = h(x^2)$ for small x.

Hint: Let $\varphi \colon \mathbf{R} \to \mathbf{R}$, $\varphi(x) := x^2$, and denote by $\mathcal{C}$ the module $\mathcal{E}_1$ over itself with respect to the outer product $\mathcal{E}_1 \times \mathcal{C} \to \mathcal{C}$, $([h], [g]) \mapsto [h] \cdot [g] := (\varphi^*[h])[g]$. Apply Corollary (3) of the Malgrange–Mather Theorem to $\mathcal{C}$ with $\mathcal{A} := \{0\}$ and a suitable finite subset B of $\mathcal{C}$.

9.2. Let $p_1, \ldots, p_n$ be the n **elementary symmetric polynomials** in n variables, i.e.,

$$p_1(x_1, \ldots, x_n) = x_1 + \cdots + x_n$$
$$p_2(x_1, \ldots, x_n) = x_1 x_2 + \cdots + x_1 x_n + \cdots + x_{n-1} x_n$$
$$\vdots$$
$$p_n(x_1, \ldots, x_n) = x_1 x_2 \cdots x_n.$$

Recall that a polynomial p is **symmetric** if $p(x_1, \ldots, x_n) = p(x_{\pi(1)}, \ldots, x_{\pi(n)})$ for all permutations π of the indices, and recall also that every symmetric polynomial p is a polynomial in the elementary symmetric polynomials $p_1, \ldots, p_n$, i.e., $p(x) = q(p_1(x), \ldots, p_n(x))$ for all $x = (x_1, \ldots, x_n)$, where q is a polynomial in n variables. This result will be generalized in this exercise.

Call $[f] \in \mathcal{E}_n$ a symmetric germ, if $f(x_1, \ldots, x_n) = f(x_{\pi(1)}, \ldots, x_{\pi(n)})$ for small x and all permutations π. Show that there is a germ $[h] \in \mathcal{E}_n$ satisfying $f(x) = h(p_1(x), \ldots, p_n(x))$ for small x.

Hint: Consider the polynomial map $\varphi \colon \mathbf{R}^n \to \mathbf{R}^n, \varphi(x) := (p_1(x), \ldots, p_n(x))$, and let $\mathcal{C} := \mathcal{E}$ be the module over $\mathcal{E}$ with respect to the outer product $\mathcal{E} \times \mathcal{C} \to \mathcal{C}, ([h], [g]) \mapsto [h] \cdot [g] := (\varphi^*[h])[g]$. Moreover, let B be the finite subset $\{[x^\nu] \colon 0 \le \nu_i < n, i = 1, \ldots, n\}$ of $\mathcal{C}$. Evaluating the polynomial

$$(\xi - x_1) \cdots (\xi - x_n) = \xi^n - p_1(x)\xi^{n-1} + \cdots + (-1)^n p_n(x)$$

at $\xi = x_i$, show that $[x_i^n]$ is an element of $\langle m \cdot \mathcal{C} \rangle$ for $i = 1, \ldots, n$. Apply Corollary (3) of the Malgrange–Mather Preparation Theorem and use the classical result about symmetric polynomials cited above.

9.3. Give a short proof of Mather's Division Theorem (10) using Corollary (3) of the Malgrange–Mather Preparation Theorem (2).

Hint: Let $\mathcal{E}_1$ be embedded in $\mathcal{E}_{n+1}$ as usual. Use the decompositions

$$\mathcal{E}_{n+1} = \mathcal{E}_1 + \langle m_s \mathcal{E}_{n+1} \rangle, \quad \mathcal{E}_1 = \langle 1, t, \ldots, t^{k-1} \rangle + m_1^k$$

following from Exercise 4.18 and from (5.10). Note that $\langle F(\cdot, 0) \rangle_{\mathcal{E}_1} = m_1^k$ holds for F as in (10).

9.4. Let h, f, and g be smooth functions of two real variables.

(i) Show that

$$\bar{\partial}(h \circ (f, g)) = (\bar{\partial}f)(D_1 h \circ (f, g)) + (\bar{\partial}g)(D_2 h \circ (f, g))$$

holds, where $\bar{\partial} := \frac{1}{2}(D_1 + iD_2)$.

(ii) Suppose that f and g are real-valued and that $F := f + ig$ is holomorphic, i.e., that $\bar{\partial}F = 0$. Prove the formula

$$\bar{\partial}(h \circ (f, g)) = \overline{\partial F}(\bar{\partial}h) \circ (f, g)$$

where $\partial := \frac{1}{2}(D_1 - iD_2)$.

(iii) Use (ii) to verify the relation

$$\frac{\partial}{\partial \bar{z}} \tilde{G}(z, x, \lambda) = (LH)(\varphi(z, x, \lambda))$$

claimed in part (ii) of the proof of the Nirenberg Extension Lemma (7).

Solutions

9.1. Let $\mathcal{C}$ be defined as in the hint. Since $[x] \cdot [1] = (\varphi^*[x])[1] = [x^2]$, the ideal $\langle m_1 \cdot \mathcal{C} \rangle$ contains m_1^2. Therefore, $\mathcal{C} \subset \langle B \rangle + \langle m_1 \cdot \mathcal{C} \rangle$ holds for $B := \{[1], [x]\}$. Clearly, $\langle B \rangle$ is contained in $\langle \mathcal{E}_1 \cdot B \rangle$ and $\mathcal{E}_1$ is a finitely generated module over itself. Thus, (3) applies to $\mathcal{C}, B$, and $\mathcal{A} := \{0\}$, yielding $\mathcal{C} \subset \langle \mathcal{E}_1 \cdot B \rangle$. This means that for every germ $[f] \in \mathcal{E}_1$ the representation

$$f(x) = h(x^2) + h_1(x^2)x$$

holds for small x with $[h], [h_1] \in \mathcal{E}_1$. If $[f]$ is even, then the odd term must vanish, and the assertion $f(x) = h(x^2)$ follows.

9.2. Refer to $\mathcal{C}$ and B as defined in the hint. Moreover, it follows from the hint that

$$[x_j^n] = \sum_{j=1}^{n} (-1)^{j+1} [p_j][x_i^{n-j}] \quad \text{for} \quad i = 1, \ldots, n.$$

Since $\varphi^*[x_j] = [p_j]$, the latter implies $[x_i^n] \in \langle m \cdot \mathcal{C} \rangle$, so that the inclusion $\mathcal{C} \subset \langle B \rangle + \langle m \cdot \mathcal{C} \rangle$ holds. Clearly, $\langle B \rangle$ is contained in $\langle \mathcal{E} \cdot B \rangle$ and $\mathcal{E}$ is a finitely generated module over itself. Thus, Corollary (3) of the Malgrange–Mather Preparation Theorem applies to $\mathcal{C}, B$, and $\mathcal{A} := \{0\}$, yielding $\mathcal{C} \subset \langle \mathcal{E} \cdot B \rangle$. This means that for every germ $[f] \in \mathcal{E}$ the representation

$$(*) \qquad f(x) = \sum_{\substack{0 \le \nu_i < n \\ 1 \le i \le n}} h_\nu(p_1(x), \ldots, p_n(x))x^\nu$$

holds for small x with some $[h_\nu] \in \mathcal{E}$.

Now assume that $[f]$ is symmetric. For every permutation π of the set $\{1, \ldots, n\}$ and every $x = (x_1, \ldots, x_n) \in \mathbf{R}^n$ let $\pi(x) := (x_{\pi(1)}, \ldots, x_{\pi(n)})$. Then symmetrization of $(*)$ yields

$$f(x) = \frac{1}{n!} \sum_\pi f(\pi(x)) = \sum_\nu h_\nu(p_1(x), \ldots, p_n(x))q_\nu(x),$$

where q_ν are the symmetric polynomials

$$\frac{1}{n!} \sum_\pi (\pi(x))^\nu.$$

It is classical that the latter are polynomials of the elementary symmetric polynomials. Thus, the assertion follows.

9.3. As noted in the hint,

$$\mathcal{E}_{n+1} = \langle F_0 \rangle \mathcal{E}_1 + \langle 1, t, \ldots, t^{k-1} \rangle + \langle m_s \mathcal{E}_{n+1} \rangle$$

holds for $F_0(t) := F(0,t)$. Decomposing $[F] \in \mathcal{E}_{n+1}$ according to $\mathcal{E}_{n+1} = \mathcal{E}_1 + \langle m_s \mathcal{E}_{n+1} \rangle$ yields

$$\langle F_0 \rangle_{\mathcal{E}_1} \subset \langle F \rangle_{\mathcal{E}_{n+1}} + \langle m_s \mathcal{E}_{n+1} \rangle.$$

Hence $\mathcal{E}_{n+1}$ is contained in

$$\langle F \rangle_{\mathcal{E}_{n+1}} + \langle 1, t, \ldots, t^{k-1} \rangle + \langle m_s \mathcal{E}_{n+1} \rangle.$$

By (3), this implies

$$\mathcal{E}_{n+1} = \langle F \rangle_{\mathcal{E}_{n+1}} + \langle 1, t, \ldots, t^{k-1} \rangle_{\mathcal{E}_s},$$

which is the assertion of Mather's Division Theorem (10).

9.4.

(i) By the chain rule, one has $D_j(g \circ (f,g)) = (D_1 g \circ (f,g))D_j f + (D_2 g \circ (f,g))D_j g$ for $j = 1,2$. This implies the assertion.

(ii) $\bar{\partial}F = 0$ means that the Cauchy–Riemann differential equations $D_1 f = D_2 g$ and $D_2 f = -D_1 g$ hold. Hence $\bar{\partial}f = \frac{1}{2}\overline{\partial F}$ and $\bar{\partial}g = \frac{i}{2}\overline{\partial F}$ follow. Then (i) yields the result.

(iii) For the proof of the assertion, the variables x and $\lambda_1, \ldots, \lambda_{k-1}$ are omitted, since they are kept constant. Then H is a function of the two complex variables z and μ_0 and $\tilde{G}(z)$ equals $H(z, P_k(z))$. Denoting the partial functions of H by H_z and H_{μ_0} one obtains from the chain rule that

$$\frac{\partial}{\partial \bar{z}}\tilde{G}(z) = \frac{\partial}{\partial \bar{z}}H_{\mu_0}(z) + \frac{\partial}{\partial \bar{z}}(H_z \circ P_k)(z)$$

with $\mu_0 = P_k(z)$. The second summand is equal to

$$\overline{P_k'(z)}(\partial H / \partial \bar{\mu}_0)(z, P_k(z))$$

by (ii), which proves the assertion.

Chapter 10

The Fundamental Theorem on Universal Unfoldings

The second major theorem in Catastrophe Theory is René Thom's Fundamental Theorem on Universal Unfoldings, which will be proved in this chapter. In addition to its theoretical importance, this theorem gives a concrete method for calculating a universal unfolding of a given germ. An important application is Table 1 of Chapter 7, which lists the universal unfoldings of the **7** elementary catastrophes. The proof of the Fundamental Theorem on Universal Unfoldings relies heavily on the so-called Main Lemma, which will be treated first. After that, just relatively simple algebraic arguments will be needed to derive the Fundamental Theorem. The proof of the Main Lemma is essentially based on the same idea used in (4.22), which was the major tool to show Mather's Sufficient Criterion for Determinacy (4.24). In particular, the problem is reduced to solving an ordinary differential equation.

Main Lemma (1). Let k and r be positive integers, and let $[f]$ be a k-determined germ in m_n^2. Then any two r-parameter k-transversal unfoldings of $[f]$ are equivalent.

Proof. The proof will be subdivided into six parts. For every r-parameter unfolding $[H]$, let $\gamma_j(H)$ be the germ in m_n of the function

$$\frac{\partial}{\partial u_j}H(x,0) - \frac{\partial}{\partial u_j}H(0,0)$$

defined for small x in $\mathbf{R}^n$; recall that the $\gamma_j(H)$, $j = 1,\ldots,r$, span the space $\mathcal{V}[H]$ introduced in Chapter 8.

(i) There is an unfolding $[H]$ in $\mathcal{E}_{n+r}$ of $[f]$ such that for every k-transversal unfolding $[F]$ of $[f]$ in $\mathcal{E}_{n+r}$ there exists a k-transversal unfolding $[F']$ of $[f]$ in $\mathcal{E}_{n+r}$ equivalent to $[H]$ and satisfying $\gamma_j(F') - \gamma_j(F) \in \mathcal{J}[f]$ for $j = 1,\ldots,r$.

Let $[F] \in \mathcal{E}_{n+r}$ be a k-transversal unfolding of $[f]$. By (8.1), $c := \mathrm{cod}[f]$ is at most r. Suppose that $[g_1] + \mathcal{J}[f], \ldots, [g_c] + \mathcal{J}[f]$ is a basis of $m_n/\mathcal{J}[f]$. Then there are real coefficients a_{jl} with

$$\gamma_j(F) + \mathcal{J}[f] = \sum_{l=1}^{c} a_{jl}([g_l] + \mathcal{J}[f]).$$

The $r \times c$ matrix (a_{jl}) has maximal rank, since $\gamma_1(F) + \mathcal{J}[f], \ldots, \gamma_r(F) + \mathcal{J}[f]$ spans $m_n/\mathcal{J}[f]$ by (8.1). Extend (a_{jl}) to an invertible $r \times r$ matrix by adding $r - c$ columns. This matrix determines a linear coordinate transformation $\psi : \mathbf{R}^r \to \mathbf{R}^r$,

$$\psi(v)_l := \sum_{j=1}^{r} v_j a_{jl}.$$

Now, for u in $\mathbf{R}^r$ and for small x in $\mathbf{R}^n$ define the function

$$H(x, u) := f(x) + u_1 g_1(x) + \cdots + u_c g_c(x)$$

whose germ in $\mathcal{E}_{n+r}$ obviously is an unfolding of $[f]$. Then $F'(x, v) := H(x, \psi(v))$ gives an unfolding $[F']$ of $[f]$ equivalent to $[H]$ satisfying

$$\gamma_j(F') = \sum_{l=1}^{c} \gamma_l(H) a_{jl} = \sum_{l=1}^{c} [g_l] a_{jl}.$$

Therefore, $\gamma_j(F') - \gamma_j(F)$ is in $\mathcal{J}[f]$ for $j = 1, \ldots, r$, which implies $\mathcal{V}[F] \subset \mathcal{J}[f] + \mathcal{V}[F']$ and hence the k-transversality of $[F']$ by (8.1).

(ii) Let $[F]$ and $[F']$ be two k-transversal unfoldings of $[f]$ in $\mathcal{E}_{n+r}$ such that $\gamma_j(F') - \gamma_j(F)$ is in $\mathcal{J}[f]$ for $j = 1, \ldots, r$. Then $[F]$ and $[F']$ are equivalent. This assertion will be proved in part (vii) using the intermediate steps developed in (ii)–(vi). Then the Main Lemma follows easily in (viii).

Let $U \subset \mathbf{R}^n$ and $V \subset \mathbf{R}^r$ be open balls around the origin such that both F and F' are defined on $U \times V$. On $U \times V$ consider the one-parameter family $F_t := (1 - t)F + tF'$ of smooth functions. Obviously, $[F_t]$ is an r-parameter unfolding of $[f]$ with $[F_0] = [F]$ and $[F_1] = [F']$. Moreover, $\gamma_j(F_t) - \gamma_j(F_s) = (t-s)(\gamma_j(F') - \gamma_j(F))$ lies in $\mathcal{J}[f]$, so that k-transversality of $[F]$ implies that of $[F_t]$ by (8.1). Therefore, to prove (ii) it suffices to show that for every $s \in [0, 1]$ there is an open interval I_s in $\mathbf{R}$ containing s such that $[F_t]$ is equivalent to $[F_s]$ for all $t \in I_s$; see (4.22i). From now on fix an $s \in [0, 1]$.

(iii) The existence of such an interval I_s will be reduced to solving an initial-value problem for an ordinary differential equation. It suffices to show the existence of open neighborhoods $A \subset U$ and $B \subset V$ of the origin, of an open interval I containing 0, and of smooth maps $\varphi : A \times B \times I \to \mathbf{R}^n$,

$\psi : B \times I \to \mathbf{R}^r$, and $\gamma : B \times I \to \mathbf{R}$ satisfying

(α) $\qquad \varphi(x, u, 0) = x, \qquad \psi(u, 0) = u,$

$\qquad\qquad \gamma(u, 0) = 0 \quad$ for all $\quad (x, u) \in A \times B,$.

(β) $\qquad\qquad \varphi(x, 0, t) = x \quad$ for all $\quad (x, t) \in A \times I;$

(γ) $\qquad\qquad B \to \mathbf{R}^r, \qquad u \mapsto \psi(u, t),$

$\qquad\qquad$ is a local diffeomorphism at the origin with

$\qquad\qquad \psi(0, t) = 0$ for all $t \in I;$

(δ) $\qquad\qquad \gamma(0, t) = 0 \quad$ for all $\quad t \in I;$

(ε) $\qquad\qquad (\varphi(x, u, t), \psi(u, t)) \in U \times V \quad$ and

$\qquad F_{s+t}(\varphi(x, u, t), \psi(u, t)) + \gamma(u, t) = F_s(x, u)$

$\qquad$ for all $\quad (x, u, t) \in A \times B \times I.$

Indeed, given the maps φ, ψ, and γ, consider the partial map $\varphi_t :$ $A \times B \to \mathbf{R}^n$ defined by $\varphi_t(x, u) := \varphi(x, u, t)$ for $t \in I$ and the analogously defined partial maps ψ_t and γ_t. Clearly, for every t in I, one has $[\varphi_t] \in m_{n+r,n}, [\psi_t] \in \mathcal{G}_r, [\gamma_t] \in m_r$ and $\varphi_t(x, 0) = x$ as well as $F_s(x, u) = F_{s+t}(\varphi_t(x, u), \psi_t(u)) + \gamma_t(u)$ for (x, u) in an open neighborhood of the origin in $\mathbf{R}^{n+r}$. This means that the unfoldings $[F_s]$ and $[F_{s+t}]$ are equivalent for all $t \in I$, and $I_s := s + I$ is an interval, as desired.

(iv) Let (α), (β), (γ), and (δ) be satisfied, and suppose that $\varphi(A \times B \times I) \subset U$ and $\psi(B \times I) \subset V$ hold. Set $\Phi(x, u, t) := F_{s+t}(x, u)$ for $(x, u, t) \in A \times B \times I$. Then (ε) follows from

$$\frac{d}{dt}\Phi(\varphi(x, u, t), \psi(u, t), t) + \frac{\partial}{\partial t}\gamma(u, t) = 0,$$

due to the initial condition

$$\Phi(\varphi(x, u, 0), \psi(u, 0), 0) + \gamma(u, 0) = \Phi(x, u, 0) = F_s(x, u).$$

Hence (ε) can be replaced by

(ε') $\qquad\qquad (\varphi(x, u, t), \psi(u, t)) \in U \times V \quad$ and

$$\frac{\partial}{\partial t}\Phi(\varphi(x, u, t), \psi(u, t), t)$$

$$+ \sum_{i=1}^{n} D_{1i}\Phi(\varphi(x, u, t), \psi(u, t), t)\frac{\partial}{\partial t}\varphi_i(x, u, t)$$

$$+ \sum_{j=1}^{r} D_{2j}\Phi(\varphi(x, u, t), \psi(u, t), t)\frac{\partial}{\partial t}\psi_j(u, t)$$

$$+ \frac{\partial}{\partial t}\gamma(u, t) = 0$$

$\qquad$ for all $\quad (x, u, t) \in A \times B \times I,$

where D_{1i} denotes the partial derivative with respect to the ith coordinate of the first argument and D_{2j} that with respect to the jth coordinate of the second argument. Furthermore, φ_i is the ith component of φ and ψ_j the jth of ψ.

(v) Conditions (α), (β), (γ), (δ), and (ε') follow if for some open interval J containing the origin there are smooth maps $\tilde{\varphi} \colon U \times V \times J \to \mathbf{R}^n$, $\tilde{\psi} \colon V \times J \to \mathbf{R}^r$, and $\tilde{\gamma} \colon V \times J \to \mathbf{R}$, satisfying

(ζ)
$$\tilde{\varphi}(x, 0, t) = 0, \qquad \tilde{\psi}(0, t) = 0,$$
$$\tilde{\gamma}(0, t) = 0 \quad \text{for all} \quad (x, t) \in U \times J;$$

(η)
$$\frac{\partial}{\partial t}\Phi(x, u, t) + \sum_{i=1}^{n} \tilde{\varphi}_i(x, u, t) \frac{\partial}{\partial x_i}\Phi(x, u, t)$$
$$+ \sum_{j=1}^{r} \tilde{\psi}_j(u, t) \frac{\partial}{\partial u_j}\Phi(x, u, t)$$
$$+ \tilde{\gamma}(u, t) = 0 \quad \text{for all} \quad (x, u, t) \in U \times V \times J,$$

where $\tilde{\varphi}_i$ and $\tilde{\psi}_j$ are the components of $\tilde{\varphi}$ and $\tilde{\psi}$.

To verify this assertion, for $(x, u) \in U \times V$ consider the system of differential equations

$$\frac{\partial}{\partial t}\varphi(x, u, t) = \tilde{\varphi}(\varphi(x, u, t), \psi(u, t), t),$$
$$\frac{\partial}{\partial t}\psi(u, t) = \tilde{\psi}(\psi(u, t), t),$$
$$\frac{\partial}{\partial t}\gamma(u, t) = \tilde{\gamma}(\gamma(u, t), t),$$

with the initial condition

$$\varphi(x, u, 0) = x, \quad \psi(u, 0) = u, \quad \gamma(u, 0) = 0.$$

By the existence theorem in Appendix A.3, there is an open set $A \times B$ in $U \times V$ containing the origin and an open interval $I \subset J$ with $0 \in I$, so that for every $(x, u) \in A \times B$ there is a unique solution

$$I \to U \times V \times \mathbf{R}, t \mapsto (\varphi(x, u, t), \psi(u, t), \gamma(u, t))$$

of the above system with

$$(\varphi(x, u, 0), \psi(u, 0), \gamma(u, 0)) = (x, u, 0)$$

having the property that the map

$$A \times B \times I \to \mathbf{R}^n \times \mathbf{R}^r \times \mathbf{R},$$
$$(x, u, t) \mapsto (\varphi(x, u, t), \psi(u, t), \gamma(u, t))$$

is smooth.

Then (α) is trivial. Also (β), (δ), and $\psi(0,t) = 0$ for $t \in I$ are obviously true, because of (ζ) and the uniqueness of the solution to the above initial-value problem. Furthermore, the first assertion in (ε') is clearly satisfied, and the second assertion follows from (η) after the variable x is replaced by $\varphi(x, u, t)$ and the parameter u by $\psi(u, t)$. What remains is to verify the first part of (γ). Since ψ is smooth with $D_1\psi(u, 0) = E_r$, it follows that $\det D_1\psi(u, t)$ does not vanish and ψ_t is a local diffeomorphism at the origin for small t. Replacing I by a sufficiently small open subinterval containing the origin, the last condition (γ) is satisfied, too.

(vi) The existence of maps $\tilde{\varphi}, \tilde{\psi}$, and $\tilde{\gamma}$ as in (v) is ensured if the equation

$$\mathcal{E}_{n+r+1} = \langle A\mathcal{E}_{n+r+1}\rangle + \langle B\mathcal{E}_{r+1}\rangle$$

holds, where

$$A := \left\{ \frac{\partial}{\partial x_1}\Phi, \ldots, \frac{\partial}{\partial x_n}\Phi \right\} \quad \text{and} \quad B := \left\{ \frac{\partial}{\partial u_1}\Phi, \ldots, \frac{\partial}{\partial u_r}\Phi, 1 \right\}.$$

To see this, notice that $\tilde{\varphi}(x, 0, t) = 0$ for small x and t is equivalent to $[\tilde{\varphi}_i] \in \langle m_r\mathcal{E}_{n+r+1}\rangle, i = 1, \ldots, n$. This is an easy consequence of (1.28), according to which germs $[g_{ik}] \in \mathcal{E}_{n+r+1}$ for $k = 1, \ldots, n$ exist with

$$\tilde{\varphi}_i(x, u, t) = \sum_{k=1}^{r} g_{ik}(x, u, t)u_k$$

for small arguments. Analogously, one has $[\tilde{\psi}_j] \in \langle m_r\mathcal{E}_{r+1}\rangle$ for $j = 1, \ldots, r$ and $[\tilde{\gamma}] \in \langle m_r\mathcal{E}_{r+1}\rangle$. Since $\partial\Phi/\partial t = F' - F$ and $F'(x, 0) - F(x, 0) = f(x) - f(x) = 0$, the germ $[\partial\Phi/\partial t]$ lies in $\langle m_r\mathcal{E}_{n+r+1}\rangle$. Therefore, conditions (ζ) and (η) are satisfied if

$$\langle m_r\mathcal{E}_{n+r+1}\rangle \subset \langle Am_r\mathcal{E}_{n+r+1}\rangle + \langle Bm_r\mathcal{E}_{r+1}\rangle$$

holds. This inclusion follows from the equation in (vi) after both sides of it are multiplied by m_r.

(vii) Now it will be shown that the equation in (vi) holds, which completes the proof of the claim in (ii). It suffices to verify $\mathcal{E}_{n+r+1} = \langle A\mathcal{E}_n\rangle + \langle B\rangle + \langle m_{r+1}\mathcal{E}_{n+r+1}\rangle$, since this obviously implies $\mathcal{E}_{n+r+1} = \langle A\mathcal{E}_{n+r+1}\rangle + \langle B\mathcal{E}_{r+1}\rangle + \langle m_{r+1}\mathcal{E}_{n+r+1}\rangle$, and then Corollary (9.3) to the Malgrange–Mather Preparation Theorem yields the assertion.

Let $[h]$ be in $\mathcal{E}_{n+r+1}$. Denote the restriction of h to $\mathbf{R}^n \times \{0\} \subset \mathbf{R}^n \times \mathbf{R}^{r+1}$ by h' and regard $[h']$ as an element of $\mathcal{E}_n$. Because $[F_s]$ is k-transversal and $[f]$ is k-determined, using (8.1) it follows that there are germs $[h_i]$ in $\mathcal{E}_n$ for $i = 1, \ldots, n$ and real numbers $b_0, b_1, \ldots, b_r$ with

$$[h'] = \sum_{i=1}^{n}[D_if][h_i] + b_0 + \sum_{j=1}^{r} b_j\gamma_j(F_s).$$

Set

$$c := b_0 - \sum_{j=1}^{r} b_j \frac{\partial}{\partial u_j} F_s(0, 0).$$

Since $[D_i f]$ is the restriction of $[\partial \Phi / \partial x_i]$ to $\mathbf{R}^n \times \{0\} \subset \mathbf{R}^n \times \mathbf{R}^{r+1}$ and $\gamma_j(F_s)$ that of

$$\left[\frac{\partial}{\partial u_j} \Phi - \frac{\partial}{\partial u_j} F_s(0, 0) \right],$$

the difference

$$[h] - \sum_{i=1}^{n} \left[\frac{\partial}{\partial x_i} \Phi \right] [h_i] - c - \sum_{j=1}^{r} b_j \left[\frac{\partial}{\partial u_j} \Phi \right]$$

vanishes on $\mathbf{R}^n \times \{0\} \subset \mathbf{R}^n \times \mathbf{R}^{r+1}$ and hence lies in $\langle m_{r+1} \mathcal{E}_{n+r+1} \rangle$. This gives the desired decomposition of $[h]$.

(viii) Let $[F]$ and $[G]$ be two k-transversal unfoldings in $\mathcal{E}_{n+r}$ of $[f]$, and let $[H]$, $[F']$, and $[G']$ be as in (i). Then $[F']$ and $[G']$ are equivalent, because they are both equivalent to $[H]$. Moreover, (ii) implies that $[F']$ is equivalent to $[F]$ and $[G']$ to $[G]$. Hence $[F]$ and $[G]$ are equivalent.

Now the investigations in Chapter 8 concerning the relationship between versal and k-transversal unfoldings can be continued. Recall that versal unfoldings are k-transversal for every $k \in \mathbf{N}$, as shown in (8.3). A partial converse is given in the following lemma.

Lemma (2). Let k be a positive integer, and consider a k-determined germ $[f]$ in m_n^2. Then a k-transversal unfolding of $[f]$ is versal.

Proof. Let $[F] \in \mathcal{E}_{n+r}$ be a k-transversal unfolding of $[f]$, and consider an arbitrary unfolding $[G] \in \mathcal{E}_{n+s}$ of $[f]$. The claim is that $[G]$ is induced by $[F]$.

Let $[H]$ be the sum of $[F]$ and $[G]$. Then $[G]$ is induced by $[H]$, according to (7.7). Furthermore, $[H]$ is k-transversal, since $\mathcal{V}[F] \subset \mathcal{V}[H]$ holds. Let $[K]$ be the $(r+s)$-parameter unfolding of $[f]$ given by $K(x, u, v) := F(x, u)$. Obviously, $[K]$ is also k-transversal.

Now the Main Lemma (1) states that $[H]$ and $[K]$ are equivalent, implying that $[H]$ is induced by $[K]$ and hence by $[F]$. By the Transitivity of Inducing (7.6), it follows that $[G]$ is induced by $[F]$.

Next it will be shown that a germ must be finitely determined if it has an unfolding that is k-transversal for all positive integers k. Furthermore, the number of parameters of such an unfolding must be at least equal to the codimension of the initial germ.

Lemma (3). Let $[F]$ be an r-parameter unfolding of a germ $[f]$ in m_n^2. Suppose that $[F]$ is k-transversal for an integer $k \geq r+1$. Then $[f]$ is $(r+2)$-determined, and the codimension of $[f]$ is at most r.

Proof. By hypothesis, $m_n = \mathcal{J}[f] + m_n^{k+1} + \mathcal{V}[F]$ holds. For the ideal $\mathcal{A} := \mathcal{J}[f] + m_n^{k+1}$, this implies $\dim(m_n/\mathcal{A}) \leq \dim \mathcal{V}[F] \leq r$. By (5.7), there is a nonnegative integer l satisfying $m_n^l \subset \mathcal{A}$ and $\dim(\mathcal{E}_n/\mathcal{A}) \geq l$. Then $r + 1 \geq l$ and $m_n^{r+1} \subset \mathcal{A}$ hold. The corollary (4.18) to Nakayama's Lemma yields $m_n^{r+1} \subset \mathcal{J}[f]$. From (4.25), it follows that $[f]$ is $(r + 2)$-determined. Now $m_n = \mathcal{A} + \mathcal{V}[F]$ and $m_n^{k+1} \subset \mathcal{J}[f]$ yield $m_n = \mathcal{J}[f] + \mathcal{V}[F]$ and hence $\mathrm{cod}[f] \leq r$.

Using the preceding results and (8.3), the main properties of versal unfoldings are immediately obtained.

Theorem (4). An r-parameter unfolding of a germ in m_n^2 is versal if and only if it is k-transversal for an integer $k > r + 1$.

Corollary (5). An unfolding of a germ in m_n^2 is versal exactly when it is k-transversal for every positive integer k.

Versality Criterion (6). An unfolding $[F]$ of a germ $[f]$ in m_n^2 is versal if and only if $m_n = \mathcal{J}[f] + \mathcal{V}[F]$ holds.

Proof. Let $[F]$ be versal. Due to (5) and (3), $[f]$ is finitely determined, and $m_n = \mathcal{J}[f] + \mathcal{V}[F]$ follows from (8.1b) and (5). The converse assertion is implied by (8.1a) and (5).

In applications of Catastrophe Theory, this criterion is crucial in determining whether or not a given unfolding is versal. As a simple application of (6), note that a germ in m_n^2 is a versal unfolding of itself if and only if it has the origin as a nondegenerate critical point.

Existence and Uniqueness for Versal Unfoldings (7). A germ in m_n^2 has a versal unfolding if and only if it is finitely determined. Any two r-parameter versal unfoldings of a germ in m_n^2 are equivalent.

Proof. By (5) a versal unfolding of a germ $[f] \in m_n^2$ is k-transversal for every k, and (3) implies that $[f]$ is finitely determined. Conversely, let $k := \det[f]$ be finite. According to (8.2), there always exists an unfolding of $[f]$ that is k-transversal, and (2) implies its versality. This proves the first assertion.

To prove the second assertion, consider two r-parameter versal unfoldings of a germ $[f]$ in m_n^2. By the first assertion, $[f]$ is finitely determined. Then the second assertion follows from (8.3) and the Main Lemma (1).

The theory culminates in the second major theorem of Catastrophe Theory that follows.

Fundamental Theorem on Universal Unfoldings (8).

(a) A germ $[f]$ in m_n^2 has a universal unfolding if and only if $[f]$ is finitely determined.

(b) If r is the codimension of a finitely determined germ $[f]$ in m_n^2, then the germ in m_{n+r}^2 of

$$F(x, u) := f(x) + \sum_{j=1}^{r} u_j g_j(x)$$

is a universal unfolding of $[f]$ for every basis $[g_1] + \mathcal{J}[f], \ldots, [g_r] + \mathcal{J}[f]$ of $m_n / \mathcal{J}[f]$.

(c) Every versal unfolding of $[f]$ in m_n^2 is equivalent to the sum of the unfolding $[F]$ given in (b) and a constant unfolding of $[f]$.

Proof.

(a) The assertion follows immediately from (7).

(b) The unfolding $[F] \in \mathcal{E}_{n+r}$ is versal by (6). Due to (5) and (3), the number r of parameters of $[F]$ is minimal, and hence, $[F]$ is universal.

(c) Let $[G]$ be an s-parameter versal unfolding of $[f]$, and set $r := \mathrm{cod}[f]$. Then $s \geq r$ follows from (b). Add to $[F]$ a constant unfolding of $[f]$ in order to get an unfolding with s parameters. Since the sum is still versal, it is equivalent to $[G]$ according to the second part of (7).

Now Table 1 at the end of Chapter 7 is verified, completing the classification of unfoldings of germs of codimension at most 4. For germs of codimension 5 and 6, the classification of their universal unfoldings is treated in Exercise 7.6. The classification for codimension 7 and corank 2 is given in Exercise 7.7.

Exercises

10.1. Some examples for the fold and the cusp catastrophes, which were mentioned in Chapter 2, will be conclusively treated. Recall the remarks at the end of Chapter 7 about unfoldings of equivalent germs at different points.

(a) Let $\alpha_0 \in {]}0, \pi/2{[}$ and consider $V(\Theta, \alpha) := \Theta \sin \alpha - \sin \alpha_0 \sin \Theta$ for small Θ and $\alpha - \alpha_0$, which is the potential of the wheel on a slope. Show that $[V]$ is an example of the fold catastrophe by verifying that the germ of the partial map $V_{\alpha_0}(\Theta) := V(\Theta, \alpha_0)$ is equivalent to $[\Theta^3]$ and that $[V]$ is a universal unfolding of $[V_{\alpha_0}]$.

(b) Let $a := (7 + \sqrt{97})/12$ and consider the potential of Zeeman's catastrophe machine

$$V(\Theta; u, v) := \left(\left[\frac{17}{4} - 2\cos\Theta \right]^{1/2} - 1 \right)^2$$

$$+ \left(\left[(u+a)^2 + v^2 + \frac{1}{4} + (u+a)\cos\Theta - v\sin\Theta \right]^{1/2} - 1 \right)^2$$

for small Θ, u, and v. Show that $[V]$ is an example of the cusp catastrophe by proving that the germ of the partial map $V_0(\Theta) := V(\Theta; 0, 0)$ is equivalent to $[\Theta^4]$ up to a constant germ and that $[V]$ is a universal unfolding of $[V_0]$.

(c) The modelling of the population growth of the spruce budworm yields the smooth function of three variables

$$F(x; u, v) := -\frac{1}{2}ux^2 + \frac{1}{3}\frac{u}{v}x^3 + x - \arctan x$$

for arbitrary x and u and $v \neq 0$. Verify that $[F]$ is not a fold catastrophe at $x = 0$. To show this, fix $v_0 > 0$, and let $f(x) := x - \arctan x$ be the partial map of F at $(u, v) = (0, v_0)$. Prove that $[f]$ is equivalent to $[x^3]$ and that $[F]$ is not a versal unfolding of $[f]$.

(d) Let F be as in (c). Denote by g the partial map of F at

$$(u, v) := \left(\frac{3}{8}\sqrt{3}, 3\sqrt{3} \right).$$

Show that as a germ at $x = \sqrt{3}$, $[g]$ is equivalent to $[(x - \sqrt{3})^4]$ up to a constant germ and that $[F]$ is a universal unfolding of $[g]$. Thus, $[F]$ is a cusp catastrophe.

10.2. Consider the Euler arch, which was introduced in Exercise 2.1. Prove that the germ of

$$V(\Theta; \alpha, \beta) := 2k\Theta^2 + \alpha\sin\Theta - 2\beta(1 - \cos\Theta)$$

at $(0; 0, 2k)$, where k is positive, is a universal unfolding of $[V(\cdot; 0, 2k)]$, which is equivalent to $[\Theta^4]$. In other words show that $[V]$ is an example of the cusp catastrophe.

10.3. For the caustic occurring in a cup of coffee (see Exercise 2.2) show that the ray length

$$F(\Theta; u, v) = \cos\Theta + [(u - \cos\Theta)^2 + (v - \sin\Theta)^2]^{1/2}$$

is a universal unfolding of the germ $[F(\cdot; \frac{1}{2}, 0)]$, and this germ is equivalent to $-[\Theta^4]$ up to a constant. Thus, $[F]$ is an example of the dual cusp catastrophe.

10.4. Consider the unfolding $[F] \in \mathcal{E}_{2+4}$ given by

$$F(x, y; u, v, w, t) := x - x \cos y - x^3 \cos y$$
$$+ \sin(ux + vx^2) + \sin(wy + ty^2)$$

of the germ $[f] = [x - x \cos y - x^3 \cos y]$. Show that $[F]$ is versal but not universal. Reduce the number of parameters of $[F]$ to obtain a universal unfolding.

Solutions

10.1.

(a)   Since

$$V_{\alpha_0}(\Theta) = (\sin \alpha_0)(\Theta - \sin \Theta) = (\tfrac{1}{6} \sin \alpha_0)\Theta^3 + \text{higher order terms},$$

it follows immediately from (1.34) that $[V_{\alpha_0}] \sim [\Theta^3]$.

Now apply the criterion for versality (6) to prove the versality of $[V]$. Clearly, $\mathcal{J}[V_{\alpha_0}] = \langle \Theta^2 \rangle_{\mathcal{E}} = m^2$ and

$$\mathcal{V}[V] = \left\langle \frac{\partial}{\partial \alpha} V(\Theta, \alpha_0) - \frac{\partial}{\partial \alpha} V(0, \alpha_0) \right\rangle = \langle \Theta \cos \alpha_0 \rangle = \langle \Theta \rangle$$

hold. Therefore, $\mathcal{J}[V_{\alpha_0}] + \mathcal{V}[V] = m$, and the versality follows. $[V]$ is even universal, since the fold catastrophe is a one-parameter unfolding.

(b)   Expanding $V_0(\Theta)$ in powers of Θ, one finds

$$V_0(\Theta) = \left(\left[\frac{17}{4} - 2 \cos \Theta \right]^{1/2} - 1 \right)^2 + \left(\left[a^2 + \frac{1}{4} + a \cos \Theta \right]^{1/2} - 1 \right)^2$$

$$= \left(\left[\frac{9}{4} + \Theta^2 - \frac{\Theta^4}{12} \right]^{1/2} - 1 \right)^2$$

$$+ \left(\left[\left(a + \frac{1}{2} \right)^2 - \frac{a}{2}\Theta^2 + \frac{a}{24}\Theta^4 \right]^{1/2} - 1 \right)^2$$

$$+ \text{powers of } \Theta \text{ greater than } 4$$

$$= \left(\frac{3}{2} \left[1 + \frac{2}{9} \Theta^2 - \frac{1}{54} \Theta^4 - \frac{2}{81} \Theta^4 \right] - 1 \right)^2$$

$$+ \left(\left(a + \frac{1}{2} \right) \left[1 - \frac{a\Theta^2}{4(a + (1/2))^2} \right. \right.$$

$$\left. + \frac{a\Theta^4}{48(a + (1/2))^2} - \frac{a\Theta^4}{128(a + (1/2))^2} \right] - 1 \right)^2$$

$+$ powers of Θ greater than 4

$$= \frac{1}{4} + \frac{1}{3} \Theta^2 + \frac{5}{108} \Theta^4 + (a - (1/2))^2$$

$$- \frac{a}{2} \frac{a - (1/2)}{a + (1/2)} \Theta^2 + c\Theta^4$$

$+$ powers of Θ greater than 4, where

$$c := \frac{a^2}{16(a + (1/2))^2} + a \frac{a - (1/2)}{a + (1/2)} \left(\frac{1}{24} - \frac{a}{16(a + (1/2))^2} \right).$$

Since $a = (7 + \sqrt{97})/12$, it follows that $V_0(\Theta) = \text{constant} + c\Theta^4 + \cdots$ with $c > 0$. Therefore, by (1.34), $[V_0]$ is equivalent to $[\Theta^4]$ up to a constant germ.

The criterion for versality (6) will be used to prove the versality of $[V]$. Obviously, $\boldsymbol{J}[V_0] = \langle \Theta^3 \rangle_\varepsilon = m^3$ holds. Moreover,

$$\frac{\partial}{\partial u} V(\Theta; 0, 0) - \frac{\partial}{\partial u} V(0)$$

$$= \left(\frac{1}{2} - \left(\frac{1}{2a+1} \right)^2 \right) \Theta^2 \text{ plus higher order terms and}$$

$$\frac{\partial}{\partial v} V(\Theta; 0, 0) = \frac{1 - 2a}{1 + 2a} \Theta \text{ plus terms of order at least 3.}$$

Therefore $\boldsymbol{J}[V_0] + \boldsymbol{V}[V] = m$, and $[V]$ is versal. Because the cusp catastrophe is a two-parameter unfolding, $[V]$ is universal.

(c) Since $f(x) = \frac{1}{3}x^3 +$ higher order terms, $[f]$ is equivalent to $[x^3]$ by (1.34). Obviously, $\boldsymbol{J}[f] = m^2$ and due to

$$\frac{\partial}{\partial u} F(x; 0, v_0) = -\frac{1}{2}x^2 + \frac{1}{3v_0}x^3 \quad \text{and} \quad \frac{\partial}{\partial v} F(x; 0, v_0) = 0,$$

one has $\boldsymbol{J}[f] + \boldsymbol{V}[F] = m^2$, showing that $[F]$ is not versal by (6).

(d)   From

$$g'(x) = \frac{\partial}{\partial x} F\left(x; \frac{3}{8}\sqrt{3}, 3\sqrt{3}\right) = \frac{x/8}{1+x^2}(x-\sqrt{3})^3$$

it follows that

$$g(x) = \text{constant} + \frac{\sqrt{3}}{128}(x-\sqrt{3})^4 + \text{higher order terms in}(x-\sqrt{3})$$

using (1.25). Therefore, $[g]$ is equivalent to $[(x-\sqrt{3})^4]$ up to a constant germ.

Then $\mathcal{J}[g] = \langle(x-\sqrt{3})^3\rangle_{\mathcal{E}}$, where $\mathcal{E}$ refers here to germs at $\sqrt{3}$. Furthermore,

$$\frac{\partial}{\partial u} F\left(x; \frac{3}{8}\sqrt{3}, 3\sqrt{3}\right) - \frac{\partial}{\partial u} F\left(\sqrt{3}; \frac{3}{8}\sqrt{3}, 3\sqrt{3}\right)$$

$$= \frac{\sqrt{3}}{27}(x-\sqrt{3})^3 - \frac{1}{6}(x-\sqrt{3})^2 - \frac{2\sqrt{3}}{3}(x-\sqrt{3})$$

and

$$\frac{\partial}{\partial v} F\left(x; \frac{3}{8}\sqrt{3}, 3\sqrt{3}\right) - \frac{\partial}{\partial v} F\left(\sqrt{3}; \frac{3}{8}\sqrt{3}, 3\sqrt{3}\right)$$

$$= -\frac{\sqrt{3}}{216}(x-\sqrt{3})^3 - \frac{1}{24}(x-\sqrt{3})^2 - \frac{\sqrt{3}}{24}(x-\sqrt{3}).$$

Therefore,

$$\mathcal{J}[g] + \mathcal{V}[F] = \langle(x-\sqrt{3})^3\rangle_{\mathcal{E}} + \langle(x-\sqrt{3})^2$$
$$+ 4\sqrt{3}(x-\sqrt{3}), (x-\sqrt{3})^2 + \sqrt{3}(x-\sqrt{3})\rangle = \langle x-\sqrt{3}\rangle_{\mathcal{E}}.$$

This shows that $[F]$ is versal by (6). $[F]$ is universal, again because the cusp catastrophe is a two-parameter unfolding.

10.2.   Since $V(\Theta; 0, 2k) = 2k\Theta^2 - 4k(1 - \cos\Theta) = \frac{k}{6}\Theta^4$ plus higher order terms in Θ, it follows immediately from (1.34) that $[V(\cdot; 0, 2k)] \sim [\Theta^4]$.

Now apply the criterion for versality (6) to prove the versality of $[V]$. Clearly, $\mathcal{J}[V(\cdot; 0, 2k)] = \langle\Theta^3\rangle_{\mathcal{E}} = m^3$ holds. Moreover,

$$\frac{\partial}{\partial\alpha} V(\Theta; 0, 2k) = \sin\Theta = \Theta \text{ plus terms of orders not less than 3,}$$

and

$$\frac{\partial}{\partial\beta} V(\Theta; 0, 2k) = 2(\cos\Theta - 1)$$

$$= -\Theta^2 \text{ plus terms of orders not less than 3.}$$

Consequently, $\mathcal{J}[V(\cdot; 0, 2k)] + \mathcal{V}[V] = m$ follows, proving the versality by (6). $[V]$ is even universal, since the cusp catastrophe is a two-parameter unfolding.

10.3. Since

$$F\left(\Theta; \frac{1}{2}, 0\right) = \cos\Theta + \left[\frac{5}{4} - \cos\Theta\right]^{1/2}$$

$$= \frac{3}{2} - \frac{1}{4}\Theta^4 \text{ plus higher order terms,}$$

the germ of $F\left(\cdot; \frac{1}{2}, 0\right) - \frac{3}{2}$ is equivalent to $-[\Theta^4]$.

Now apply the criterion for versality (6) to prove the versality of $[F]$. Clearly, $\mathcal{J}[F(\cdot; \frac{1}{2}, 0)] = \langle\Theta^3\rangle_{\mathcal{E}} = m^3$ holds. Moreover,

$$\frac{\partial}{\partial u}F\left(\Theta; \frac{1}{2}, 0\right) = \frac{(1/2) - \cos\Theta}{[(5/4) - \cos\Theta]^{1/2}}$$

$$= -1 + 2\Theta^2 \text{ plus terms of orders at least 4,}$$

and

$$\frac{\partial}{\partial v}F\left(\Theta; \frac{1}{2}, 0\right) = \frac{-\sin\Theta}{[(5/4) - \cos\Theta]^{1/2}}$$

$$= -2\Theta \text{ plus terms of orders not less than 3.}$$

Therefore, $m = \mathcal{J}[F(\cdot; \frac{1}{2}, 0)] + \mathcal{V}[F]$ is valid, and $[F]$ is versal. $[F]$ is even universal, since the dual cusp catastrophe is a two-parameter unfolding.

10.4. The Jacobi ideal of $[f]$ is generated by the germs of $1 - \cos y - 3x^2\cos y$ and $(x\sin y)(1+3x^2)$. The second function has a germ equal to $[xy][h(x,y)]$ with

$$h(x,y) := \frac{\sin y}{y}(1+3x^2)$$

for $y \neq 0$ and $h(x,0) := 1 + 3x^2$. Since $[h]$ is invertible, it can be disregarded. Dividing the germ of the first function by $[-\cos y]$ yields

$$\left[3x^2 - \frac{1}{2}y^2k(y)\right] \quad \text{with} \quad k(y) := \frac{2(1 - \cos y)}{y^2\cos y}$$

for $y \neq 0$ and $k(0) := 1$. Hence $\mathcal{J}[f]$ is equal to

$$\left\langle 3x^2 - \frac{1}{2}y^2k(y), xy\right\rangle_{\mathcal{E}}.$$

A further simplification is achieved by decomposing the germ

$$[3x^2] = [3x^2k(y) + 3x^2(1 - k(y))] \in [3x^2k(y)] + \langle xy\rangle_{\mathcal{E}}.$$

Therefore,

$$\mathcal{J}[f] = \left\langle 3x^2 - \frac{1}{2}y^2, xy\right\rangle_{\mathcal{E}}$$

follows.

From this one sees that $[x^3]$ and $[y^3]$ lie in $\mathcal{J}[f]$, since

$$3x^3 = \left(3x^2 - \frac{1}{2}y^2\right)x + (xy)\frac{1}{2}y.$$

Consulting the following diagram (cf. Exercise 5.4), it is apparent that the cosets $[x] + \mathcal{J}[f], [y] + \mathcal{J}[f]$, and $[x^2] + \mathcal{J}[f]$ form a basis of $m/\mathcal{J}[f]$. In particular, the codimension of $[f]$ is 3.

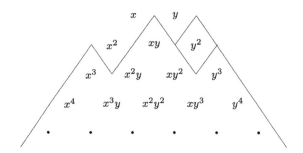

To compute $\mathcal{V}[F]$, note that

$$\frac{\partial}{\partial u}F(x,y;0) = x, \quad \frac{\partial}{\partial v}F(x,y;0) = x^2,$$

$$\frac{\partial}{\partial w}F(x,y;0) = y, \quad \frac{\partial}{\partial t}F(x,y;0) = y^2$$

holds, giving $\mathcal{V}[F] = \langle x, x^2, y, y^2 \rangle$.

From both results it follows that $\mathcal{J}[f] + \mathcal{V}[F] = m$, which implies the versality of $[F]$ by (6).

If the parameter t is dropped—i.e., consider the 3-parameter unfolding $[G]$ given by

$$G(x,y;u,v,w) := f(x,y) + \sin(ux + vx^2) + \sin(wy)$$

—then $[G]$ is still versal for the same reasons. In addition, the number of parameters is minimal, because it is equal to the codimension of $[f]$; see (8). Hence, $[G]$ is universal.

Chapter 11

Genericity

An unfolding is called **elementary** if it is a versal unfolding of a germ whose codimension is at most 4. This notion of an **elementary unfolding** also includes unfoldings of germs for which the origin is not critical. Such unfoldings are automatically versal (see Exercise 7.12).

A main issue in the preceding chapters is the finite classification of elementary unfoldings: Every elementary unfolding of a degenerate critical point is equivalent to one of the **7** plus 3 universal unfoldings of the elementary catastrophes in Table 1 of Chapter 7 up to the addition of a constant unfolding and a nondegenerate normal quadratic form in the remaining variables. If the critical point is not degenerate, then the unfolding is equivalent to the sum of a constant unfolding and a nondegenerate normal quadratic form of all variables. Every unfolding of a noncritical point is equivalent to the germ of a one-dimensional projection plus a constant unfolding of the latter.

More explicitly, this result means that every elementary unfolding $[G]$ in m_{n+s} of a germ $[g]$ in m_n has a representative of the form

$$G(z, w) = F\big(\omega(z, w), \psi(w)\big) + \gamma(w)$$

for small z and w, where $[\psi] \in \mathcal{G}_s$, $[\gamma] \in m_s$, and $[\omega] \in m_{n+s,n}$ is such that $z \mapsto \omega(z, 0)$ is a local diffeomorphism at the origin, cf. Equation (7.9). If the origin is a degenerate critical point the unfolding $[F]$ is determined by a sum

$$F(x, y; u, v) = H(x, u) + q(y)$$

where q is a nondegenerate quadratic form and $[H]$ is a universal unfolding listed in Table 1 of Chapter 7. If the origin is a critical point but nondegenerate, then the first summand H is missing and q depends on all variables. In the case that the origin is not critical, then simply $[F] = [x_1]$ holds.

The classification of elementary unfoldings is a beautiful accomplishment of catastrophe theory and its methods. But it would lose some of its significance if the occurrence of elementary unfoldings were rare. Fortunately, the contrary is true, and the proof of this fact is a further achievement of the theory. The main result of this chapter shows that most unfoldings with at most 4 parameters are elementary. All other unfoldings with at most 4 parameters are exceptional.

To be precise, the set of all smooth functions F in n variables and r parameters for $r \geq 0$ is equipped with the **Whitney C^∞-topology**. At every point, F is considered to be a representative of an r-parameter unfolding. The assertion is that for r at most 4 those F that represent an elementary unfolding at every point are **generic**. In other words, such functions F form an open and dense subset of the set of all smooth functions.

By a main result of the following chapter, it is also true that no elementary unfolding is exceptional (see Exercise 12.2).

Recall that a finite classification of germs is achieved up to codimension 6 in Chapter 6 (see Exercises 6.2 and 6.3). Consequently, the classification of versal unfoldings of germs whose codimension is at most 6 is still finite. In this chapter the proof of theorem (35) on **Measure Theoretical Genericity** can be extended to the case of codimension 5 and most probably also to the case of codimension 6. This is briefly discussed before (27).

To introduce the specifics, let $F \in C^\infty(\mathbf{R}^{n+r})$ be a real valued smooth function of n variables and r parameters with $r \geq 0$. For every state x in $\mathbf{R}^n$ and every value of the parameters u in $\mathbf{R}^r$, the function F determines by

$$F^{x,u}(x', u') := F(x' + x, u' + u) - F(x, u)$$

an r-parameter unfolding

$$[F^{x,u}] \in m_{n+r}$$

of the germ

$$[f^{x,u}] \in m_n, \quad f^{x,u}(x') := F^{x,u}(x', 0).$$

The focus is on the subset of smooth functions

(1) $\mathcal{F} := \{F \in C^\infty(\mathbf{R}^{n+r}) : [F^{x,u}] \text{ is versal for all } (x, u) \in \mathbf{R}^{n+r}\}.$

If F is in $\mathcal{F}$, the codimension of $[f^{x,u}]$ is at most r, since by the Fundamental Theorem on Universal Unfoldings (10.8) the number of parameters of a versal unfolding is at least equal to the codimension of its organization center. Therefore, if $r \in \{0, 1, ..., 4\}$, $\mathcal{F}$ is just the subset of smooth functions F for which $[F^{x,u}]$ is an elementary unfolding for every $(x, u) \in \mathbf{R}^{n+r}$.

A main result of this chapter, theorem (38) on **Topological Genericity**, says that with respect to the Whitney C^∞-topology, $\mathcal{F}$ is open and dense in $C^\infty(\mathbf{R}^{n+r})$ for $r \in \{0, 1, ..., 4\}$; this topology is introduced after (7).

The following considerations rely upon the criterion (10.4). According to the geometrical version of k-transversality in Exercise 8.1, the unfolding $[F^{x,u}]$ is versal if and only if for some integer k such that

$$k \geq r + 2$$

(and hence, by (10.5), for every positive integer k) the equation

(2) $j^k(m) = T_z(z\mathcal{G}^k) + \operatorname{Im} Dj_e^k[F^{x,u}](0)$

holds, where z denotes the k-jet $j^k[f^{x,u}]$. Note that (2) holds even if $[f^{x,u}]$ is in $m \setminus m^2$, in which case $\mathcal{J}[f^{x,u}] = \mathcal{E}$ and $[F^{x,u}]$ is automatically versal by Exercise 7.12. Indeed, it suffices to recall Exercise 4.16, according to which

$$T_z(z\mathcal{G}^k) = j^k(< m\mathcal{J}[f^{x,u}] >)$$

is valid.

It is convenient to identify $j^k(m)$ with $\mathbf{R}^{N-1}$ for $N = \binom{n+k}{k}$ by taking Taylor coefficients as done after (4.6). In this representation it is easy to verify that the entities appearing in (2) depend smoothly on the derivatives of F up to order $k + 1$. Set

$$F^k : \mathbf{R}^{n+r} \longrightarrow \mathbf{R}^{N-1}, \quad F^k(x, u) := \left(\frac{1}{\nu!} D_1^\nu F(x, u) \right)_{0 < |\nu| \le k}.$$

Because of $D^\nu f^{x,u}(0) = D_1^\nu F(x, u)$, z in (2) becomes

(3) $z = F^k(x, u)$.

In part (iv) of the proof of the Linearization Lemma (4.37), the following expression is obtained for the tangent space $T_z(z\mathcal{G}^k)$ at z to the orbit $z\mathcal{G}^k$ in $j^k(m)$:

(4) $T_z(z\mathcal{G}^k) = \langle zD_{j\kappa}M(\epsilon) : 1 \le j \le n, 0 < |\kappa| \le k \rangle.$

Finally, Exercise 8.1 shows that

(5) $\text{Im } Dj_e^k[F^{x,u}](0) = DF^k(x, u)(\mathbf{R}^{n+r}) = < z_{11}, ..., z_{1n}, z_{21}, ..., z_{2r} >$

holds, using the abbreviations

$$z_{1i} := \frac{\partial}{\partial x_i} F^k(x, u) \text{ and } z_{2j} := \frac{\partial}{\partial u_j} F^k(x, u) \text{ for } i = 1, ..., n \text{ and } j = 1, ..., r.$$

Equation (2) now reads

(6) $\mathbf{R}^{N-1} = \langle zD_{j\kappa}M(\epsilon) : 1 \le j \le n, 0 < |\kappa| \le k \rangle + < z_{11}, ..., z_{2r} >.$

All the z-vectors z, z_{1i}, and z_{2j} depend smoothly on (x, u).

In this paragraph one must forget the dependence on (x, u), and one considers the right side of (6) as a subspace of $\mathbf{R}^{N-1}$ depending on the vector $\omega = (z, z_{11}, ..., z_{1n}, z_{21}, ..., z_{2r})$ in $\mathbf{R}^{(1+n+r)(N-1)}$. If ω is such that equation (6) holds, then (6) is satisfied by all vectors ω' in a neighborhood of ω. In order to see this, recall that the rank of a matrix Z is equal to the maximal dimension of the quadratic submatrices with nonvanishing determinant. Since the determinant is a continuous function of the entries, it follows that the rank of the matrices Z' in some neighborhood of Z is at least equal to the rank of Z. In other words the rank of a matrix is a **lower semicontinuous** function of the entries. Consider now the matrix Z whose rows are the vectors $zD_{j\kappa}M(\epsilon)$ for $1 \leq j \leq n$, $0 < |\kappa| \leq k$, the vectors z_{1i} for $1 \leq i \leq n$, and the vectors z_{2j} for $1 \leq j \leq r$. Note that for every j and κ the map $z \mapsto zD_{j\kappa}M(\epsilon)$ is continuous. It follows that the rank of Z is a lower semicontinuous function of ω. This implies

Lemma (7). If (6) holds for $\omega \in \mathbf{R}^{(1+n+r)(N-1)}$, then there exists a $\tau > 0$ such that (6) holds for all $\omega' \in \mathbf{R}^{(1+n+r)(N-1)}$ satisfying

$$|\omega' - \omega|_{max} < \tau$$

in the maximum norm.

It is clear that the topology with which the space of all smooth functions on $\mathbf{R}^{n+r}$ will be equipped should control the derivatives, optimally for every order and on all of $\mathbf{R}^{n+r}$. The set $\mathcal{F}$ in (1) should belong to a coarse topology, so that $\mathcal{F}$ being open is a strong property, and at the same time $\mathcal{F}$ should be dense with respect to a fine topology, so that denseness is also a strong result. Both requirements can be fulfilled by Whitney's C^l- and C^∞-topologies. Put $m := n + r$.

Definition. Let l be a nonnegative integer, and let F be in $C^\infty(\mathbf{R}^m)$. A basis of open neighborhoods of F with respect to the **Whitney C^l-topology** is given by the sets

$$\mathcal{V}^l_\delta(F) := \{G \in C^\infty(\mathbf{R}^m) : |D^\nu(G+F)(p)| < \delta(p) \text{ for all}$$

$$\nu \in \mathbf{N}^m_0 \text{ with } 0 \leq |\nu| \leq l \text{ and } p \in \mathbf{R}^m\},$$

where δ runs through all continuous functions on $\mathbf{R}^m$ for which $\delta(p) > 0$ holds for all $p \in \mathbf{R}^m$. A basis of open neighborhoods of F with respect to the **Whitney C^∞-topology** is given by the sets $\mathcal{V}^l_\delta(F)$ for all δ and all l. In the following, the prefixes C^l- and C^∞- refer to the corresponding Whitney topologies.

In Exercise 11.1 it will be verified that the sets $\mathcal{V}^l_\delta(F)$ are bases of open C^l-neighborhoods, i.e.,

(i) $F \in \mathcal{V}^l_\delta(F)$

(ii) For $\mathcal{V}^l_\delta(F)$ and $\mathcal{V}^l_{\delta'}(F)$ there is $\mathcal{V}^l_{\delta''}(F) \subset \mathcal{V}^l_\delta(F) \cap \mathcal{V}^l_{\delta'}(F)$

(iii) For $G \in \mathcal{V}^l_\delta(F)$ there is $\mathcal{V}^l_{\delta'}(G) \subset \mathcal{V}^l_\delta(F)$.

Regarding the Whitney C^∞-topology, note that the sets $\mathcal{V}^l_\delta(F)$, where l as well as δ varies, satisfy the axioms for bases of open neighborhoods. Indeed, since $\mathcal{V}^{l'}_\delta(F) \subset \mathcal{V}^l_\delta(F)$ holds for $l' \geq l$, for $\mathcal{V}^l_\delta(F)$ and $\mathcal{V}^{l'}_{\delta'}(F)$ there is $\mathcal{V}^{l''}_{\delta''}(F) \subset \mathcal{V}^l_\delta(F) \cap \mathcal{V}^{l'}_{\delta'}(F)$.

A subset $\boldsymbol{\mathcal{X}}$ of $C^\infty(\mathbf{R}^m)$ is C^∞-open, if and only if there is a $\boldsymbol{\mathcal{V}}^l_\delta(F) \subset \boldsymbol{\mathcal{X}}$ for every $F \in \boldsymbol{\mathcal{X}}$. Recall also that $\boldsymbol{\mathcal{X}}$ is C^∞-dense if and only if $\boldsymbol{\mathcal{V}}^l_\delta(F) \cap \boldsymbol{\mathcal{X}} \neq \emptyset$ for every $\boldsymbol{\mathcal{V}}^l_\delta(F)$. A general fact of topology will be used in the sequel, namely that the intersection $\boldsymbol{\mathcal{X}}_1 \cap \boldsymbol{\mathcal{X}}_2$ of two open and dense sets $\boldsymbol{\mathcal{X}}_1$ and $\boldsymbol{\mathcal{X}}_2$ is still open and dense. Indeed $\boldsymbol{\mathcal{X}}_1 \cap \boldsymbol{\mathcal{X}}_2$ is dense, since for every nonempty open set $\boldsymbol{\mathcal{U}}$ the open intersections $\boldsymbol{\mathcal{U}} \cap \boldsymbol{\mathcal{X}}_1$ and $\boldsymbol{\mathcal{U}} \cap (\boldsymbol{\mathcal{X}}_1 \cap \boldsymbol{\mathcal{X}}_2) = (\boldsymbol{\mathcal{U}} \cap \boldsymbol{\mathcal{X}}_1) \cap \boldsymbol{\mathcal{X}}_2$ are nonempty due to the denseness of $\boldsymbol{\mathcal{X}}_1$ and $\boldsymbol{\mathcal{X}}_2$, respectively.

A special type of C^l-open set often occurs: For every $\tau > 0$ and for every closed subset A of $\mathbf{R}^m$, let

$$\boldsymbol{\mathcal{V}}^l_{\tau,A}(F) := \{G \in C^\infty(\mathbf{R}^m) : |D^\nu(G - F)(a)| < \tau \text{ for all}$$
$$0 \leq |\nu| \leq l \text{ and } a \in A\}.$$

Lemma (8). The equality

$$\boldsymbol{\mathcal{V}}^l_{\tau,A}(F) = \bigcup \{\boldsymbol{\mathcal{V}}^l_\delta(F) : \ \delta \text{ satisfies } A \subset \delta^{-1}(]0,\tau[)\}$$

holds. In particular, $\boldsymbol{\mathcal{V}}^l_{\tau,A}(F)$ is C^l-open.

Proof. For the inclusion "$\supset$", consider G in $\boldsymbol{\mathcal{V}}^l_\delta(F)$ with $\delta^{-1}(]0,\tau[) \supset A$. Then
$$|D^\nu(G - F)(a)| < \delta(a) < \tau$$
holds for $0 \leq |\nu| \leq l$ and $a \in A$. Hence, G lies in $\boldsymbol{\mathcal{V}}^l_{\tau,A}(F)$.

For the "$\subset$" inclusion, consider G in $\boldsymbol{\mathcal{V}}^l_{\tau,A}(F)$. For $p \in \mathbf{R}^m$ let
$$\gamma(p) := \max\{|D^\nu(G - F)(p)| : 0 \leq |\nu| \leq l\},$$
$$\delta(p) := \gamma(p) + \frac{1}{2}|\tau - \gamma(p)| + d(p, A),$$
where $d(p, A) := \inf\{|p - a| : a \in A\}$ denotes the distance of p to A. Obviously, δ is continuous and positive. If a is in A, then $\delta(a) < \tau$ holds, since $d(a, A) = 0$ and $\gamma(a) < \tau$ by the definition of $\boldsymbol{\mathcal{V}}^l_{\tau,A}(F)$.

G lies in $\boldsymbol{\mathcal{V}}^l_\delta(F)$ if δ is greater than γ everywhere. Clearly $\delta \geq \gamma$. Suppose that equality $\delta(p) = \gamma(p)$ holds for some $p \in \mathbf{R}^m$. Then $\gamma(p) = \tau$ and $d(p, A) = 0$ follow. However, the latter implies $p \in A$, since A is closed. This contradicts the fact that $\gamma(a) < \tau$ for $a \in A$.

These preliminary remarks will be concluded by the definition of the sets
$$\boldsymbol{\mathcal{F}}(A) := \{F \in C^\infty(\mathbf{R}^{n+r}) : [F^{x,u}] \text{ is versal for all } (x, u) \in A\}$$
for $A \subset \mathbf{R}^{n+r}$. They all contain $\boldsymbol{\mathcal{F}}(\mathbf{R}^{n+r})$, which is $\boldsymbol{\mathcal{F}}$ in (1). A main idea of the following proofs is to express $\boldsymbol{\mathcal{F}}$ in terms of the sets $\boldsymbol{\mathcal{F}}(A)$ for suitable $A \subset \mathbf{R}^{n+r}$.

First the openness of $\boldsymbol{\mathcal{F}}$ will be proved. This is much easier than the proof of the denseness of $\boldsymbol{\mathcal{F}}$ and will be achieved in Corollary (12). Openness holds without any restriction on the number r of parameters.

Lemma (9). Let A be a compact subset of $\mathbf{R}^{n+r}$, and let F be in $\mathcal{F}(A)$. Then for $k \geq r + 2$ there is a $\tau > 0$ such that $\mathcal{V}_{\tau,A}^{k+1}(F) \subset \mathcal{F}(A)$. In particular, $\mathcal{F}(A)$ is C^{r+3}-open.

Proof. Let a be in A. Then (6) holds for $(x, u) = a$. Since the derivatives $D^\nu F$ are continuous, it follows from (7) that there is a neighborhood W of a, such that (6) is satisfied for all $(x, u) \in W$. Set $B := \overline{W} \cap A$, where $\overline{W}$ denotes the closure of W. Then B is compact. It follows again from (7) that there is a $\tau > 0$ such that

$$\mathcal{V}_{\tau,B}^{k+1}(F) \subset \mathcal{F}(B).$$

As A is compact there are finitely many sets B of this kind, say, $B_1, \ldots, B_s$, which cover A, i.e.,

$$A = \bigcup_{\sigma=1}^{s} B_\sigma.$$

Now denote by τ the minimum of the corresponding $\tau_1, \ldots, \tau_s$. Then τ is positive and one obtains:

$$\mathcal{V}_{\tau,A}^{k+1}(A) \subset \bigcap_{\sigma=1}^{s} \mathcal{V}_{\tau_\sigma,B_\sigma}^{k+1}(F) \subset \bigcap_{\sigma=1}^{s} \mathcal{F}(B_\sigma) = \mathcal{F}(A).$$

In the next step, unions

(10)
$$A = \bigcup_{\sigma=1}^{\infty} A_\sigma$$

of countably many compact sets $A_\sigma \subset \mathbf{R}^{n+r}$ are considered, where the A_σ are contained in open, mutually disjoint neighborhoods W_σ. Choose continuous functions $\beta_\sigma : \mathbf{R}^{n+r} \to [0, 1]$ such that $\beta_\sigma | A_\sigma = 1$ and $\beta_\sigma | (\mathbf{R}^{n+r} \setminus W_\sigma) = 0$ (see Exercise 11.2).

Lemma (11). Let $A \subset \mathbf{R}^{n+r}$ be a set as in (10). Then $\mathcal{F}(A)$ is C^{r+3}-open.

Proof. Choose β_σ as after (10). Then obviously

$$\beta := 1 - \sum_{\sigma=1}^{\infty} \beta_\sigma$$

is well-defined, nonnegative, and continuous.

Now let F be in $\mathcal{F}(A)$. Then F lies in $\mathcal{F}(A_\sigma)$ for every σ. According to (9), there is a $\tau_\sigma > 0$ such that

$$\mathcal{V}_{\tau_\sigma,A_\sigma}^{k+1}(F) \subset \mathcal{F}(A_\sigma).$$

Put $\delta := \beta + \sum_{\sigma=1}^{\infty} \tau_\sigma \beta_\sigma$. Obviously δ is a well-defined, nonnegative, and continuous function on $\mathbf{R}^{n+r}$. It is even positive, since $\delta \geq \beta$ and because

$\beta(p) = 0$ implies $p \in A_\sigma$ for some σ and hence $\delta(p) = \tau_\sigma > 0$. Because of $\delta | A_\sigma = \tau_\sigma$,

$$\mathcal{V}_\delta^{k+1}(F) \subset \bigcap_{\sigma=1}^{\infty} \mathcal{V}_{\tau_\sigma, A_\sigma}^{k+1}(F) \subset \bigcap_{\sigma=1}^{\infty} \mathcal{F}(A_\sigma) = \mathcal{F}(A).$$

The whole space $\mathbf{R}^{n+r}$ can be written as the union of two sets A and A' as in (10). For example, for $\sigma \in \mathbf{N}$ choose

$$A_\sigma := \left\{ p \in \mathbf{R}^{n+r} : \sigma - 1 \leq |p| \leq \sigma + 1 \right\}$$

and

$$A'_\sigma := \left\{ p \in \mathbf{R}^{n+r} : \sigma \leq |p| \leq \sigma + 2 \right\}.$$

Since $\mathcal{F} = \mathcal{F}(A) \cap \mathcal{F}(A')$ holds, an immediate consequence of (11) is

Corollary (12). $\mathcal{F}$ is C^{r+3}-open for all integers $n \geq 1$ and $r \geq 0$.

The remainder of this chapter is devoted to the proof of the denseness of $\mathcal{F}$. Some elements of differential geometry will be needed for this. They are developed to the extent of being able to prove the **Fundamental Lemma on Transversality** (18). Actually, the proof of (18) is a skillful application of **Sard's Theorem**. It is beyond the scope of this book to present a proof of this measure theoretical result, which is only needed there. Appendix A.6 contains a brief discussion of Sard's Theorem.

From (18) it is easy to show the transversality result (20) on which the theorem on **Measure Theoretical Genericity** (35) is based. Corollary (20) applies to $\mathcal{F} \subset C^\infty(\mathbf{R}^{n+r})$ if and only if the codimension of the real algebraic subset R_{r+1}^{r+2} of $j^{r+2}(m)$ defined in (22) exceeds $n+r$. A detailed analysis in (27) up to (34) will show that this condition is satisfied for r less than 5. Finally, theorem (38) on **Topological Genericity** will be derived from (35) concluding this chapter.

Definition. Let d and n be integers satisfying $0 \leq d \leq n$. A nonvoid subset S of $\mathbf{R}^n$ is called a d-**dimensional regular submanifold** of $\mathbf{R}^n$ if for every $p \in S$ there is an open neighborhood U of p in $\mathbf{R}^n$ and a diffeomorphism $\varphi : U \to V$ onto an open subset V of $\mathbf{R}^n$ satisfying

$$\varphi(S \cap U) = \{ y \in \mathbf{R}^n : y_{d+1} = \cdots = y_n = 0 \} \cap V.$$

In the literature often the term **embedded** is used instead of regular.

It is easy to argue with the help of the Inverse Function Theorem (1.29) that the dimension d is uniquely determined. It is denoted by $\dim S$. The integer $\mathrm{cod} S := n - d$ is called the **codimension** of S. Regular submanifolds occur quite naturally.

Lemma (13). The **graph** of a smooth function $f : W \to \mathbf{R}^m$ defined on an open subset $W \subset \mathbf{R}^d$ is a regular submanifold of $\mathbf{R}^d \times \mathbf{R}^m$ with dimension d.

Proof. The diffeomorphism

$$\varphi : W \times \mathbf{R}^m \to W \times \mathbf{R}^m, \ \varphi(x, y) := (x, f(x) - y)$$

maps the graph $\{(x, f(x)) : x \in W\}$ of f onto $W \times \{0\}$.

In (33) the set S_7 and parts of S_5 are described locally as graphs of smooth functions and hence are regular submanifolds. A regular submanifold can also be determined by equations:

Lemma (14). Let $X \subset \mathbf{R}^n$ be open and $f : X \to \mathbf{R}^m$ a smooth map such that

$$S := f^{-1}(\{0\})$$

is nonempty and the derivatives $Df(p)$ are surjective for all $p \in S$. Then S is a regular submanifold of $\mathbf{R}^n$ with codimension m.

Proof. Let p be in S. Since the rank is lower semicontinuous (see after (6)), $Df(x)$ is surjective for all x in some neighborhood of p in $\mathbf{R}^n$. According to the Rank Theorem in A.4, there is a diffeomorphism $\varphi : U \to V$ defined on an open neighborhood U of p in X onto an open set $V \subset \mathbf{R}^n$ and a diffeomorphism v defined on some open neighborhood of the origin in $\mathbf{R}^m$ onto another such neighborhood such that $v(0) = 0$ and

$$v \circ f(x) = (\varphi_1(x), \dots, \varphi_m(x))$$

holds for all x in U. This implies

$$\varphi(S \cap U) = \{y \in V : y_1 = \cdots = y_m = 0\}.$$

The foregoing result, which is also referred to as the **Lemma on the Regular Value**, will be generalized in (17).

Definition. The **tangent space** T_pS **at the point** $p \in S$ **to the regular submanifold** S of $\mathbf{R}^n$ consists of all vectors in $\mathbf{R}^n$, which occur as tangent vectors $c'(0)$ to curves c in $\mathbf{R}^n$ based at p, whose image lies in S.

Let $\varphi : U \to V$ be a diffeomorphism as considered in the definition of a regular submanifold, and set $\psi := (\varphi_{d+1}, \dots, \varphi_n)$. Then obviously

(15) $$T_pS = (D\varphi(p))^{-1} \left(\mathbf{R}^d \times \{0\} \right) = (D\psi(p))^{-1} (\{0\})$$

holds. In particular T_pS is a linear subspace of $\mathbf{R}^n$ with dimension $d = \dim S$. In (14), where S is given by $f^{-1}(\{0\})$, ψ is of the form $v \circ f$ with a diffeomorphism v so that $T_pS = (Df(p))^{-1}(\{0\})$ holds by (15).

A notion fundamental to the following considerations will be introduced now.

Definition. Let m and n be nonnegative integers. Let $f : \mathbf{R}^m \to \mathbf{R}^n$ be a smooth map and let S be a regular submanifold of $\mathbf{R}^n$. Then f is said to be **transverse to S at** $p \in \mathbf{R}^m$ if either $f(p) \notin S$ or

$$f(p) \in S \quad \text{and} \quad \mathbf{R}^n = T_{f(p)}S + \mathrm{Im}Df(p)$$

holds. The map f is called **transverse to S** if it is transverse to S at every point $p \in \mathbf{R}^m$.

The dimension of the image $\mathrm{Im}Df(p) = Df(p)(\mathbf{R}^m)$ is at most m. In the case that m is smaller than $\mathrm{cod}S$, the map f is transverse to S if and only if

$$(16) \qquad\qquad\qquad f(\mathbf{R}^m) \cap S = \emptyset$$

holds. This fact will be of great importance in (23).

More generally than by equations (cf. (14)), regular submanifolds arise as preimages of regular submanifolds under transverse maps:

Theorem (17). Let f be transverse to S, and let s be the codimension of S. Suppose that $R := f^{-1}(S)$ is not empty. Then R is a regular submanifold of $\mathbf{R}^m$ with codimension s. For the tangent space to R at $p \in R$ one obtains

$$T_pR = (Df(p))^{-1}T_{f(p)}S.$$

Proof. Let p be in R. There is an open neighborhood $U \subset \mathbf{R}^n$ of $f(p)$ and a diffeomorphism $\varphi : U \to V$ onto an open set $V \subset \mathbf{R}^n$ such that $\varphi(S \cap U) = \{y \in V : y_1 = \cdots = y_s = 0\}$. Put $\psi := (\varphi_1, \ldots, \varphi_s)$. On the open neighborhood $W := f^{-1}(U)$ of p define the map $g : W \to \mathbf{R}^s$ by $g(x) = \psi(f(x))$. Then $g^{-1}(\{0\}) = R \cap W$ holds.

It will be shown now that $Dg(x)$ is surjective for all x in $R \cap W$. According to (14) this will imply that $R \cap W$, and hence R itself, is a regular submanifold of $\mathbf{R}^m$ with $\mathrm{cod}R = s$.

Because of $T_{f(x)}S = (D\varphi(f(x)))^{-1}(\{0\} \times \mathbf{R}^{n-s})$, it follows from the transversality of f to S that $\mathbf{R}^n = (\{0\} \times \mathbf{R}^{n-s}) + D\varphi(f(x))(\mathrm{Im}Df(x))$ by (15). This implies $\mathbf{R}^s = D\psi(f(x))(\mathrm{Im}Df(x))$, resulting in the surjectivity of $Dg(x)$.

Finally, by (15) fol.,

$$T_pR = (Dg(p))^{-1}(\{0\}) = (Df(p))^{-1}(D\psi(f(p)))^{-1}(\{0\}) =$$

$$(Df(p))^{-1}T_{f(p)}S$$

holds for $p \in R$.

The next result needs Sard's Theorem. An appropriate version of it is explained in Appendix A.6.

Fundamental Lemma on Transversality (18). Let m, n, and l be nonnegative integers. Let $F : \mathbf{R}^m \times \mathbf{R}^l \to \mathbf{R}^n$ be smooth, and let S be a regular submanifold of R^n. Suppose that F is transverse to S. Then there is a **Lebesgue null set** $\Lambda \subset \mathbf{R}^l$ such that for all $\lambda \in \mathbf{R}^l \setminus \Lambda$ the partial map

$$F_\lambda : \mathbf{R}^m \to \mathbf{R}^n, \ F_\lambda(p) := F(p, \lambda)$$

is transverse to S.

Proof. If $R := F^{-1}(S)$ is empty, then $F_\lambda^{-1}(S)$ is empty and hence, F_λ is transverse to S for all λ. Therefore, suppose that $R \neq \emptyset$. By (17), R is a regular submanifold. Let $\pi : R \to \mathbf{R}^l$, $\pi(p, \lambda) := \lambda$, be the restriction to R of the projection of $\mathbf{R}^m \times \mathbf{R}^l$ onto $\mathbf{R}^l$.

According to Sard's Theorem A.6, the set

$$\left\{ \lambda \in \mathbf{R}^l : \mathrm{Im} D\pi(p, \lambda) = \mathbf{R}^l \text{ for all } p \in \pi^{-1}(\{\lambda\}) \right\}$$

is the complement of a Lebesgue null set. Obviously, $\pi^{-1}(\{\lambda\}) = \emptyset$ means that $F_\lambda(\mathbf{R}^m) \cap S = \emptyset$. Therefore, it suffices to show that $\mathrm{Im} D\pi(p, \lambda) = \mathbf{R}^l$ implies the transversality of F_λ to S at $p \in F_\lambda^{-1}(S)$, which will be done now.

Fix (p, λ) in R satisfying $\mathrm{Im} D\pi(p, \lambda) = \mathbf{R}^l$. Then for every $\xi \in \mathbf{R}^l$ there is a $u \in \mathbf{R}^m$ such that $(u, \xi) \in T_{(p,\lambda)}R$. Indeed, since there is a curve $c = (c_1, c_2)$ in $\mathbf{R}^m \times \mathbf{R}^l$ based at (p, λ), whose image lies in R and which satisfies $\xi = (\pi \circ c)'(0) = c_2'(0)$, it suffices to take $u = c_1'(0)$. This implies

(i) $$\mathbf{R}^m \times \mathbf{R}^l = T_{(p,\lambda)}R + \mathbf{R}^m \times \{0\},$$

since for $(u, \xi) \in \mathbf{R}^m \times \mathbf{R}^l$ there is a $u' \in \mathbf{R}^m$ such that $(u', \xi) \in T_{(p,\lambda)}R$, as stated above, and hence, $(u, \xi) = (u', \xi) + (u - u', 0)$ holds.

From (i) and the transversality of F to S at $(p, \lambda) \in F^{-1}(S)$, i.e.,

(ii) $$\mathbf{R}^n = T_{F(p,\lambda)}S + \mathrm{Im} DF(p, \lambda),$$

the transversality

(iii) $$\mathbf{R}^n = T_{F_\lambda(p)}S + \mathrm{Im} DF_\lambda(p)$$

of F_λ to S at $p \in F_\lambda^{-1}(S)$ will follow:

Fix v in $\mathbf{R}^n$. By (ii) there are $v' \in T_{F(p,\lambda)}S = T_{F_\lambda(p)}S$ and $(u, \xi) \in \mathbf{R}^m \times \mathbf{R}^l$ such that $v = v' + DF(p, \lambda)(u, \xi)$. By (i) there are $(u', \xi') \in T_{(p,\lambda)}R$ and $u'' \in \mathbf{R}^m$ such that $(u, \xi) = (u', \xi') + (u'', 0)$. Therefore, $v = v' + DF(p, \lambda)(u', \xi') + DF(p, \lambda)(u'', 0)$ holds. This yields (iii), since $DF(p, \lambda)(u', \xi') \in T_{F(p,\lambda)}S = T_{F_\lambda(p)}S$ by (17) and since $DF(p, \lambda)(u'', 0) = DF_\lambda(p)u''$.

Now the decisive application of this lemma can be initiated. Let $P^k \setminus \mathbf{R}$ be the set of all polynomials in n variables of degree at most k without the constants, which will be identified with $\mathbf{R}^{N-1}$ for $N = \binom{n+k}{k}$ in the usual way by taking coefficients (cf. (4.6)). In particular, subsets of $P^k \setminus \mathbf{R}$ with Lebesgue measure zero are defined. For $F \in C^\infty(\mathbf{R}^{n+r})$ and $P \in P^k \setminus \mathbf{R}$ denote by $F + P$ the function $(x, u) \to F(x, u) + P(x)$ on $\mathbf{R}^{n+r}$. Recall the definition of F^k before (3).

Theorem (19). Let S be a regular submanifold of $P^k \setminus \mathbf{R}$, and let F be in $C^\infty(\mathbf{R}^{n+r})$. Then $(F + P)^k$ is transverse to S for all P in the complement of some Lebesgue null set of $P^k \setminus \mathbf{R}$.

Proof. For $a \in \mathbf{R}^{N-1}$, let $P := \sum_{0 < |\alpha| \le k} a_\alpha x^\alpha$. Consider the smooth map
$\Psi : \mathbf{R}^{n+r} \times \mathbf{R}^{N-1} \to \mathbf{R}^{N-1}$ given by $\Psi(x, u, a) := (F + P)^k(x, u)$.
The image of $D\Psi(x, u, a)$ contains the $N - 1$ vectors

$$\frac{\partial}{\partial a_\beta} \left(\sum_{0 < |\alpha| \le k} a_\alpha \left(\frac{1}{\nu!} D^\nu x^\alpha \right) \right)_{0 < |\nu| \le k} = \left(\frac{1}{\nu!} D^\nu x^\beta \right)_{0 < |\nu| \le k}$$

for $0 < |\beta| \le k$, which are linearly independent for every fixed value of x. To show this, consider $\mathbf{R}^{N-1}$ as a subspace of the n-fold product $\mathbf{R}^{k+1} \times \cdots \times \mathbf{R}^{k+1}$. The ith factor $\mathbf{R}^{k+1}$ is spanned by the linearly independent vectors $(\frac{1}{\nu_i!} D^{\nu_i} x^{\beta_i})_{0 \le \nu_i \le k} = (*, \ldots, *, 1, 0, \ldots, 0)$ for $\beta_i = 0, \ldots, k$, where 1 stands in the β_ith position and where $*, \ldots, *$ denotes unspecified entries.

Therefore, $D\Psi(x, u, a)$ is surjective for every (x, u, a). In particular, Ψ is transverse to S. This implies that $\Psi_a = (F + P)^k$ is transverse to S for all a in the complement of some Lebesgue null set of $\mathbf{R}^{N-1}$ by the preceding Fundamental Lemma.

Denote by $\Lambda(F, S) \subset P^k \setminus \mathbf{R}$ the exceptional Lebesgue null set occurring in the previous theorem. Suppose that $\mathcal{M}$ is a countable set of regular submanifolds of $\mathbf{R}^{N-1}$. Then

$$\Lambda(F) := \bigcup \{\Lambda(F, S) : S \in \mathcal{M}\}$$

is still a Lebesgue null set.

Corollary (20). Let F be in $C^\infty(\mathbf{R}^{n+r})$. Then $(F + P)^k$ is transverse to all $S \in \mathcal{M}$ for every P in the complement of the Lebesgue null set $\Lambda(F)$ in $P^k \setminus \mathbf{R}$.

The goal now is to prove the Measure Theoretical Genericity (35), which states that for every $F \in C^\infty(\mathbf{R}^{n+r})$ there exists an exceptional Lebesgue null set $\Lambda(F)$ of $P^k \setminus \mathbf{R}$ such that $F + P \in \mathcal{F}$ holds for all P not in $\Lambda(F)$. The set $P^k \setminus \mathbf{R}$ is naturally identified with the jet space $j^k(m)$ (see the comments preceding (4.6)). Therefore, in order to show (35), the criterion (10.4) in its geometrical version (2) suggests applying the above transversality result (20) to the set $\mathcal{M}$ of orbits $z\mathcal{G}^k$ in $j^k(m)$. But there are two difficulties. The first is technical and is overcome in (21). The second one, however, occurs for $n \geq 2$ and is fundamental. It can be solved by uniting orbits to higher dimensional invariant regular submanifolds. This is explained in (22).

Orbits and Regular Submanifolds (21).   The orbits $z\mathcal{G}^k$ are regular submanifolds of $\mathbf{R}^{N-1}$. As mentioned in the remark after (4.32), this is a deep result on semialgebraic group actions. The lengthy proof cannot be reproduced in this book. Fortunately, (20) can be applied to orbits without referring to this result.

First it will be shown that an orbit $S \subset \mathbf{R}^{N-1}$ is the union of countably many regular submanifolds. For $z \in S$ one has $S = z\mathcal{G}^k$. As noted at the beginning of the proof of (4.37), it follows from the Rank Theorem A.4 that there is an open neighborhood W_z of the identity in $\mathcal{G}^k$ and a diffeomorphism v_z on an open neighborhood V_z of z in $\mathbf{R}^{N-1}$ onto the open cube

$$L = \left\{ x \in \mathbf{R}^{N-1} : |x_l| < 1,\ 1 \leq l \leq N-1 \right\}$$

in $\mathbf{R}^{N-1}$, such that

$$S_z := zW_z$$

satisfies $S_z \subset V_z$ and

$$v_z(S_z) = \{x \in L : x_{d+1} = \cdots = x_{N-1} = 0\}.$$

Here d is the rank of the orbital map. It is equal to the dimension of the tangent space at z to the orbit, see part (iv) of the proof of (4.37). In particular, S_z is a d-dimensional regular submanifold of $\mathbf{R}^{N-1}$.

There are countably many such S_z that cover the orbit S. Indeed, fix z_0 in S. For every ξ in $\mathcal{G}^k$ the open set $\xi W_{z_0\xi}$ is a neighborhood of ξ. Since $\mathcal{G}^k$ is the union of countably many compact subsets (see part (iii) of the proof of (4.37)), there is a countable subset X of $\mathcal{G}^k$ such that $\mathcal{G}^k = \bigcup \{\xi W_{z_0\xi} : \xi \in X\}$. Hence, one has

$$S = z_0\mathcal{G}^k = \bigcup \{S_z : z \in z_0 X\}.$$

Now, for the application of (2), it suffices to consider these countably many regular submanifolds S_z instead of the orbit S, since by the Linearization Lemma (4.37)

$$T_{\bar{z}}\left(\bar{z}\mathcal{G}^k\right) = T_{\bar{z}}(S_z)$$

holds for every $\bar{z}$ in S_z.

In the sequel, the corank $(N-1) - d$ of the orbital map is also called the **codimension of the orbit**, and it is denoted by cod$S$.

The invariant sets R_c^k (22). The second problem mentioned above is that $\mathbf{R}^{N-1}$ consists of uncountably many orbits under $\mathcal{G}^k$ if n and k are at least 3 or if k is at least 4 in the case of $n = 2$ (see Exercises 6.6 and 6.7). Thus, the transversality result (20) cannot be applied to the set of all orbits in $\mathbf{R}^{N-1}$ in the case $n \geq 2$. The solution to this problem is based on the fact that the set of k-jets of germs with codimension at least c

$$R_c^k := \left\{ z = j^k[f] \in \mathbf{R}^{N-1} : [f] \in m^2,\ \mathrm{cod}[f] \geq c \right\},$$

where $0 \leq c \leq k - 1$ and $k \geq 2$, is a finite union of orbits and other regular submanifolds.

There is an intrinsic reason for this, namely that R_c^k is a real algebraic set; see Exercise 11.4. No explicit use of this property is made here, but for the relevant set R_5^6 such a finite union is shown in (28)–(34).

In the case of only one variable, $\mathbf{R}^{N-1}$ consists of finitely many orbits under $\mathcal{G}^k$. Therefore, (20) applies directly to this case without restriction on the number r of parameters. It follows that the Measure Theoretical Genericity (35) and the Topological Genericity (38) hold for $n = 1$ and $r \geq 0$. See Exercise 11.8.

Lemma (23). Suppose that for some $k \geq r + 2$ the hypotheses (a) and (b) hold:

(a) R_{r+1}^k is a finite union $\bigcup S$ of orbits and other regular submanifolds S such that every S satisfies

$$n + r < \mathrm{cod}S.$$

(b) $\mathbf{R}^{N-1} \setminus R_{r+1}^k$ consists of finitely many orbits.

Then for every $F \in C^\infty(\mathbf{R}^{n+r})$ there is a Lebesgue null set $\Lambda(F) \subset P^k \setminus \mathbf{R}$ such that $F + P \in \mathcal{F}$ for every $P \notin \Lambda(F)$.

Proof. Denote by $\mathcal{M}_{r+1}^k$ the set of orbits and the other regular submanifolds in (a) and (b). Because of (21), Corollary (20) holds for $\mathcal{M} = \mathcal{M}_{r+1}^k$. Consider now $G := F + P$ with G^k transverse to all $S \in \mathcal{M}_{r+1}^k$. Since $\dim \mathrm{Im}DG^k(p) \leq n + r$ for all $p \in \mathbf{R}^{n+r}$,

$$G^k\left(\mathbf{R}^{n+r}\right) \cap R_{r+1}^k = \emptyset$$

follows (cf. (16)). Hence, for every $p \in \mathbf{R}^{n+r}$, $G^k(p)$ lies in some of the orbits in $\mathbf{R}^{N-1} \setminus R_{r+1}^k$ to which G^k is transverse. From (2) and (5), it follows that $G \in \mathcal{F}$.

The important question now is, of course, for which k and r with $k \geq r + 2$ the hypotheses (a) **and** (b) in (23) hold. For $r = 7$ the hypothesis (b) is not satisfied. This is to be expected since a finite classification of germs of codimension 7 is no longer possible, and indeed, in the case of 3 variables, $\mathbf{R}^{N-1} \setminus R_8^k$ contains the uncountably many different orbits $[x^3 + y^3 + z^3 + 3cxyz]\mathcal{G}^k$ for $c \in \mathbf{R} \setminus \{-1\}$ by Exercises 5.11 and 6.6. For $r \leq 6$ the hypothesis (b) is satisfied since the classification of germs with codimension up to 6 is finite.

The hypotheses (a) and (b) in (23) are valid for r' and k if they hold for r and k with $0 \leq r' < r \leq k - 2$. This will follow easily from the next result (26), which is needed in the sequel. For $[f] \in m^2$ and $k \geq 2$ let

$$(24) \qquad \mathrm{cod}^k[f] := \dim \left(j^k \left(m^2 \right) / T_z \left(z\mathcal{G}^k \right) \right)$$

where $z = j^k[f]$. Note that because of $T_z(z\mathcal{G}^k) = j^k(\langle m\mathcal{J}[f]\rangle) \subset j^k(m^2)$ by Exercise 4.16, the right side of (24) is well defined. Since $\dim(j^k(m)/j^k(m^2))$ is equal to n, the formula

$$(25) \qquad \mathrm{cod} \left(z\mathcal{G}^k \right) = n + \mathrm{cod}^k[f]$$

holds for the codimension of the orbit $z\mathcal{G}^k$ in $j^k(m)$. For a result closely related to (26) see Exercise 11.3.

> **Codimensions of the $\mathcal{G}^k$-orbits (26).** Let $[f]$ be in m^2 and let $k \geq 2$. Then $\mathrm{cod}^k[f] \leq \mathrm{cod}[f]$ holds, and the following implications are valid:
>
> $$\mathrm{cod}[f] \leq k - 2 \Rightarrow \det[f] \leq k \Rightarrow m^{k+1} \subset \langle m\mathcal{J}[f]\rangle \Leftrightarrow \mathrm{cod}^k[f] = \mathrm{cod}[f].$$
>
> **Proof.** By (4.8), m^2/m^{k+1} is isomorphic with $j^k(m^2)$, and therefore $(\langle m\mathcal{J}[f]\rangle + m^{k+1})/m^{k+1}$ is isomorphic with $j^k(\langle m\mathcal{J}[f]\rangle)$. Moreover, $m^{k+1} \subset \langle m\mathcal{J}[f]\rangle + m^{k+1} \subset m^2$ holds. Therefore,
>
> $$\dim \left(j^k \left(m^2 \right) / j^k \left(\langle m\mathcal{J}[f]\rangle \right) \right) \leq \dim \left(m^2 / \langle m\mathcal{J}[f]\rangle \right)$$
>
> is valid, and obviously equality holds if and only if m^{k+1} is contained in $\langle m\mathcal{J}[f]\rangle$. By (4.38) the latter is the case if $\det[f] \leq k$, and this holds by (5.9) if $\mathrm{cod}[f] \leq k - 2$.
>
> The left side of the above inequality is $\mathrm{cod}^k[f]$, and the right side is equal to $\mathrm{cod}[f]$ by (5.17).

Regarding the assertion preceding (24), note that $R_{r'+1}^k \supset R_{r+1}^k$ and that $R_{r'+1}^k \setminus R_{r+1}^k$ consists of finitely many orbits S in $j^k(m^2)$ whose codimensions satisfy $n + r' < \mathrm{cod}S \ (\leq n + r)$ by (26).

The subsequent investigations up to (34) are devoted to the proof of (a) and (b) in (23) for R_5^6. By the above argument, this implies the validity of the conclusion of (23) for r less than 5, which is the content of (35).

An analogous lengthy analysis would show that condition (a) in (23) on the codimensions is still satisfied for $r = 5$. Therefore (35) and, as a consequence, (38) hold for $r = 5$, too. Most probably these results remain valid even in the case $r = 6$.

Theorem (27). Let $n \geq 2$. Then $\mathbf{R}^{N-1} \setminus R_5^6$ consists of the following $11n - 2$ orbits:

(a) 1 orbit $j^6[x]\mathcal{G}^6 = j^6(m \setminus m^2)$ due to the noncritical point,

(b) $n + 1$ orbits $j^6[q_{s,n}]\mathcal{G}^6$, $0 \leq s \leq n$, due to the nondegenerated critical points,

(c) $6n$ orbits $j^6[q_{s,n-1} + \eta^{l+1}x^l]\mathcal{G}^6$, $\eta \in \{-1,1\}$, $3 \leq l \leq 6$, $0 \leq s \leq n-1$, due to the germs of corank 1,

(d) $4(n-1)$ orbits $j^6[q_{s,n-2} + f]\mathcal{G}^6$ for $[f]$ equal to $[x^3 - xy^2]$, $[x^3 + y^3]$, $[x^2y + y^4]$, $[-x^2y - y^4]$, and $0 \leq s \leq n-2$, due to the germs of corank 2.

Proof. The list is complete by Table 1 in Chapter 6.

The $11n - 2$ orbits in (27) are distinct. This follows from (4.48), but this result is not needed.

Theorem (28). Let $n \geq 2$. Then R_5^6 is the union of the following $6n - 5$ orbits and $n + 2$ other regular submanifolds:

(a) n orbits $j^6[q_{s,n-1}]\mathcal{G}^6$, $0 \leq s \leq n-1$,

(b) $5(n-1)$ orbits $j^6[q_{s,n-2} + x^2y]\mathcal{G}^6$, $j^6[q_{s,n-2} + x^2y + \eta y^l]\mathcal{G}^6$, $\eta \in \{-1,1\}$, $l \in \{5,6\}$, $0 \leq s \leq n-2$,

(c) 3 regular submanifolds whose union is
$$S_5 := \bigcup_{0 \leq s \leq n-2} j^6([q_{s,n-2} + x^3] + m_2^4)\mathcal{G}^6,$$

(d) 1 regular submanifold $S_7 := \bigcup_{0 \leq s \leq n-2} j^6([q_{s,n-2}] + m_2^4)\mathcal{G}^6$,

(e) $n - 2$ regular submanifolds $\bigcup_{0 \leq s \leq t} j^6([q_{st}] + m_{n-t}^3)\mathcal{G}^6$, $0 \leq t \leq n-3$.

Proof. (i) First it will be verified that the sets in (a)–(e) lie in R_5^6.

By (5.3) and (6.6), the codimensions of $[q_{s,n-1}]$ and of $[x^2y]$ are infinite. Further, $\text{cod}[x^2y + y^5] = \text{cod}[x^2y - y^5] = 5$ holds by Exercise 6.2. Finally, note that $[x^2y - y^6]$ is equivalent to $-[x^2y + y^6]$ and that $\text{cod}(\pm[x^2y + y^6]) = 6$ as shown in Exercise 6.3. Thus all the orbits in (a) and (b) lie in R_5^6.

Since the classifications achieved in (6.1) and in Exercises 6.2, 6.3, and 6.4 are complete, it follows that $\text{cod}[f] \geq 8$ and $\text{cod}([x^3] + [f]) \geq 5$ hold for $[f]$ in m_2^4. Hence, the sets in (c) and (d) are subsets of R_5^6.

The corank of the germs arising from the sets in (e) is at least 3, so that their codimension is at least 6 by (5.19). Therefore, the sets in (e) also lie in R_5^6.

(ii) Now it must be shown that there are no further jets in R_5^6 in addition to those in (a)–(e). For this, consider $[f]$ in m^2 with $\text{cod}[f]$ at least 5. Apply the Reduction Lemma (3.2) to $[f]$. Because of (5.6), it follows that $\text{cor}[f] \geq 1$. An immediate consequence is also that $j^6[f]$ lies in some of the sets in (e) if $\text{cor}[f] \geq 3$ holds.

If $\text{cor}[f]$ is equal to 1, then $[f]$ is equivalent to $[q_{s,n-1}] + [g]$ with $[g] \in m_1^3$. According to (1.34), it follows that $[g]$ is either zero or equivalent to $\pm[x^l]$.

In the latter case l is at least 7 by (5.2). Thus $j^6[f]$ lies in some of the orbits in (a).

The case $\mathrm{cor}[f] = 2$ remains to be treated. By (3.2), the germ $[f]$ is equivalent to $[q_{s,n-2}] + [g]$ with $[g] \in m_2^3$. According to the Classification of Cubic Forms on $\mathbf{R}^2$ (6.4), one may choose $[g] \in [k] + m_2^4$ with $[k]$ equal to 0, $[x^3]$, $[x^2 y]$, $[x^3 - xy^2]$, or $[x^3 + y^3]$. But since $\mathrm{cod}[k] = \det[k] = 3$ holds for $[k] = [x^3 - xy^2]$ and $[k] = [x^3 + y^3]$ by (6.6), only the cases $[k] = 0$, $[x^3]$, and $[x^2 y]$ can occur. The cases $[k] = 0$ and $[k] = [x^3]$ are listed in (d) and (c).

For the last case $[k] = [x^2 y]$, Exercise 6.1(b) shows that the germ $[g]$ is either equivalent to one of the 6 germs $[x^2 y \pm y^4]$, $[x^2 y \pm y^5]$, $[x^2 y \pm y^6]$ or lies in $[x^2 y] + m_2^7$. The cases $[x^2 y \pm y^4]$ are discarded, since $[x^2 y - y^4]$ is equivalent to $-[x^2 y + y^4]$ and, by Table 1 in Chapter 6, $\mathrm{cod}(\pm[x^2 y + y^4]) = 4 < 5$. The remaining cases are listed in (b).

(iii) Finally, the assertion has to be proved that the sets in (d) and (e) are regular submanifolds and that S_5 in (c) is the union of 3 regular submanifolds. The following theorems (29) and (33) will include this.

The $6n - 5$ orbits and $n + 2$ regular submanifolds in (28) are $7n - 3$ mutually disjoint sets. This follows from (4.48) but is not needed.

Theorem (29). Let k and t be integers satisfying $k \geq 2$ and $0 \leq t \leq n$. Then

$$\bigcup_{0 \leq s \leq t} j^k \left([q_{st}] + m_{n-t}^3 \right) \mathcal{G}^k$$

is a regular submanifold of $\mathbf{R}^{N-1}$ with codimension

$$n + \frac{1}{2}(n - t + 1)(n - t).$$

Proof. By the Reduction Lemma (3.2), the set in question becomes

$$\left\{ z \in \mathbf{R}^{N-1} : z_\nu = 0 \text{ for } |\nu| = 1, \ C \text{ has rank } t \right\},$$

where C denotes the symmetric $n \times n$ matrix determined by the quadratic form (see the definition after (1.15))

$$x \mapsto \sum_{|\nu|=2} z_\nu x^\nu.$$

Let $\mathcal{S}$ be the $\frac{1}{2}n(n+1)$-dimensional vector space of the symmetric $n \times n$ matrices (see the proof of (1.36)), and denote by $\mathcal{S}^t$ the subset of matrices in $\mathcal{S}$ with rank t. Obviously, one has to show that $\mathcal{S}^t$ is a $\frac{1}{2}(n-t+1)(n-t)$-codimensional regular submanifold of $\mathcal{S}$. The cases $t = 0$ and $t = n$ are easy, since $\mathcal{S}^0 = \{0\}$ and $\mathcal{S}^n = \det^{-1}(\mathbf{R} \setminus \{0\})$ is open in $\mathcal{S}$. So let $0 < t < n$.

Note first that $\boldsymbol{S}^t = \bigcup_{o \le s \le t} \{T^t E_{st} T : T \text{ invertible}\}$ holds by (1.24). Therefore by homogeneity, for every s, it suffices to consider a suitable neighborhood of E_{st} in $\boldsymbol{S}$. For $C \in \boldsymbol{S}$ use the notation

$$C = \begin{pmatrix} A & B \\ B^t & H \end{pmatrix}$$

where A denotes a symmetric $t \times t$ matrix. Then

$$\mathcal{U} := \{C \in \boldsymbol{S} : \det A \ne 0\}$$

is an open neighborhood of E_{st} in $\boldsymbol{S}$. For $C \in \mathcal{U}$ the factorization

$$C = \begin{pmatrix} A & 0 \\ B^t & E_{n-t} \end{pmatrix} \begin{pmatrix} E_t & A^{-1}B \\ 0 & H - B^t A^{-1} B \end{pmatrix}$$

is valid. Therefore, C is an element of $\boldsymbol{S}^t$ if and only if $H - B^t A^{-1} B = 0$. Cramer's rule shows that the map $A \mapsto A^{-1}$ is a rational diffeomorphism. Therefore,

$$\mathcal{U} \to \mathcal{U}, \; C \mapsto \begin{pmatrix} A & B \\ B^t & H - B^t A^{-1} B \end{pmatrix}$$

is a diffeomorphism, which maps $\mathcal{U} \cap \boldsymbol{S}^t$ onto $\mathcal{U} \cap \{C \in \boldsymbol{S} : H = 0\}$. Note that the inverse map simply replaces the minus sign by the plus sign in the matrix.

Hence the sets in part (e) of (28) for $n \ge 3$ and $0 \le t \le n - 3$ are regular submanifolds with codimension at least $n + \frac{1}{2}(3+1) \cdot 3 = n + 6$. The treatment of the sets S_5 and S_7 in (28) needs a little more preparation.

For $0 < t < n$ the matrices C in $\mathcal{U} \cap \boldsymbol{S}^t$, where $\mathcal{U}$ is defined in the proof of (29), can also be described in a different way using (1.24). Let T be an invertible matrix such that $C = T^t E_{st} T$. For the submatrices in

$$C = \begin{pmatrix} A & B \\ B^t & H \end{pmatrix}, \; T = \begin{pmatrix} X & Y \\ Z & S \end{pmatrix}, \; E_{st} = \begin{pmatrix} D & 0 \\ 0 & 0 \end{pmatrix},$$

the equations

$$(30) \qquad A = X^t DX, \; B = X^t DY, \; H = B^t A^{-1} B$$

follow. The first implies that X is invertible. Conversely, if X is any invertible matrix and Y is arbitrary, then A, B, and H as defined by (30) determine a matrix C in $\mathcal{U} \cap \boldsymbol{S}^t$. The matrices Z and S must satisfy only the condition that T is invertible. This means that

$$(31) \qquad V := S - ZX^{-1}Y$$

is invertible, which is easily seen from the factorization

$$T = \begin{pmatrix} X & 0 \\ Z & E_{n-t} \end{pmatrix} \begin{pmatrix} E_t & X^{-1}Y \\ 0 & V \end{pmatrix}.$$

In the sequel, the relation $X^{-1}Y = A^{-1}B$ is used frequently.

The next lemma is used for the analysis of the subsets S_5 and S_7 of R_5^6 in (28). The variables in $\mathbf{R}^n$ are denoted by $(x, y) \in \mathbf{R}^{n-2} \times \mathbf{R}^2$ according to the Reduction Lemma (3.2).

Lemma (32). Let $n \geq 2$, and let $[k_0] \in m_2^3$ be a cubic form. Then

$$\left([q_{s,n-2} + k_0] + m_2^4\right) \mathcal{G}_n = \{[q_{s,n-2} + k][\tau] : [\tau] \in \mathcal{G}_n \text{ is linear and}$$
$$[k] \in m_n^3 \text{ a cubic form satisfying } \overline{k} = k_0\}$$
$$+ m_n^4$$

holds, where $\overline{k}(y) := k(0, y)$ is that part of k that depends only on the **essential variables** y.

Proof. In order to show the inclusion "$\subset$", consider a germ $[f] = ([q_{s,n-2} + k_0] + [g])[\varphi]$ with $[g]$ in m_2^4 and $[\varphi]$ in $\mathcal{G}_n$. Then $[f]$ is equal to $[q_{s,n-2} + k_0][\varphi] + [r_1]$ with $[r_1] \in m_n^4$, where φ may be assumed to be a polynomial map of order at most 2. Let τ be the linear part of φ. Then $\psi := \varphi \circ \tau^{-1}$ satisfies $[\varphi] = [\psi][\tau]$ and $\psi(x, y) = (x, y) + Q(x, y)$, where Q is zero or a homogeneous polynomial map of degree 2. Therefore, one has $(q_{s,n-2} + k_0)(\psi(x, y)) = q_{s,n-2}(x) + k(x, y)$ plus terms of order ≥ 4 with $k(x, y) := 2\sum_{i=1}^{n-2} \varepsilon_i x_i Q_i(x, y) + k_0(y)$, where $\varepsilon_i \in \{-1, 1\}$ are the signs of $q_{s,n-2}$. Obviously, the cubic term k satisfies $\overline{k} = k_0$ and $[f] = [q_{s,n-2} + k][\tau] + [r]$ holds for some $[r] \in m_n^4$.

As to the proof of the reverse inclusion "$\supset$", consider a germ $[q_{s,n-2} + k][\tau] + [r]$ as indicated. It is equivalent to $[f] := [q_{s,n-2} + k] + [r_1]$ for $[r_1] := [r][\tau]^{-1} \in m_n^4$. By the Reduction Lemma (3.2), the germ $[f]$ is equivalent to $[q_{s,n-2}] + [f_R]$ for some $[f_R] \in m_2^3$. Moreover, by Exercise 3.3, the cubic term of $[f_R]$ is equal to the cubic part of $[f]$, which depends only on the essential variables. The latter is $[k_0]$ by assumption.

Now the proof of (28) is accomplished, and the codimensions of the remaining submanifolds are calculated.

Theorem (33). Let $n \geq 2$. The set S_5 in (28) is the union of three regular submanifolds of $\mathbf{R}^{N-1}$, one of codimension $n + 5$ and two of codimension $n + 6$. The set S_7 is a regular submanifold of codimension $n + 7$.

Proof. The image under the 6-jet map of the invariant set at the left side of the equality sign in (32) is equal to S_5 for $k_0 = y_1^3$ and S_7 for $k_0 = 0$. In what follows, the image of the right side is analysed. First let $n > 2$.

Let T be the invertible matrix representing τ, and let C be the matrix in $\mathcal{S}^{n-2}$ which represents $q := q_{s,n-2} \circ \tau$. Put $K := k \circ \tau$. As in the proof of (29), it suffices to consider C in $\mathcal{U} \cap \mathcal{S}^{n-2}$. Therefore, (30) applies to C. The condition $\overline{k} = k_0$ means that

$$K\left(-A^{-1}By, y\right) = k_0\left(Vy\right) \text{ for } y \in \mathbf{R}^2,$$

since $K(x,y) = k(\tau(x,y)) = k(Xx + Yy, Zx + Sy)$. Therefore, by (30) and (31), $Xx + Yy = 0$ holds if and only if $x = -X^{-1}Yy = -A^{-1}By$. By the latter, $Zx + Sy$ becomes $(S - ZX^{-1}Y)y = Vy$.

The above condition on K is equivalent to certain relations among the coefficients of K. These follow by differentiating with respect to the essential variables. Put $Q := -A^{-1}B$. Then the relations are

$$K_{\alpha\beta\gamma} + Q_{i\alpha}Q_{j\beta}Q_{l\gamma}K_{ijl} +$$
$$Q_{i\alpha}Q_{j\beta}K_{ij\gamma} + Q_{i\alpha}Q_{l\gamma}K_{il\beta} + Q_{j\beta}Q_{l\gamma}K_{jl\alpha} +$$
$$Q_{i\alpha}K_{i\beta\gamma} + Q_{j\beta}K_{j\alpha\gamma} + Q_{l\gamma}K_{l\alpha\beta} =$$
$$V_{\xi\alpha}V_{\eta\beta}V_{\zeta\gamma}k_{0,\xi\eta\zeta}.$$

Here, Greek indices $\alpha, \beta, \gamma, \xi, \eta, \zeta$ designate partial derivatives with respect to the essential variables y_1, y_2 and Latin indices i, j, l with respect to the remaining variables $x_1, \ldots, x_{n-2}$. Moreover, whenever an index appears twice in a product the sum is to be taken over this index.

In an abbreviated form these equations read

(*) $$K_{\alpha\beta\gamma} + r_{\alpha\beta\gamma} = V_{\xi\alpha}V_{\eta\beta}V_{\zeta\gamma}k_{0,\xi\eta\zeta}$$

where the $r_{\alpha\beta\gamma}$ are rational functions of the coefficients of the terms in q and K, which depend not solely on the essential variables.

(i) Consider now the case $k_0 = 0$, which yields the set S_7. Denote by $\bar{q}$ and $\overline{K}$ the essential parts of q and K, i.e., $\bar{q}(y) = q(0,y)$ and $\overline{K}(y) = K(0,y)$. Being the coefficients of H in (30), the 3 coefficients of the quadratic form $\bar{q}$ are rational functions of the remaining coefficients of q. Similarly (*) shows that the 4 coefficients of the cubic form $\overline{K}$ are rational functions of the remaining coefficients of q and K. The n coefficients of the linear term in S_7 are zero. This shows the assertion about S_7.

(ii) The case $k_0 = y_1^3$ yields S_5. Write $(Vy)_1 = v_1 y_1 + v_2 y_2$, where $(v_1, v_2) \neq (0,0)$, since V is invertible by (31). Then the equations (*) read

$$K_{\alpha\beta\gamma} + r_{\alpha\beta\gamma} = 6v_\alpha v_\beta v_\gamma.$$

The free coefficients v_1, v_2 can be replaced by $v_\alpha = 6^{-\frac{1}{3}}(K_{\alpha\alpha\alpha} + r_{\alpha\alpha\alpha})^{\frac{1}{3}}$, and the remaining 2 essential coefficients $K_{112}(= K_{121} = K_{211})$ and $K_{122}(= K_{212} = K_{221})$ are determined by

$$K_{\alpha\beta\gamma} = (K_{\alpha\alpha\alpha} + r_{\alpha\alpha\alpha})^{\frac{1}{3}} (K_{\beta\beta\beta} + r_{\beta\beta\beta})^{\frac{1}{3}} (K_{\gamma\gamma\gamma} + r_{\gamma\gamma\gamma})^{\frac{1}{3}} - r_{\alpha\beta\gamma}.$$

Since the cube root is a non-differentiable function at the origin, three cases have to be distinguished. First the case where K_{111} and K_{222} are free variables with $K_{111} \neq -r_{111}$ and $K_{222} \neq -r_{222}$, determines a subset of S_5, which is a regular submanifold with codimension diminished by 2 with respect to the case (i). The case that K_{111} is a free variable with $K_{111} \neq$

$-r_{111}$, whereas $K_{222} = -r_{222}$ holds, determines a regular submanifold with codimension diminished by 1. Interchanging the indices 1 and 2 in the previous case yields the last case.

The case $n = 2$ is even simpler because the functions $r_{\alpha\beta\gamma}$ are missing.

Codimension of the orbits in R_5^6 (34).

(a) $\mathrm{cod}^6 [q_{s,n-1}] = 5$ for $0 \le s \le n - 1$.

(b) $\mathrm{cod}^6 [q_{s,n-2} + x^2 y] = 7$ for $0 \le s \le n - 2$.

(c) $\mathrm{cod}^6 [q_{s,n-2} + x^2 y + \eta y^l] = l$ for $l \in \{5,6\}, \eta \in \{-1,1\}, 0 \le s \le n-2$.

Proof. As to (a) and (b) the formula $\mathrm{cod}^6[f] = \dim(j^6(m^2)/j^6(\langle m\mathcal{J}[f]\rangle))$ (cf. after (24)) is used.

(a) Note that $j^6(\langle m\mathcal{J}[q_{s,n-1}]\rangle)$ is identified with the linear space of all polynomials of degree at most 6 without constant and linear terms and without any of the monomials

$$x_n^2, x_n^3, \ldots, x_n^6.$$

(b) Here the monomials x_{n-1}^2 and $x_{n-1}x_n$ are also missing, since $\langle m\mathcal{J}[x^2 y]\rangle = \langle xy^2, x^2 y, x^3 \rangle_{\mathcal{E}} = [x]m^2$ in $\mathcal{E}_2$.

(c) By Exercises 6.2 and 6.3, $\det[x^2 y + \eta y^l] = \mathrm{cod}[x^2 y + \eta y^l] = l$ holds for $l \in \{5,6\}, \eta \in \{-1,1\}$. Hence, the assertion follows from (26).

This accomplishes the verification of the hypotheses of (23) for R_5^6. Thus, as argued before (27), the following theorem is proved.

Measure Theoretical Genericity (35). Let $r \in \{0,1,\ldots,4\}$. Then for every $F \in C^\infty(\mathbf{R}^{n+r})$ there is a Lebesgue null set $\Lambda(F) \subset P^6 \setminus \mathbf{R}$, such that

$$F + P \in \mathcal{F}$$

for all $P \notin \Lambda(F)$.

From this measure theoretical result the topological result (38) is derived in just a few additional steps. Recall that $r \le 4$.

Lemma (36). If $A \subset \mathbf{R}^{n+r}$ is compact, then $\mathcal{F}(A)$ is C^∞-dense in $C^\infty(\mathbf{R}^{n+r})$.

Proof. Let F be in $C^\infty(\mathbf{R}^{n+r})$, and let $\mathcal{V}_\delta^l(F)$ be a C^∞-neighborhood of F. In order to show $\mathcal{F}(A) \cap \mathcal{V}_\delta^l(F) \ne \emptyset$, choose a smooth function β with compact support K such that $\beta(p) = 1$ holds for all $p \in A$ (see Exercise 11.2). Then

$$b := \sup\left\{\sum_{\lambda \le \mu} \binom{\mu}{\lambda} |D^\lambda \beta(p)| : 0 \le |\mu| \le l, \, p \in \mathbf{R}^{n+r}\right\}$$

and

$$c := \sup \{\nu! \, |x^\nu| : 0 \le |\nu| \le 6, \, (x, u) \in K\}$$

are finite.

Now let $\varepsilon > 0$ be given. Since open sets are not Lebesgue null sets, according to (35) there exists a polynomial $P \in P^6 \backslash \mathbf{R}$ such that $F + P \in \mathcal{F}$ and such that the sum of the absolute values of the coefficients of P is smaller than ε.

Clearly $F + \beta P \in \mathcal{F}(A)$ holds. If $\varepsilon > 0$ is small enough, then $F + \beta P \in \mathcal{V}^l_\delta(F)$ also holds. Indeed, note that $D^\mu(\beta P)(p) = 0$ for all $p \notin K$ and that for all $p = (x, u) \in K$

$$|D^\mu(\beta P)(p)| \le \sum\nolimits_{\lambda \le \mu} \binom{\mu}{\lambda} |D^\lambda \beta(p) D^{\mu - \lambda} P(x)| \le bc\varepsilon$$

is valid by Leibniz's rule of derivation. Therefore,

$$\varepsilon := (bc)^{-1} \min\{\delta(p) : p \in K\}$$

yields the assertion.

Lemma (37). Let A be as in (10). Then $\mathcal{F}(A)$ is C^∞-dense in $C^\infty(\mathbf{R}^{n+r})$.

Proof. In order to show $\mathcal{F}(A) \cap \mathcal{V}^l_\delta(F) \ne \emptyset$, let β_σ be as in the remark following (10). In addition, let β_σ be smooth with compact support K_σ, which is possible by Exercise 11.2. Then

$$b_\sigma := \sup \left\{ \sum\nolimits_{\lambda \le \mu} \binom{\mu}{\lambda} |D^\lambda \beta_\sigma(p)| : 0 \le |\mu| \le l, \, p \in \mathbf{R}^{n+r} \right\}$$

is finite and nonzero. For

$$\tau_\sigma := b_\sigma^{-1} \min\{\delta(p) : p \in K_\sigma\} > 0$$

it follows that $\beta_\sigma H$ is in $\mathcal{V}^l_\delta(0)$ for all $H \in \mathcal{V}^l_{\tau_\sigma, K_\sigma}(0)$. Indeed, $D^\mu(\beta_\sigma H)(p) = 0$ for $p \notin K_\sigma$ and $|D^\mu(\beta_\sigma H)(p)| \le \sum_{\lambda \le \mu} \binom{\mu}{\lambda} |D^\lambda \beta_\sigma(p) D^{\mu - \lambda} H(p)| < b_\sigma \tau_\sigma \le \delta(p)$ for $p \in K_\sigma$.

According to (36), there is a $G_\sigma \in \mathcal{F}(A_\sigma) \cap \mathcal{V}^l_{\tau_\sigma, K_\sigma}(F)$. Put

$$G := \beta F + \sum_{\sigma=1}^\infty \beta_\sigma G_\sigma \text{ with } \beta := 1 - \sum_{\sigma=1}^\infty \beta_\sigma.$$

Note that because of the disjointness of the supports K_σ, both functions β and G are well defined and smooth.

If p is in A_σ, then $G(p) = G_\sigma(p)$ holds. Therefore, $G \in \bigcap_{\sigma=1}^\infty \mathcal{F}(A_\sigma) = \mathcal{F}(A)$.

If $p \notin \bigcup_\sigma K_\sigma$, then $G(p) - F(p) = 0$ is valid. If $p \in K_\sigma$, then $G(p) - F(p) = \beta_\sigma(p)(G_\sigma(p) - F(p))$ holds. Since $G_\sigma - F \in \mathcal{V}^l_{\tau_\sigma, K_\sigma}(0)$, it follows that $\beta_\sigma(G_\sigma - F) \in \mathcal{V}_\delta^l(0)$ as shown above. Therefore, $G - F \in \mathcal{V}_\delta^l(0)$ is true, implying $G \in \mathcal{V}_\delta^l(F)$.

This chapter concludes with the main genericity result on elementary unfoldings:

Topological Genericity (38).   Let $r \in \{0, 1, \ldots, 4\}$. Then $\mathcal{F}$ is open and dense in $C^\infty(\mathbf{R}^{n+r})$ with respect to the Whitney C^∞-topology.

Proof. Choose A and A' in $\mathbf{R}^{n+r}$ as in the remark following (11). Then $\mathcal{F}(A)$ and $\mathcal{F}(A')$ are C^∞-open by (11). They are C^∞-dense by (37). There-fore, as noted just before (8), $\mathcal{F} = \mathcal{F}(A) \cap \mathcal{F}(A')$ is C^∞-open and C^∞-dense.

Recall that $\mathcal{F}$ is even C^{r+3}-open by (12).

Exercises

11.1.   Fix a nonnegative integer l. For $F \in C^\infty(\mathbf{R}^m)$ and δ a strictly positive continuous function on $\mathbf{R}^m$, show that the sets $\mathcal{V}_\delta^l(F)$ satisfy the axioms (i)–(iii) for bases of open neighborhoods (see after (7)).

Moreover, verify that the set τ of subsets of $C^\infty(\mathbf{R}^m)$ consisting of arbitrary unions of sets $\mathcal{V}_\delta^l(F)$ is indeed a **topology** on $C^\infty(\mathbf{R}^m)$, i.e., that τ contains the void set and the whole space and that τ is closed under finite intersections and arbitrary unions of sets from τ.

11.2.   Let $A \subset \mathbf{R}^m$ be compact and let $W \subset \mathbf{R}^m$ be open such that $A \subset W$. Show that there is a smooth function $\beta : \mathbf{R}^m \to \mathbf{R}$ with compact **support** satisfying $\beta(\mathbf{R}^m) \subset [0,1]$, $\beta|A = 1$, and $\beta|(\mathbf{R}^m \setminus W) = 0$. The support of β is the closure of $\{x \in \mathbf{R}^m : \beta(x) \neq 0\}$ and is denoted by $\mathrm{supp}\beta$.
Hint: Let $\varepsilon > 0$ and $A_\varepsilon := \{y \in \mathbf{R}^m : |y - a| \leq \varepsilon$ for all $a \in A\}$. Then β is ob-tained by **Friedrichs' smoothing** of the **indicator function** χ_ε of A_ε, where χ_ε is defined by $\chi_\varepsilon|A_\varepsilon = 1$ and $\chi_\varepsilon|(\mathbf{R}^m \setminus A_\varepsilon) = 0$. Let ρ be any nonnegative smooth function on $\mathbf{R}^m$ satisfying $\rho(x) = 0$ for $|x| \geq 1$ and $\int \rho(x)dx = 1$. For instance, take $\exp((|x|^2 - 1)^{-1})$ for $|x| < 1$ multiplied by a normalizing constant. Set

$$\beta(x) := \int \chi_\varepsilon(y)\rho_\varepsilon(x - y)dy$$

for $\rho_\varepsilon(x) := \varepsilon^{-m}\rho(x/\varepsilon)$.

11.3.   Let $[f]$ be in m^2 and let $k \geq 2$. Show that the formula

$$\mathrm{cod}^k[f] = \dim\left(m^2/(\langle m\mathcal{J}[f]\rangle + m^{k+1})\right)$$

holds, and the implication

$$\mathrm{cod}^k[f] \leq k - 2 \Rightarrow \mathrm{cod}^k[f] = \mathrm{cod}[f]$$

is valid. Compare this result with (26).

11.4. Let k and c be integers such that $k \geq 2$ and $c \in \{0, 1, \ldots, N - 1 - n\}$. Recall that $N = \binom{n+k}{k}$ and that as usual $j^k(m)$ is identified with $\mathbf{R}^{N-1}$. Set

$$\tilde{R}^k_c := \left\{ z = j^k[f] \in \mathbf{R}^{N-1} : [f] \in m^2, \ \mathrm{cod}^k[f] \geq c \right\}.$$

Show that

(a) $\tilde{R}^k_c \subset R^k_c$ (cf. (22)).

(b) $\tilde{R}^k_c = R^k_c$ for $0 \leq c \leq k - 1$.

(c) $\tilde{R}^k_c$ is a real algebraic set, i.e., there is a finite set D_c of polynomials p in $N - 1$ real variables $z = (z_1, \ldots, z_{N-1})$ such that $\tilde{R}^k_c$ is equal to the set of common zeros

$$N(D_c) := \bigcap \left\{ p^{-1}(\{0\}) : p \in D_c \right\}.$$

Hint for (c): According to (4), the dimension of the tangent space at $z = j^k[f]$ to the orbit $z\mathcal{G}^k$ in $\mathbf{R}^{N-1}$ for $[f] \in m^2$ is equal to the rank of the linear map

$$\mathbf{R}^{n(N-1)} \to \mathbf{R}^{N-1}, \ \zeta \mapsto \sum_{j,\kappa} z D_{j\kappa} M(\varepsilon) \zeta_{j\kappa}.$$

11.5. Let n, r, and k be integers satisfying $n \geq 1$, $r \geq 0$, and $k \geq 1$. Let $V \subset \mathbf{R}^{N-1}$ be open, e.g., $V = \mathbf{R}^{N-1} \setminus R^k_{r+1}$ with $k \geq r + 2$ (cf. (23)). Show that

$$\mathcal{V} := \left\{ F \in C^\infty \left(\mathbf{R}^{n+r} \right) : F^k(p) \in V \text{ for all } p \in \mathbf{R}^{n+r} \right\}$$

is C^k-open.

11.6. Recall that a smooth function is a Morse function if all its critical points are nondegenerate. Show that the set of Morse functions on $\mathbf{R}^n$ is a C^3-open and C^∞-dense subset of $C^\infty(\mathbf{R}^n)$.

11.7. Show that the set $\mathcal{R}$ of smooth functions on $\mathbf{R}^n$ without critical points is C^1-open. For $n = 1$ show that $\mathcal{R}$ is not C^1-dense in $C^\infty(\mathbf{R})$; for general n see Exercise 12.2.

11.8. Show that in the case of one variable, i.e., $n = 1$, the Measure Theoretical Genericity (35) and the Topological Genericity (38) hold for any number r of parameters.
Hint: If $n = 1$, then $\mathbf{R}^{N-1}$ consists of only finitely many orbits under $\mathcal{G}^k$. Therefore, (20) applies directly without restriction on r.

Solutions

11.1. Axiom (i) is trivially satisfied. As to axiom (ii), let $F \in C^\infty(\mathbf{R}^m)$ and δ, δ' be strictly positive, continuous functions on $\mathbf{R}^m$. Then the point-wise minimum δ'' of δ and δ' is still strictly positive and continuous. Clearly, $\mathcal{V}_{\delta''}^l(F) \subset \mathcal{V}_\delta^l(F) \cap \mathcal{V}_{\delta'}^l(F)$. Also axiom (iii) is satisfied. Indeed, for $G \in \mathcal{V}_\delta^l(F)$, the function $\delta' : \mathbf{R}^m \to \mathbf{R}$,

$$\delta'(p) := \delta(p) - \max\{|D^\nu(G - F)(p)| : 0 \leq |\nu| \leq l\}$$

is strictly positive and continuous. By the triangle inequality, $\mathcal{V}_{\delta'}^l(G) \subset \mathcal{V}_\delta^l(F)$ follows easily.

Obviously, $\emptyset$ and $C^\infty(\mathbf{R}^m)$ belong to τ. Since the formula

$$\left(\bigcup_\iota \mathcal{V}_{1,\iota}\right) \cap \left(\bigcup_\kappa \mathcal{V}_{2,\kappa}\right) = \bigcup_{\iota,\kappa}(\mathcal{V}_{1,\iota} \cap \mathcal{V}_{2,\kappa})$$

holds for the intersection of two arbitrary unions, τ is a topology if every intersection $\mathcal{V} := \mathcal{V}_\delta^l(F) \cap \mathcal{V}_{\delta'}^l(F')$ belongs to τ. For the proof let G be in $\mathcal{V}$. According to axiom (iii), $\mathcal{V}_\gamma^l(G) \subset \mathcal{V}_\delta^l(F)$ and $\mathcal{V}_{\gamma'}^l(G) \subset \mathcal{V}_{\delta'}^l(F')$ hold for suitable γ and γ'. By axiom (ii), there is a suitable γ'' satisfying $\mathcal{V}_{\gamma''}^l(G) \subset \mathcal{V}$.

11.2. Since A_ε is compact and $(x, y) \mapsto \rho_\varepsilon(x - y)$ is smooth, β is smooth with

$$D^\nu \beta(x) = \int \chi_\varepsilon(y) D^\nu \rho_\varepsilon(x - y)\, dy.$$

Clearly, β is nonnegative and $\beta(x) \leq \int \rho_\varepsilon(x - y)\,dy = \int \rho_\varepsilon(y)\,dy = 1$ for all x. If $a \in A$ then $\beta(a) = \int \rho_\varepsilon(a - y)\,dy = 1$, since $\rho_\varepsilon(x) = 0$ for $|x| \geq \varepsilon$ and $|a - y| > \varepsilon$ for $y \notin A_\varepsilon$. The support of β is contained in $A_{2\varepsilon}$ and hence is compact. Indeed, for $x \in \mathbf{R}^m \setminus A_{2\varepsilon}$ and $y \in A_\varepsilon$, it follows that $|x - y| \geq |x - a| - |a - y| > 2\varepsilon - \varepsilon = \varepsilon$, whence $\beta(x) = 0$.

It remains to choose $\varepsilon > 0$ such that $A_{2\varepsilon} \subset W$. Since $B := \mathbf{R}^m \setminus W$ is closed and disjoint from the compact set A, the distance $d := \inf\{|a - b| : a \in A, b \in B\}$ is positive. Choose $\varepsilon := d/3$. Then for $x \in A_{2\varepsilon}$, $b \in B$, and $a \in A$, it follows that $|x - b| \geq |b - a| - |a - x| \geq d - 2\varepsilon = \varepsilon$, implying $x \notin B$.

11.3. The formula for $\mathrm{cod}^k[f]$ follows immediately from the first two lines of the proof of (26).

Since the dimensions

$$d_j := \dim\left(m^2 / (\langle m\mathcal{J}[f]\rangle + m^{j+1})\right)$$

increase from $d_1 = 0$ to $d_k = \mathrm{cod}^k[f]$ by $k - 1$ steps, $\mathrm{cod}^k[f] \leq k - 2$ implies that $\langle m\mathcal{J}[f]\rangle + m^l = \langle m\mathcal{J}[f]\rangle + m^{l+1}$ for some $l \in \{2, ..., k\}$. By (4.18), this is equivalent to $m^l \subset \langle m\mathcal{J}[f]\rangle$, and hence $\mathrm{cod}^k[f] = \dim\left(m^2/\langle m\mathcal{J}[f]\rangle\right)$ holds. The latter is equal to $\mathrm{cod}[f]$ by (5.17).

11.4. By (26), $\mathrm{cod}^k[f] \leq \mathrm{cod}[f]$, which implies (a).

Now assume $c \leq k - 1$. Then

$$\mathbf{R}^{N-1} \setminus R_c^k = j^k(m \setminus m^2) \cup \left\{ j^k[f] : [f] \in m^2, \ \mathrm{cod}[f] < c \right\}.$$

In this formula $\mathrm{cod}[f]$ can be replaced by $\mathrm{cod}^k[f]$ due to (26) and Exercise 11.3, since either $\mathrm{cod}[f] \leq k - 2$ or $\mathrm{cod}^k[f] \leq k - 2$ holds. Hence, $\mathbf{R}^{N-1} \setminus R_c^k = \mathbf{R}^{N-1} \setminus \tilde{R}_c^k$ implying (b).

For the proof of (c), let $z \in \tilde{R}_c^k$. Let the matrix $Z(z)$ represent the linear map given in the hint, and let $D_c(z)$ denote the set of determinants of the quadratic submatrices of $Z(z)$ of dimension $N - n - c$. Denote by D_c the set of polynomials in z corresponding to the determinants.

Then $N(D_c)$ is the set of all $z \in \mathbf{R}^{N-1}$ such that the rank of $Z(z)$ is smaller than $N - n - c$, i.e., the codimension of the orbit $z\mathcal{G}^k$ in $\mathbf{R}^{N-1}$ is at least $n + c$. This means by (25) that $z \in N(D_c)$ if and only if $\mathrm{cod}^k[f] \geq c$ for $[f] \in m^2$ with $z = j^k[f]$. Hence, $N(D_c) = \tilde{R}_c^k$ holds.

11.5. If $V = \mathbf{R}^{N-1}$, then $\mathcal{V} = C^\infty(\mathbf{R}^{n+r})$ is C^k-open. Hence, let $V \neq \mathbf{R}^{N-1}$ and take $F \in \mathcal{V}$. For $p \in \mathbf{R}^{n+r}$ there is a $\delta(p) > 0$ such that

$$B_p := \left\{ z \in \mathbf{R}^{N-1} : \left| z_\nu - \frac{1}{\nu!} D_1^\nu F(p) \right| < \delta(p) \text{ for } 0 < |\nu| \leq k \right\} \subset V.$$

Obviously, there is a maximal $\delta(p) < \infty$ with this property. Take this. It suffices to show that $p \mapsto \delta(p)$ is continuous, since then $\mathcal{V}_\delta^k(F) \subset \mathcal{V}$ holds. Indeed, if $G \in \mathcal{V}_\delta^k(F)$, then $|D^\nu(G - F)(p)| < \delta(p)$ for $0 < |\nu| \leq k$ and hence $G^k(p) \in B_p$ holds for all $p \in \mathbf{R}^{n+r}$. This implies that $G^k(p) \in V$ for all p. Thus $G \in \mathcal{V}$ and therefore $\mathcal{V}_\delta^k(F) \subset \mathcal{V}$.

Turn to the proof of the continuity of δ. For $z \in V$, let $B(z; r(z)) \subset V$ be the open ball with respect to the maximum norm on $\mathbf{R}^{N-1}$ with center z and maximal radius $r(z)$. Let $z' \in B(z; r(z))$.

(i) Obviously, $r(z) - |z - z'|_{\max} > 0$. Then for every $w \in B(z'; r(z) - |z - z'|_{\max})$, it follows that $|w - z'|_{\max} + |z' - z|_{\max} < r(z)$. Hence $|w - z|_{\max} < r(z)$, i.e., $w \in B(z; r(z))$. Since $r(z')$ is maximal, this implies $r(z') \geq r(z) - |z - z'|_{\max}$.

(ii) For every $w \in B(z; r(z))$, it follows that $|w - z'|_{\max} \leq |w - z|_{\max} + |z - z'|_{\max} \leq r(z) + |z - z'|_{\max}$. Hence, $B(z; r(z)) \subset B(z'; r(z) + |z - z'|_{\max})$. Since $r(z)$ is maximal, this implies $r(z') \leq r(z) + |z - z'|_{\max}$.

The results in (i) and (ii) yield $|r(z') - r(z)| \leq |z' - z|$. Since $\delta(p) = r(F^k(p))$ and F^k is continuous, the assertion follows.

11.6. If the number r of parameters is zero, then $\mathcal{F}$ in (1) is just the set of Morse functions on $\mathbf{R}^n$ by Exercise 7.12 and by (10.6) (see also Exercise 7.10). The assertion follows from (12) and (38).

11.7. Note that $\mathcal{R} = \{F \in C^\infty(\mathbf{R}^n) : F^1(x) \in \mathbf{R}^n \setminus \{0\} \text{ for all } x \in \mathbf{R}^n\}$. Since $\mathbf{R}^n \setminus \{0\}$ is open in $\mathbf{R}^n$, $\mathcal{R}$ is C^1-open by Exercise 11.5.

Now let $n = 1$ and $q : \mathbf{R} \to \mathbf{R}$, $q(x) = x^2$. Obviously $q \notin \mathcal{R}$. Even $\mathcal{V}^1_{1,[-1,1]}(q) \cap \mathcal{R} = \emptyset$ holds. Indeed, if $G \in \mathcal{V}^1_{1,[-1,1]}(q)$, then G' satisfies $|G'(a) - 2a| < 1$ for all $a \in [-1, 1]$. In particular, $|G'(-1) + 2| < 1$ and $|G'(1) - 2| < 1$, whence $G'(-1) < 0$ and $G'(1) > 0$. Hence, there is an $x \in]-1, 1[$ such that $G'(x) = 0$. Therefore, G is not in $\mathcal{R}$.

Since $\mathcal{V}^1_{1,[-1,1]}$ is C^1-open, this proves that $\mathcal{R}$ is not C^1-dense in $C^\infty(\mathbf{R})$.

11.8. By (1.34), $m_1 \setminus m_1^\infty = \bigcup \{ [\varepsilon^{l-1} x^l] \mathcal{G}_1 : l \in \mathbf{N}, \ \varepsilon \in \{1, -1\} \}$ holds. This implies

$$j^k(m_1) = \bigcup \left\{ j^k [\varepsilon^{l-1} x^l] \mathcal{G}_1^k : l \in \{1, ..., k\}, \ \varepsilon \in \{1, -1\} \right\}$$

for every $k \geq 1$. Hence, $\mathbf{R}^{N-1}$ is the union of only finitely many orbits under $\mathcal{G}_1^k$. Therefore, as argued in (21), the hypothesis of (20) is satisfied for $k \geq r+2$. By (20), for every $F \in C^\infty(\mathbf{R}^{1+r})$ there is a null set $\Lambda(F) \subset P^k \setminus \mathbf{R}$ such that $(F + P)^k$ is transverse to all orbits for all $P \notin \Lambda(F)$. Because of (5) and (2), this proves the Measure Theoretical Genericity (35) for $n = 1$ and $k \geq r + 2$. From this the Topological Genericity (38) follows quite analogously for $n = 1$ and arbitrary $r \geq 0$ as for $r \leq 4$ and arbitrary $n \geq 1$.

Chapter 12

Stability

Every degenerate critical point of a smooth function is unstable when subjected to a perturbation. The addition of an arbitrarily small term as a perturbation diminishes the degree of degeneracy of the critical point and creates new critical points of smaller degeneracy in the neighborhood. This behavior of degenerate critical points has been studied in detail in Examples (7.2) and (7.3) of Chapter 7.

The notion of an unfolding of a germ has been introduced in Chapter 7 as the basic concept to describe perturbations. A main point of the Fundamental Theorem on Universal Unfoldings (10.8) is that all possible perturbations of a finitely determined germ are included in a single unfolding, which is called a versal unfolding. The theory is completed in this chapter by considering the stability of unfoldings. The culmination is the result (12), which says that for unfoldings the properties **versality** and **stability** imply each other. At this point, a very satisfactory conclusion of the theory is obtained. Thus, Catastrophe Theory reveals in a mathematically rigorous manner the true complementary nature of the seemingly unreconcilable notions of versality and stability. This is of great importance for a theory of cognition, and is thoroughly discussed by René Thom in his book *Stabilité Structurelle et Morphogénèse* [T1].

In this chapter the number r of parameters of an unfolding is not restricted. The notations of Chapter (11) are frequently used.

In a first step toward an appropriate definition of stability for an unfolding, it is shown that versality is a persistent property.

> **Lemma (1).** Let $[F] \in \mathcal{E}_{n+r}$ be a versal unfolding of a germ in $\mathcal{E}_n$. It is no restriction to assume that F is defined on $\mathbf{R}^{n+r}$ (see before (4.2)). Then there is a compact neighborhood A of the origin in $\mathbf{R}^{n+r}$ and a $\tau > 0$ such that
> $$\mathcal{V}_{\tau,A}^{r+3}(F) \subset \mathcal{F}(A).$$

This means that $[G^{x,u}]$ is versal for all (x, u) in some neighborhood of the origin in $\mathbf{R}^{n+r}$ and for all G in some C^{r+3}-neighborhood of F.

Proof. Because of (11.2)–(11.7) it follows immediately by continuity that there is a compact neighborhood A of $0 \in \mathbf{R}^{n+r}$ such that $F \in \mathcal{F}(A)$ holds. The assertion derives from (11.9).

Therefore, a slightly perturbed versal unfolding $[F]$ of a germ $[f]$ stays versal at all points of a neighborhood of the origin. However, if $[G]$ denotes the unfolding of the germ $[g] := [G(\cdot, 0)]$, which results from $[F]$ by a perturbation, then the germs $[f^{0,0}](= [f - f(0)])$ and $[g^{0,0}](= [g - g(0)])$ need not be equal nor even equivalent. This is demonstrated in the following simple example.

Let $[F] \in \mathcal{E}_{1+1}$, $F(x, u) := x^3 + ux$, be the well-known universal unfolding of the fold catastrophe. Perturbing this by the addition of the term

$$t[x^4 + x^2 + u + 1] \in \mathcal{E}_{1+1}$$

for arbitrarily small $t \neq 0$, one obtains $G(x, u) = x^3 + ux + t(x^4 + x^2 + u + 1)$ and

$$g^{0,0}(x) = x^3 + t\left(x^4 + x^2\right).$$

Since $0 \in \mathbf{R}$ is a nondegenerate critical point of $[g^{0,0}]$, clearly $[g^{0,0}]$ is not equivalent to $[x^3] = [f^{0,0}]$. This confirms the above statement.

However, what about the germs $[g^{b,c}]$ for $(b, c) \in \mathbf{R}^{1+1}$? It is easy to verify that

$$g^{b,c}(x) = (1 + tx + 4tb)x^3 + (6tb^2 + 3b + t)x^2 + (4tb^3 + 3b^2 + 2tb + c)x$$

and $G^{b,c}(x, u) = g^{b,c}(x) + u(x + b + t)$ hold. Choosing

$$b = -2/3t \left(1 + \sqrt{1 - 8t^2/3}\right)^{-1}, \; c = -4tb^3 - 3b^2 - 2tb$$

for $|t| < \sqrt{3/8}$, then $g^{b,c}(x) = (\sqrt{1 - 8t^2/3} + tx)x^3$ and clearly

$$[g^{b,c}] \sim [f^{0,0}].$$

Hence, one finds that the organization center of the perturbed unfolding is equivalent to the original germ—not at the origin but at a slightly shifted position.

This implies that $[G^{b,c}]$ is an unfolding of a germ that is equivalent to $[f^{0,0}]$. Such a situation is considered at the end of Chapter 7. It follows that the relation (7.9) holds for $[G^{b,c}]$ in place of $[G]$. The parameter transformation is given here by the identity map. Therefore, the unfoldings $[F]$ and $[G^{b,c}]$ with b and c defined above are isomorphic according to the following

Definition. Let $[F]$ and $[G]$ be two unfoldings in $\mathcal{E}_{n+r}$. They are called **isomorphic**, denoted by

$$[G] \approx [F],$$

if there are germs $[\omega] \in m_{n+r,n}$, $[\psi] \in \mathcal{G}_r$, and $[\gamma] \in \mathcal{E}_r$ such that

(i) $x \mapsto \omega_0(x) := \omega(x, 0)$ is a local diffeomorphism of the origin, and

(ii) $G(y, v) = F(\omega(y, v), \psi(v)) + \gamma(v)$ holds for small (y, v) in $\mathbf{R}^{n+r}$.

Obviously, equivalent unfoldings are isomorphic. In Exercise 12.1 it is shown that the isomorphism of unfoldings is an equivalence relation and that two isomorphic unfoldings are versal when one of them is versal. If $[f]$ and $[g]$ are the organization centers of the isomorphic unfoldings $[F]$ and $[G]$ then $[g^{0,0}] = [f^{0,0}][\omega_0]$ holds with $[\omega_0] \in \mathcal{G}_n$ from (i) in the above definition.

Now the central notion of a stable unfolding will be introduced. The foregoing example shows that an unfolding may merge into an inequivalent unfolding under the influence of an arbitrarily small perturbation. However, one may expect of a stable unfolding that there are neighboring points at which the perturbed unfolding is isomorphic to the unperturbed one.

Definition. Let $[F] \in \mathcal{E}_{n+r}$ be an unfolding, and let the representative F be defined on $\mathbf{R}^{n+r}$ (cf. before (4.2)). Then the unfolding $[F]$ is **stable** if for every neighborhood of W of $0 \in \mathbf{R}^{n+r}$ and every $H \in C^\infty(\mathbf{R}^{n+r})$ there is an $\varepsilon > 0$ such that for every $t \in]-\varepsilon, \varepsilon[$ there is a $(b, c) \in W$ satisfying

$$(2) \qquad [(F + tH)^{b,c}] \approx [F].$$

If condition (2) holds for some neighborhood W, then it is satisfied for any $W' \supset W$. Therefore the above definition of stability does not depend on the choice of the representative F.

A heuristic argument will be used to deduce an algebraic property of stable unfoldings called **infinitesimal stability**, which will later turn out to be equivalent to stability. The argument is similar to that given for (4.30).

Writing the relation (2) indicating explicitly the dependence on t of the translation (b, c), one obtains

$$F(x, u) + tH(x, u) = F\left(\omega_t(x - b_t, u - c_t), \psi_t(u - c_t)\right) + \gamma_t(u - c_t).$$

Defining $\omega(x, u, t) := \omega_t(x - b_t, u - c_t)$ and $\psi(u, t)$, $\gamma(u, t)$ analogously, this becomes

$$F(x, u) + tH(x, u) = F(\omega(x, u, t), \psi(u, t)) + \gamma(u, t).$$

Without restriction $\omega(x, u, 0) = x$ and $\psi(u, 0) = u$ hold. Now the plausible assumption is made that ω, ψ, and γ depend smoothly on all three variables. Then differentiating the above equation with respect to t yields at $t = 0$ the relation

$$H(x, u) = \sum_{i=1}^{n} \frac{\partial F}{\partial x_i}(x, u)\partial_t\omega(x, u, 0) + \sum_{j=1}^{r} \frac{\partial F}{\partial u_j}(x, u)\partial_t\psi(u, 0) + \partial_t\gamma(u, 0).$$

Since $H \in C^\infty(\mathbf{R}^{n+r})$ is arbitrary,

$$(3) \qquad \mathcal{E}_{n+r} = \left\langle \frac{\partial F}{\partial x_1}, \ldots, \frac{\partial F}{\partial x_n} \right\rangle_{\mathcal{E}_{n+r}} + \left\langle \frac{\partial F}{\partial u_1}, \ldots, \frac{\partial F}{\partial u_r}, 1 \right\rangle_{\mathcal{E}_r}$$

follows.

Definition.   An unfolding $[F]$ in $\mathcal{E}_{n+r}$ is said to be **infinitesimally stable** if it satisfies (3).

The next result already points to a close relationship between stability and versality.

Theorem (4).   Let $[F] \in \mathcal{E}_{n+r}$ be an unfolding of the germ $[f] \in \mathcal{E}_n$. Then the statements (a), (b), and (c) are equivalent.

(a) $[F]$ is versal.

(b) $\mathcal{E}_n = \mathcal{J}[f] + \mathcal{V}[F] + \mathbf{R}[1]$.

(c) $[F]$ is infinitesimally stable.

Proof.   As to the equivalence of (a) and (b), it may be assumed that $[f] \in m^2$ (see Exercise 7.12). Then $\mathcal{J}[f] + \mathcal{V}[F] \subset m$ holds, and the equivalence follows from (10.6).

For the proof of the equivalence of (b) and (c) the embedding $\varphi : \mathbf{R}^n \to \mathbf{R}^{n+r}$, $\varphi(x) := (x, 0)$ is used. The pullback $\varphi^* : \mathcal{E}_{n+r} \to \mathcal{E}_n$, $\varphi^*[G] := [G \circ \varphi] = [G(\cdot, 0)]$ is a surjective linear ring homomorphism preserving the identity (see before (9.2)).

Now assume (b). Let $[H]$ be in $\mathcal{E}_{n+r}$. According to (b) there are germs $[g_i] \in \mathcal{E}_n$ for $i = 1, \ldots, n$ and real constants $c_1, \ldots, c_r, c$ such that

$$\varphi^*[H] = \sum_{i=1}^{n} [g_i][D_i f] + \sum_{j=1}^{r} c_j \left[\frac{\partial F}{\partial u_j}(\cdot, 0) \right] + c.$$

This defines a new unfolding $[H'] \in \mathcal{E}_{n+r}$ by setting

$$[H'] := [H] - \sum_{i=1}^{n} [g_i] \left[\frac{\partial F}{\partial x_i} \right] - \sum_{j=1}^{r} c_j \left[\frac{\partial F}{\partial u_j} \right] - c.$$

It satisfies $\varphi^*[H'] = 0$, which means that $H'(x, 0) = 0$ for small $x \in \mathbf{R}^n$. Therefore, by (1.28), there are germs $[H_j] \in \mathcal{E}_{n+r}$ for $j = 1, \ldots, r$ such that $H'(x, u) = \sum_{j=1}^{r} H_j(x, u) u_j$ in some neighborhood of the origin of $\mathbf{R}^{n+r}$. Hence, $[H'] \in \langle m_r \mathcal{E}_{n+r} \rangle$ holds. Since this is true for any $[H] \in \mathcal{E}_{n+r}$,

$$\mathcal{E}_{n+r} = \left\langle \frac{\partial F}{\partial x_1}, \ldots, \frac{\partial F}{\partial x_n} \right\rangle_{\mathcal{E}_n} + \left\langle \frac{\partial F}{\partial u_1}, \ldots, \frac{\partial F}{\partial u_r}, 1 \right\rangle + \langle m_r \mathcal{E}_{n+r} \rangle$$

follows. The first summand is contained in $\mathcal{A} := \left\langle \frac{\partial F}{\partial x_1}, \ldots, \frac{\partial F}{\partial x_n} \right\rangle_{\mathcal{E}_{n+r}}$ and the second one is a subset of $\langle \mathcal{E}_r B \rangle$ with $B := \left\{ \left[\frac{\partial F}{\partial u_1} \right], \ldots, \left[\frac{\partial F}{\partial u_r} \right], [1] \right\}$. Therefore, $\mathcal{E}_{n+r} = \mathcal{A} + \langle \mathcal{E}_r B \rangle + \langle m_r \mathcal{E}_{n+r} \rangle$ holds. Now (9.1) can be applied to this equation yielding (3) and hence (c).

Finally, assume (c). In order to derive (b) apply φ^* to equation (3). The left side becomes $\mathcal{E}_n$. The first summand on the right side yields $\mathcal{J}[f]$, since $\varphi^* \left[\frac{\partial F}{\partial x_i} \right]$ equals $\left[\frac{\partial f}{\partial x_i} \right]$ for $i = 1, \ldots, n$. The image of the second summand under φ^* is $\mathcal{V}[F] + \mathbf{R}[1]$. Indeed, note that $\varphi^*(\mathcal{E}_r) = \mathbf{R}[1]$, since for $[K] \in \mathcal{E}_r$ the representative K does not depend on x according to the usual embedding of $\mathcal{E}_r$ into $\mathcal{E}_{n+r}$ (see the first paragraph of Chapter 9), so that $(K \circ \varphi)(x) = K(x, 0) = K(0)$ is constant. Moreover,

$$\left(\frac{\partial F}{\partial u_j} \circ \varphi \right)(x) = \left(\frac{\partial F}{\partial u_j}(x, 0) - \frac{\partial F}{\partial u_j}(0, 0) \right) + \frac{\partial F}{\partial u_j}(0, 0)$$

holds.

The main result of this chapter is the equivalence of versality and stability for unfoldings. The equivalence to versality persists if stability is replaced by one of the following apparently stronger properties (a), (b), or (c) from (5). These differ from each other in the topology on $C^\infty(\mathbf{R}^{n+r})$, which is used for the mathematical description of a weak perturbation.

Lemma (5). Let $[F] \in \mathcal{E}_{n+r}$ be an unfolding, and let F be defined on $\mathbf{R}^{n+r}$. Consider the following properties for $[F]$.

(a) For every neighborhood W of $0 \in \mathbf{R}^{n+r}$ there is a neighborhood $\mathcal{V}_{\tau, A}^l(F)$ with some compact $A \subset W$ (see before (11.8)) such that for every $G \in \mathcal{V}_{\tau, A}^l(F)$ there is a $(b, c) \in W$ satisfying $[G^{b,c}] \approx [F]$.

(b) In (a), replace $\mathcal{V}_{\tau, A}^l(F)$ by a general C^l-neighborhood $\mathcal{V}_\delta^l(F)$ (see the definition after (11.7)).

(c) For every neighborhood W of $0 \in \mathbf{R}^{n+r}$ and for finitely many arbitrary $H_1, \ldots, H_m$ in $C^\infty(\mathbf{R}^{n+r})$ there is an $\varepsilon > 0$ such that for every $t_1, \ldots, t_m$ in $]-\varepsilon, \varepsilon[$ there is a $(b, c) \in W$ satisfying $[(F + t_1 H_1 + \cdots + t_m H_m)^{b,c}] \approx [F]$.

(d) $[F]$ is stable.

Then the implications (a)$\Rightarrow$(b)$\Rightarrow$(c)$\Rightarrow$(d) hold.

Proof. The implication (a)$\Rightarrow$(b) holds because of (11.8). The implication (c)$\Rightarrow$(d) is obvious. It remains to show that (b) implies (c). Let an open neighborhood W of $0 \in \mathbf{R}^{n+r}$ and smooth functions $H_1, \ldots, H_m$ on $\mathbf{R}^{n+r}$ be given. Assume without restriction that the closure $\overline{W}$ is compact. Choose a C^l-neighborhood $\mathcal{V}_\delta^l(F)$ of F according to (b). Let β be a smooth function with compact support A such that $\beta|W = 1$ holds (see Exercise 11.2). Now for $\varepsilon > 0$ and $|t_j| < \varepsilon$, $j = 1, \ldots, m$, set

$$G := F + \beta \left(t_1 H_1 + \cdots + t_m H_m \right).$$

Then $(G - F)(p) = 0$ holds for $p \notin A$ and

$$|D^\nu(G - F)(p)| \le \varepsilon \sup \left\{ \sum_{j=1}^m |D^\mu \left(\beta H_j \right)(q)| : 0 \le |\mu| \le l, \ q \in A \right\} < \infty$$

holds for $0 \leq |\nu| \leq l$ and $p \in A$. Therefore, and since δ assumes on A a positive minimum, it follows for sufficiently small $\varepsilon > 0$ that $|D^{\nu}(G - F)(p)| < \delta(p)$ is valid for $0 \leq |\nu| \leq l$ and for all p. Hence

$$G \in \mathcal{V}_{\delta}^{l}(F)$$

follows. Since (b) is assumed, there exists a $(b, c) \in W$ such that $[G^{b,c}] \approx [F]$. Finally, $[G^{b,c}] = [(F + t_1 H_1 + \cdots + t_m H_m)^{b,c}]$ holds because of $\beta | W = 1$.

The further line of reasoning goes as follows. First it is shown that versality implies the property (a) from (5) (see (10)) and then that versality is implied by stability (see (11)). Because of the previous results (4) and (5), the final result is achieved in (12), namely that **versality** and **stability** as well as **infinitesimal stability** and the **properties** (a), (b), (c) in (5) are all equivalent properties of unfoldings. The considerations begin with a technical lemma and a supplement to it.

Lemma (6). Let $[K]$ be in m_{n+r+1}, and suppose that there are germs $[\alpha] \in \mathcal{E}_{n+r+1,n}$, $[\beta] \in \mathcal{E}_{r+1,r}$, and $[\eta] \in \mathcal{E}_{r+1}$ such that

$$\frac{\partial K}{\partial t}(x, u, t) = \sum_{i=1}^{n} \alpha_i(x, u, t)\frac{\partial K}{\partial x_i}(x, u, t) + \sum_{j=1}^{r} \beta_j(x, u, t)\frac{\partial K}{\partial u_j}(x, u, t) + \eta(u, t)$$

holds for small $(x, u, t) \in \mathbf{R}^{n+r+1}$.

Denote by $\Phi = (\varphi, \psi)$ with $[\varphi] \in \mathcal{E}_{n+r+1,n}$ and $[\psi] \in \mathcal{E}_{r+1,r}$ the unique local solution of the system of ordinary differential equations.

$$\frac{\partial \phi}{\partial t}(x, u, t) = -\alpha(\varphi(x, u, t), \psi(u, t), t)$$

$$\frac{\partial \psi}{\partial t}(u, t) = -\beta(\psi(u, t), t)$$

satisfying the initial condition

$$\Phi(x, u, 0) = (x, u).$$

Define $[\gamma] \in \mathcal{E}_{r+1}$ by

$$\gamma(u, t) := -\int_0^t \eta(\psi(u, s), s)ds.$$

Then

(7) $$K(\Phi(x, u, t), t) + \gamma(u, t) = K(x, u, 0)$$

holds for small x, u, t.

Now the notations $p := (x, u)$ and $K_t(p) := K(p, t)$ are used. Then there are open neighborhoods W and W_1 of $0 \in \mathbf{R}^{n+r}$ and $\delta > 0$ such that

Φ is defined on $W \times] - 2\delta, 2\delta[$ and such that for every $p_1 \in W_1$, $t_1 \in]-\delta, \delta[$, and $t_2 \in \mathbf{R}$ with $|t_2 - t_1| < \delta$, there exists a $p_0 \in W$ satisfying $p_1 = \Phi(p_0, t_1)$ and

$$(8) \qquad\qquad [(K_{t_1})^{p_1}] \approx [(K_{t_2})^{p_2}],$$

where $p_2 := \Phi(p_0, t_2)$.

Proof. By A.3 the initial value problem $y' = A(y, t)$ with $y(0) = (x, u)$, where $y = (y_1, y_2)$ and $A(y, t) := -(\alpha(y_1, y_2, t), \beta(y_2, t))$, has a unique local solution $\Phi = (\varphi, \psi)$, which depends smoothly on (x, u, t). In general, ψ would depend on x. But ψ solves the initial value problem $y_2' = -\beta(y_2, t)$, which is independent of x. Therefore, by A.3, its local solution, which is unique, is independent of x.

Because of $\Phi(x, u, 0) = (x, u)$, it follows from the assumption on $[K]$ that

$$\frac{\partial K}{\partial t}(\Phi(x, u, t), t) =$$

$$\sum_{i=1}^{n} \alpha_i (\Phi(x, u, t), t) \frac{\partial K}{\partial x_i}(\Phi(x, u, t), t) +$$

$$\sum_{j=1}^{r} \beta_j (\psi(u, t), t) \frac{\partial K}{\partial u_j}(\Phi(x, u, t), t) + \eta(\psi(u, t), t) =$$

$$-\sum_{i=1}^{n} \frac{\partial \varphi_i}{\partial t}(x, u, t) \frac{\partial K}{\partial x_i}(\Phi(x, u, t), t) -$$

$$\sum_{j=1}^{r} \frac{\partial \psi_j}{\partial t}(u, t) \frac{\partial K}{\partial u_j}(\Phi(x, u, t), t) - \frac{\partial \gamma}{\partial t}(u, t)$$

holds for small x, u, t. On the other hand, one has

$$\frac{d}{dt} K(\Phi(x, u, t), t) =$$

$$\frac{\partial K}{\partial t}(\Phi(x, u, t), t) + \sum_{i=1}^{n} \frac{\partial K}{\partial x_i}(\Phi(x, u, t), t) \frac{\partial \varphi_i}{\partial t}(x, u, t) +$$

$$\sum_{j=1}^{r} \frac{\partial K}{\partial u_j}(\Phi(x, u, t), t) \frac{\partial \psi_j}{\partial t}(x, u, t).$$

Combining these two equations yields $\frac{d}{dt} K(\Phi(x, u, t), t) + \frac{\partial}{\partial t}\gamma(u, t) = 0$. Therefore, $K(\Phi(x, u, t), t) + \gamma(u, t) = c(x, u)$ holds for some germ $[c]$ in $\mathcal{E}_{n+r}$. At $t = 0$ one gets $K(x, u, 0) = c(x, u)$. This proves (7).

As to (8), consider the following expressions, which are well defined for small p, p_0, and t. It is easy to verify by (7) that

$$(K_t)^{\Phi(p_0, t)} (\Phi(p + p_0, t) - \Phi(p_0, t)) = K_t (\Phi(p + p_0, t)) - K_t (\Phi(p_0, t)) =$$

$$K_0(p + p_0) - \gamma(u + u_0, t) - K_0(p_0) + \gamma(u_0, t) =$$
$$(K_0)^{p_0}(p) - \gamma(u + u_0, t) + \gamma(u_0, t)$$

holds. For small t the map $p \mapsto \tilde{\Phi}_t(p) := \Phi(p + p_0, t) - \Phi(p_0, t)$ is a local diffeomorphism at the origin with $\tilde{\Phi}_t(0) = 0$, since $\tilde{\Phi}_0(p) = p$. Hence, the above relations show that

(*) $$\left[(K_t)^{\Phi(p_0, t)}\right] \approx [(K_0)^{p_0}]$$

is valid for p_0 in some open neighborhood W of $0 \in \mathbf{R}^{n+r}$ and for t in some open interval I around $0 \in \mathbf{R}$.

The smooth map $\Psi : W \times I \to \mathbf{R}^{n+r} \times \mathbf{R}$ defined by $\Psi(p, t) := (\Phi(p, t), t)$ is a local diffeomorphism at the origin with $\Psi(0, 0) = (0, 0)$, since

$$D\Psi(0,0) = \begin{pmatrix} E_{n+r} & D_2\Phi(0,0) \\ 0 & 1 \end{pmatrix}$$

is invertible.

Without restriction let $W \times I$ be so small that Ψ is a diffeomorphism onto the open image $\Psi(W \times I)$. Choose an open neighborhood W_1 of $0 \in \mathbf{R}^{n+r}$ and an open interval I_1 around $0 \in \mathbf{R}$ such that $W_1 \times I_1$ is contained in $\Psi(W \times I)$. Moreover, let $I_1 =]-\delta, \delta[$ for some $\delta > 0$, and choose δ so small that $]-2\delta, 2\delta[$ is contained in I. This ends the preliminaries.

Now let $p_1 \in W_1$, $|t_1| < \delta$, and $|t_2 - t_1| < \delta$. There exists a unique element $(p_0, t_0) \in W \times I$ such that $\Psi(p_0, t_0) = (p_1, t_1)$. Then $t_0 = t_1$ and $p_1 = \Phi(p_0, t_1)$ follow. Hence, (*) yields

$$[(K_{t_1})^{p_1}] \approx [(K_0)^{p_0}].$$

Since $|t_2| \leq |t_2 - t_1| + |t_1| < 2\delta$, t_2 is in I. This time (*) yields

$$[(K_0)^{p_0}] \approx [(K_{t_2})^{p_2}],$$

where $p_2 := \Phi(p_0, t_2)$. Combining both results shows (8).

For $p \in \mathbf{R}^{n+r}$ denote by $|p|_{\max}$ the maximum norm of p, i.e., the maximum of the absolute values of the coordinates of p.

Supplement (9). Let $M > 0$ be a constant satisfying $M > |\alpha_i(0)|$ and $M > |\beta_j(0)|$ for all $i = 1, \ldots, n$ and all $j = 1, \ldots, r$. Then there exists an open neighborhood W of $0 \in \mathbf{R}^{n+r}$ and an open interval I around $0 \in \mathbf{R}$ such that

(*) $$|\Phi(p, t_2) - \Phi(p, t_1)|_{\max} < M |t_2 - t_1|$$

for all $p \in W$ and all $t_1, t_2 \in I$. Furthermore, there is an open neighborhood W_1 of $0 \in \mathbf{R}^{n+r}$ and $\delta > 0$ such that for every $p_1 \in W_1$, $t_1 \in]-\delta, \delta[$ and $t_2 \in]t_1 - \delta, t_1 + \delta[$ there exists $p_2 \in \mathbf{R}^{n+r}$ satisfying $|p_2 - p_1|_{\max} < M|t_2 - t_1|$ such that (8) holds.

Proof. It suffices to prove (*), since the remaining part of the assertion follows from (*) for $p = p_0$ and the result following (7).

Since $\Phi(0) = 0$, there is an open neighborhood $X \times U \times I \subset \mathbf{R}^n \times \mathbf{R}^r \times \mathbf{R}$ of the origin such that

$$|\alpha_i(\Phi(p,t),t)| < M \text{ and } |\beta_j(\psi(u,t),t)| < M$$

for $1 \leq i \leq n$, $1 \leq j \leq r$ and for all $(x,u,t) \in X \times U \times I$.

By integrating the differential equations in (6) from t_1 to t_2 one obtains immediately the estimate

$$|\alpha_i(p,t_2) - \alpha_i(p,t_1)| < M|t_2 - t_1|$$

for $i = 1, \ldots, n$ and the analogous estimate for β_j, $j = 1, \ldots, r$. These imply (*) for $W := X \times U$.

Theorem (10). Let $[F] \in \mathcal{E}_{n+r}$ be a versal unfolding. Then $[F]$ satisfies the property (a) in (5).

Proof. Without restriction let F be defined on the whole of $\mathbf{R}^{n+r}$ and let $F(0) = 0$ hold. Fix $k \geq r + 2$.

(i) Since $[F]$ satisfies (b) in (4),

$$J^k = j^k(\mathcal{J}[f]) + j^k(\mathcal{V}[F]) + \mathbf{R}j^k[1]$$

follows. Therefore, the N-dimensional linear space J^k is spanned by the $Nn + r + 1$ k-jets

$$j^k\left[x^\kappa \frac{\partial F}{\partial x_i}(\cdot, 0)\right] \text{ for } 0 \leq |\kappa| \leq k,\ 1 \leq i \leq n,$$

$$j^k\left[\frac{\partial F}{\partial u_j}(\cdot, 0)\right] \text{ for } 1 \leq j \leq r, \text{ and}$$

$$j^k[1].$$

Choose N linearly independent k-jets from these, and denote them by $e_\iota(F)$ for $\iota = 1, \ldots, N$. Then there is a compact neighborhood A of $0 \in \mathbf{R}^{n+r}$ such that for every $(b,c) \in A$

$$\{e_\iota(F^{b,c}) : \iota = 1, \ldots, N\}$$

is a basis of J^k. This is proved by the continuity arguments given in (11.2) up to (11.7).

(ii) Let $M > 0$ be a constant such that the cube $[-M, M]^{n+r}$ is contained in A. Now the following is claimed.

There is a $\tau > 0$ such that for every $G \in \mathcal{V}_{\tau,A}^{k+1}(F)$ and every $(b,c) \in A$

$$\{e_\iota \left(G^{b,c}\right) : \iota = 1, \ldots, N\}$$

is a basis of J^k. Furthermore, if $H \in \mathcal{V}_{\tau,A}^{k+1}(0)$ and if $[h^{b,c}]$ is the organization center of $[H^{b,c}]$ for $(b,c) \in A$, then the coefficients c_ι of the expansion

$$j^k \left[h^{b,c}\right] = \sum_\iota c_\iota e_\iota \left(G^{b,c}\right)$$

are bounded by M, i.e., $|c_\iota| < M$ for $\iota = 1, \ldots, N$.

As to the proof, denote by S the sphere in J^k around the origin with radius M. Then the map

$$\gamma : A \times S \to J^k, \quad \gamma(b,c;s) := \sum_\iota s_\iota e_\iota \left(F^{b,c}\right)$$

is continuous, and it does not vanish because of the linear independence of the basis vectors. Since $A \times S$ is compact, it follows that $|\gamma| \geq \delta$ for some $\delta > 0$. Once more by the continuity arguments at the beginning of Chapter 11 there is a $\tau > 0$ such that

$$\left|\sum_\iota s_\iota e_\iota \left(G^{b,c}\right)\right| \geq \frac{\delta}{2}$$

for all $(b,c) \in A$ and $G \in \mathcal{V}_{\tau,A}^{k+1}(F)$. If τ is chosen small enough, then also

$$\left|j^k \left[h^{b,c}\right]\right| < \frac{\delta}{2}$$

can be fullfilled for all $(b,c) \in A$ and $H \in \mathcal{V}_{\tau,A}^{k+1}(0)$. Then

$$\frac{\delta}{2} > \frac{|c|}{M} \left|\sum_\iota M \frac{c_\iota}{|c|} e_\iota (G^{b,c})\right| \geq \frac{|c|}{M} \frac{\delta}{2}$$

holds for $|c| := \left(\sum_\iota |c_\iota|^2\right)^{\frac{1}{2}}$. This shows $M > |c| \geq c_\iota$ for all ι.

(iii) The next steps will show that for every $H \in \mathcal{V}_{\tau,A}^{k+1}(0)$ there is a $(b,c) \in A$ such that

$$[(F+H)^{b,c}] \approx [F]$$

holds. Obviously, this will prove the assertion of the theorem.

(iv) Fix H in $\mathcal{V}_{\tau,A}^{k+1}(0)$. For $s \in [0,1]$, $(b,c) \in A$ define $K \in C^\infty(\mathbf{R}^{n+r+1})$ by

$$K(x,u,t) := (F + (t+s)H)(x+b, u+c) - (F + (t+s)H)(b,c).$$

Obviously $K(0,0,0) = 0$. Since $G_s := F + sH$ is in $\boldsymbol{V}_{\tau,A}^{k+1}(F)$, by (ii) the set $\{e_\iota(G_s^{b,c})\}$ is a basis of J^k. By the particular choice of the jets of the basis it follows that

$$\mathcal{E}_n = \boldsymbol{\mathcal{J}}\left[g_s^{b,c}\right] + \boldsymbol{\mathcal{V}}\left[G_s^{b,c}\right] + \mathbf{R}[1] + m^{k+1}$$

holds for $g_s^{b,c} := G_s^{b,c}(\cdot, 0)$. If $\boldsymbol{\mathcal{J}}[g_s^{b,c}] \subset m$, then clearly $[G_s^{b,c}]$ is k-transversal and hence versal by (10.4). If not, then $[G_s^{b,c}]$ is versal by Exercise 7.12.

Since $[K(\cdot,\cdot,0)] = [G_s^{b,c}]$, it follows that $[K]$ is an $(r+1)$-parameter versal unfolding. By (4) it is infinitesimally stable. Multiplying (3) for $[K]$ by m_{r+1} one obtains

$$m_{r+1}\mathcal{E}_{n+r+1} = \left\langle \frac{\partial K}{\partial x_1}, \dots, \frac{\partial K}{\partial x_n} \right\rangle_{m_{r+1}\mathcal{E}_{n+r+1}} + \left\langle \frac{\partial K}{\partial u_1}, \dots, \frac{\partial K}{\partial u_r}, \frac{\partial K}{\partial t}, 1 \right\rangle_{m_{r+1}}.$$

(v) By (ii) the expansion

$$j^k\left[h^{b,c}\right] = \sum_\iota c_\iota e_\iota\left(G_s^{b,c}\right)$$

holds with $|c_\iota| < M$ for all ι. Because of the particular choice of the jets of the basis this implies that there are germs $[k_i] \in \mathcal{E}_n$ with $|k_i(0)| < M$ and constants a and b_j with $|a| < M$, $|b_j| < M$, such that the equation

$$[h^{b,c}] = \sum_{\iota=1}^n [k_i]\left[\frac{\partial}{\partial x_i}G_s^{b,c}(\cdot,0)\right] + \sum_{j=1}^r b_j\left[\frac{\partial}{\partial u_j}G_s^{b,c}(\cdot,0)\right] + a + [w]$$

holds for some germ $[w]$ in m^{k+1}. By (10.3) the germ $[g_s^{b,c}]$ is k-determined. Hence, by (4.38) there are germs $[w_i]$ in m satisfying

$$[w] = \sum_{i=1}^n [w_i]\left[\frac{\partial}{\partial x_i}g_s^{b,c}\right].$$

Note the identities

$$h^{b,c} = \frac{\partial}{\partial t}K(\cdot,0,0),$$

$$\frac{\partial}{\partial x_i}g_s^{b,c} = \frac{\partial}{\partial x_i}G_s^{b,c}(\cdot,0) = \frac{\partial}{\partial x_i}K(\cdot,0,0),$$

$$\frac{\partial}{\partial u_j}G_s^{b,c}(\cdot,0) = \frac{\partial}{\partial u_j}K(\cdot,0,0).$$

Put $[k_i'] := [k_i + w_i]$. Then it follows

$$\left[\frac{\partial}{\partial t}K(\cdot,0,0)\right] = \sum_{i=1}^n [k_i']\left[\frac{\partial}{\partial x_i}K(\cdot,0,0)\right] + \sum_{j=1}^r b_j\left[\frac{\partial}{\partial u_j}K(\cdot,0,0)\right] + a,$$

where $|k_i'(0)| < M$ still holds since $w_i(0) = 0$.

(vi) The results from (iv) and (v) will be combined now. By (v) the germ $[L]$ in $\mathcal{E}_{n+r+1}$ defined by

$$[L] := \left[\frac{\partial K}{\partial t}\right] - \sum_{i=1}^{n} [k_i'] \left[\frac{\partial K}{\partial x_i}\right] - \sum_{j=1}^{r} b_j \left[\frac{\partial K}{\partial u_j}\right] - a$$

satisfies $L(x,0,0) = 0$ for small x. This means by (1.28) that $[L]$ actually is in $m_{r+1}\mathcal{E}_{n+r+1}$. Hence, according to the result in (iv) there are germs $[\alpha_i']$ in $m_{r+1}\mathcal{E}_{n+r+1}$ and $[\beta_j']$, $[\beta]$, $[\eta']$ in m_{r+1} such that

$$[L] = \sum_{i=1}^{n} [\alpha_i'] \left[\frac{\partial K}{\partial x_i}\right] + \sum_{j=1}^{r} [\beta_j'] \left[\frac{\partial K}{\partial u_j}\right] + [\beta] \left[\frac{\partial K}{\partial t}\right] + [\eta'].$$

Comparing this expression for $[L]$ with its definition one finds

$$\left[\frac{\partial K}{\partial t}\right] = \sum_{i=1}^{n} [\alpha_i] \left[\frac{\partial K}{\partial x_i}\right] + \sum_{j=1}^{r} [\beta_j] \left[\frac{\partial K}{\partial u_j}\right] + [\eta]$$

with $[\alpha_i] := [\alpha_i' + k_i']/[1-\beta]$, $[\beta_j] := [\beta_j' + b_j]/[1-\beta]$, and $[\eta] := [\eta' + a]/[1-\beta]$. Note that $1/[1-\beta]$ is well defined, since $[\beta]$ is in m_{r+1}. Moreover, it follows that $[\alpha_i] \in \mathcal{E}_{n+r+1}$ and $[\beta_j], [\eta] \in \mathcal{E}_{r+1}$ satisfy $|\alpha_i(0)| = |k_i'(0)| < M$, $|\beta_j(0)| = |b_j| < M$, and $|\eta(0)| = |a| < M$.

(vii) By the last result, Lemma (6) and its Supplement (9) apply to $[K]$. Accordingly there are an open neighborhood W_1 of $0 \in \mathbf{R}^{n+r}$ and a $\delta > 0$ such that for every $p_1 \in W_1$, $t_1 \in]-\delta, \delta[$, and $t_2 \in]t_1 - \delta, t_1 + \delta[$ there exists $p_2 \in \mathbf{R}^{n+r}$ with $|p_2 - p_1|_{\max} < M|t_2 - t_1|$ satisfying

$$[(K_{t_1})^{p_1}] \approx [(K_{t_2})^{p_2}].$$

Hence, for every $a := (b,c) \in A$ and every $s \in [0,1]$ one has

$$\left[G_{s+t_1}^{a+p_1}\right] \approx \left[G_{s+t_2}^{a+p_2}\right].$$

Of course, W_1 and δ depend on (a,s). The open sets $(a,s) + W_1 \times]-\delta, \delta[$ cover the compact set $A \times [0,1]$. Thus, there are finitely many of them that cover it. Denoting by Δ the minimum of the occurring δ values, the following result is proven:

If $(p,t) \in A \times [0,1]$ and $t' \in \mathbf{R}$ with $|t'-t| < \Delta$, then there is a $p' \in \mathbf{R}^{n+r}$ with $|p' - p|_{\max} < M|t' - t|$ satisfying

$$[G_t^p] \approx \left[G_{t'}^{p'}\right].$$

(viii) Let L be a positive integer with $1/L < \Delta$. Then, by (vii), there is an $a_1 \in \mathbf{R}^{n+r}$ with $|a_1|_{\max} < M/L$ satisfying $[G_0^0] \approx [G_{1/L}^{a_1}]$. Recall that $]-M, M[^{n+r} \subset A$, whence $a_1 \in A$. Then, by (vii), there is $a_2 \in \mathbf{R}^{n+r}$

with $|a_2 - a_1|_{\max} < M/L$ satisfying $[G_{1/L}^{a_1}] \approx [G_{2/L}^{a_2}]$. Then $[G_0^0] \approx [G_{2/L}^{a_2}]$ follows. Since $|a_2|_{\max} \leq |a_2 - a_1|_{\max} + |a_1|_{\max} < 2M/L$, also $a_2 \in A$ holds.

After L steps, $a_L \in \mathbf{R}^{n+r}$ is reached satisfying $|a_L|_{\max} < M$, whence $a_L \in A$, and

$$[F] = [G_0^0] \approx [G_1^{a_L}] = [(F + H)^{a_L}].$$

Thus (iii) holds, and the proof is finished.

Theorem (11). Let $[F] \in \mathcal{E}_{n+r}$ be a stable unfolding. Then $[F]$ is versal.

Proof. Without restriction let $F \in C^\infty(\mathbf{R}^{n+r})$ and $F(0) = 0$. By (11.21) the orbit $z\mathcal{G}^k$ through $z := j^k[f]$ with $f := F(\cdot, 0)$ is the union of countably many regular submanifolds. Therefore, according to (20), $(F + P)^k$ is transverse to $z\mathcal{G}^k$ for all P in $P^k \setminus \mathbf{R}$ that do not lie in a Lebesgue null set $\Lambda(F)$.

Let $\sigma \in \mathbf{N}$. Then $\left(F + \frac{1}{\sigma}P\right)^k$ is transverse to $z\mathcal{G}^k$ for all $P \notin \sigma\Lambda(F)$. Since $\sigma\Lambda(F)$ is a Lebesgue null set, the countable union $\Lambda := \bigcup_\sigma \sigma\Lambda(F)$ is a Lebesgue null set, too.

Take P_0 in $P^k \setminus \mathbf{R}$ but not in Λ. Then $\left(F + \frac{1}{\sigma}P_0\right)^k$ is transverse to $z\mathcal{G}^k$ for all $\sigma \in \mathbf{N}$. Since $[F]$ is stable, there is a $\sigma \in \mathbf{N}$ and a $p \in \mathbf{R}^{n+r}$ such that $[F]$ and $[G] := \left[\left(F + \frac{1}{\sigma}P_0\right)^p\right]$ are isomorphic. This implies that $[f]$ and $[g]$ are equivalent for $g := G(\cdot, 0)$.

Hence, $[g] \in z\mathcal{G}^k$, and $[G]$ is k-transversal when $[g] \in m^2$. It follows from (11.2) that $[G]$ is versal. Thus, $[F]$ is versal, too, by Exercise 12.1.

As announced, the results (4), (5), (10), and (11) yield immediately

Corollary (12). Let $n \in \mathbf{N}$ and $r \in \mathbf{N}_0$. Then for any unfolding in $\mathcal{E}_{n+r}$, versality, infinitesimal stability, and each of the properties (a), (b), (c) in (5) is equivalent to stability.

Exercises

12.1. Show that the relation of being isomorphic is an equivalence relation for unfoldings. Show also that two isomorphic unfoldings are versal if one of them is versal.

12.2. Let n and r be integers satisfying $n \geq 1$ and $r \geq 0$. Let $W \subset \mathbf{R}^{n+r}$ be open containing the origin, and let $[H] \in \mathcal{E}_{n+r}$ be a versal unfolding. Show that the set

$$\mathcal{H} := \left\{F \in C^\infty(\mathbf{R}^{n+r}) : [F^{x,u}] \not\approx [H] \text{ for all } (x, u) \in W\right\}$$

is not C^∞-dense in $C^\infty(\mathbf{R}^{n+r})$.

Speaking qualitatively as in the introduction to Chapter 11, this result implies that none of the elementary unfoldings is exceptional for $r \leq 4$.

Solutions

12.1. Let $[F]$ and $[G]$ be unfoldings in $\mathcal{E}_{n+r}$. In what follows, the definition of isomorphic unfoldings will be tacitly referred to.

Obviously, reflexivity $[F] \approx [F]$ holds choosing $\omega(y, v) := y$, $\psi(v) := v$, and $\gamma(v) := 0$ for all (y, v).

As to symmetry, let $[G] \approx [F]$. This is equivalent to $[G - G(0,0)] \approx [F - F(0,0)]$, since $\gamma(0)$ can be chosen accordingly. Hence, it is no restriction to assume $[F], [G] \in m_{n+r}$. Then $[g] = [f][\omega_0]$ holds for the organization centers $[f]$ and $[g]$ of $[F]$ and $[G]$, respectively. Using the notation at the end of Chapter 7,

$$\chi^*[G](y, v) = G(\chi(y), v) = F(\omega(\chi(y), v), \psi(v)) + \gamma(v)$$

follows, where $[\chi] := [\omega_0]^{-1}$, $\omega(\chi(y), 0) = \omega_0(\chi(y)) = y$ for small y, $[\gamma] \in m_r$, and $[\psi] \in \mathcal{G}_r$. Hence, $\chi^*[G]$ and $[F]$ are equivalent. Therefore, $[F]$ is induced by $\chi^*[G]$ so that the parameter transformation is a local diffeomorphism. This immediately implies $[F] \approx [G]$.

Transitivity is verified analogously to (7.6).

Finally, it is shown in Chapter 7 that equivalence as well as χ^* preserve versality of unfoldings.

12.2. It is no restriction to assume that H is defined on the whole of $\mathbf{R}^{n+r}$. Since property (a) of Lemma (5) holds by Theorem (10), there is a compact set $A \subset W$ and a $\tau > 0$ such that for every $G \in \mathcal{V}_{\tau,A}^l(H)$ there is a $(b, c) \in W$ satisfying $[G^{b,c}] \approx [H]$. Therefore $\mathcal{V}_{\tau,A}^l(H) \cap \mathcal{H} = \emptyset$ holds. Since $\mathcal{V}_{\tau,A}^l(H)$ is C^∞-open by (11.8), this proves the assertion.

Appendix

Some standard calculus theorems used in this book are given in A.1 through A.5. In A.6 a version of Sard's Theorem is presented.

A.1 Implicit Function Theorem

Let $U \subset \mathbf{R}^n$ and $V \subset \mathbf{R}^m$ be open and let $(a, b) \in U \times V$. Suppose that $F : U \times V \to \mathbf{R}^m$ has continuously differentiable components, that $F(a, b) = 0$, and that the Jacobian at b of the partial map

$$V \to \mathbf{R}^m, \quad y \mapsto F(a, y)$$

is not zero. Then there are open balls $U' \subset U$ and $V' \subset V$ with centers a and b, such that there is a unique map $\psi : U' \to V'$ satisfying $F(x, \psi(x)) = 0$ for all x in U'. It follows that $\psi(a) = b$ and that the components of ψ are continuously differentiable. The Jacobi matrix of ψ at x satisfies

$$D\psi(x) = -[D_2 F(x, \psi(x))]^{-1} D_1 F(x, \psi(x)),$$

where $D_1 F(x, y)$ denotes the Jacobi matrix at x of the partial map $U \to \mathbf{R}^m, x \mapsto F(x, y)$ for $y \in V$ fixed; $D_2 F(x, y)$ is defined analogously. Finally, if the components of F are k times continuously differentiable for $k \in \mathbf{N}$, then the same is true for those of ψ.

Proof. See the proof of [Die, (10.2.1)].

A.2 Derivative of a Limiting Function

Let $U \subset \mathbf{R}^n$ be open and connected and let (f_k) be a sequence of differentiable functions $f_k : U \to \mathbf{R}$. Suppose that (i) there is an $a \in U$ such that $(f_k(a))$ is convergent and that (ii) for every $x \in U$ there is a ball $B(x)$ around x contained in U such that (Df_k) converges uniformly on $B(x)$. Then (f_k) converges uniformly on $B(x)$ for every $x \in U$ and, if

$$f(x) := \lim_{k \to \infty} f_k(x),$$

251

then f is differentiable with

$$Df(x) = \lim_{k \to \infty} Df_k(x) \quad \text{for} \quad x \in U.$$

Proof. See [Die, (8.6.3)].

A.3 Existence of Solutions of a Differential Equation

Let $U \subset \mathbf{R}^n$ and $J \subset \mathbf{R}$ be open, and let $F \colon U \times J \to \mathbf{R}^n$ be smooth. Then, for any $(a, s) \in U \times J$:

(a) There is an open set V in $\mathbf{R}^n$ with $a \in V \subset U$ and an open interval I with $s \in I \subset J$, such that for every $(x_0, t_0) \in V \times I$ there exists a unique solution $I \to U, t \mapsto u(x_0, t_0; t)$, of

$$x' = F(x, t)$$

satisfying $u(x_0, t_0; t_0) = x_0$. The map $V \times I \times I \to \mathbf{R}^n, (x_0, t_0, t) \mapsto u(x_0, t_0; t)$, is smooth.

(b) There is an open set $W \subset \mathbf{R}^n$ with $a \in W \subset V$ such that for any $(x_0, t_0, t) \in W \times I \times I$ the equation $x_0 = u(x, t; t_0)$ has a unique solution $x = u(x_0, t_0; t)$ in V.

Proof. See [Die, (10.8.1), (10.8.2)].

A.4 Rank Theorem

Let $U \subset \mathbf{R}^n$ be open, let $p \in U$, and let $F \colon U \to \mathbf{R}^m$ be smooth such that the rank r of the Jacobi matrix $DF(x)$ is constant for $x \in U$. Then there exists

(a) a diffeomorphism u of an open set $W \subset \mathbf{R}^n$ with $p \in W \subset U$ onto the open cube $K := \{x \in \mathbf{R}^n \colon |x_i| < 1, i = 1, \ldots, n\}$ with $u(p) = 0$, and

(b) a diffeomorphism v of an open set $V \subset \mathbf{R}^m$ with $F(p) \in V$ onto the open cube $L := \{x \in \mathbf{R}^m \colon |x_i| < 1, i = 1, \ldots, m\}$ with $v(F(p)) = 0$,

such that

$$F(x) = v^{-1}(F_0(u(x))) \quad \text{for all} \quad x \in W,$$

where

$$F_0 \colon K \to L, \quad (x_1, \ldots, x_n) \mapsto (x_1, \ldots, x_r, 0, \ldots, 0).$$

Proof. See [Die, (10.3.1)].

A.5 Green's Theorem

The following theorem is also referred to as **Stokes' Theorem** in the plane. Let U be an open set in $\mathbf{R}^2$. Suppose that D is a compact subset of U such that its boundary ∂D is formed by finitely many continuously differentiable curves. If f and g are continuously differentiable functions on U, then

$$\int_{\partial D} f\,dx + g\,dy = \iint_D \left(\frac{\partial}{\partial x} g - \frac{\partial}{\partial y} f \right) dx\,dy$$

holds when the interior of D lies to the left of ∂D. Considering U as a subset of the complex plane and using the notation

$$z := x + iy, \quad dz := dx + idy, \quad \frac{\partial}{\partial \bar{z}} := \frac{1}{2}\left(\frac{\partial}{\partial x} + i\frac{\partial}{\partial y} \right),$$

and $dz\,d\bar{z} := -2i\,dx\,dy$, the above formula becomes

$$\int_{\partial D} F\,dz + \iint_D \frac{\partial}{\partial \bar{z}} F\,dz\,d\bar{z} = 0$$

for every complex-valued function $F(z) = f(x,y) + ig(x,y)$ on U whose real part f and imaginary part g are continuously differentiable.

Proof. See [MW, 1.8.4].

A.6 Sard's Theorem

Let S be a regular submanifold of $\mathbf{R}^n$ and $f : S \to \mathbf{R}^m$ a smooth map. Then

$$D := \left\{ y \in \mathbf{R}^m : \operatorname{Im}Df(x) = \mathbf{R}^m \text{ for all } x \in f^{-1}(\{y\}) \right\}$$

is the complement of a Lebesgue null set.

Proof. See [M,VII.2].

The following remarks serve to explain the foregoing theorem.

(i) By definition, the set D contains $\mathbf{R}^m \setminus f(S)$.

(ii) The smoothness of f is defined as follows. For every $p \in S$ let $\varphi : U \to V$ be a diffeomorphism on an open neighborhood U of p in $\mathbf{R}^n$ onto an open set $V \subset \mathbf{R}^n$ such that $\varphi(U \cap S) = \{ y \in V : y_{d+1} = \cdots = y_n = 0 \}$, where d denotes the dimension of S. Let $\beta := (\varphi_1, \dots, \varphi_d)|(U \cap S)$. Then $\beta(U \cap S)$ is open in $\mathbf{R}^d$, and f is said to be **smooth** if

$$f \circ \beta^{-1} : \beta(U \cap S) \to \mathbf{R}^m$$

is smooth. It is easy to verify that this property does not depend on the particular choice of φ.

(iii) Let $c : I \to \mathbf{R}^n$ be a curve on $\mathbf{R}^n$ based at p (see before (4.37)) satisfying $c(I) \subset S$. Then $f \circ c$ is a curve in $\mathbf{R}^m$, and the **differential of f at p** is defined by

$$Df(p) : T_pS \to \mathbf{R}^m, \quad Df(p)c'(0) := (f \circ c)'(0).$$

This definition does not depend on the particular choice of c, since $f \circ c = (f \circ \beta^{-1}) \circ (\varphi_1, \ldots, \varphi_d) \circ c$. Therefore, $(f \circ c)'(0)$ depends only on $c'(0)$ due to the chain rule.

(iv) A set $\Lambda \subset \mathbf{R}^m$ is a **Lebesgue null set** if for every $\varepsilon > 0$ there is a cover of Λ by countably many open cubes such that the sum of their volumes is less than ε.

(v) Unions of countably many Lebesgue null sets obviously are Lebesgue null sets. It is also clear that open sets are not Lebesgue null sets. Therefore, the complement of a Lebesgue null set, as e.g., the set D in Sard's Theorem, is dense.

References

[A]   Arnol'd, V.I., Critical Points of Smooth Functions and Their Normal Forms, Russian Math. Surveys *30*, 1975, 1–75.

[BCR]   Bochnak, J., M. Coste and M.-F. Roy, Géométrie algébrique réelle., Springer, Berlin 1987.

[Be]   Berry, M.V., Beyond Rainbows, Current Science *59*, 1990, 1175–1191.

[BL]   Bröcker, T. and L.C. Lander, Differentiable Germs and Catastrophes, London Math. Society Lecture Note Series *17*, Cambridge University Press, Cambridge 1975.

[BM]   Bierstone, E. and M. Milman, Semianalytic and Subanalytic Sets, IHES, 1988; or: Publications Mathematiques *67*, 1988, 5–42.

[BR]   Benedetti, R. and J.-J. Risler, Real algebraic and semi-algebraic sets, Hermann, Paris 1990.

[Bra]   Braun, M., Differential Equations and their Applications, Springer, New York 1978.

[Bro]   Brown, C., Chaos and Catastrophe Theories: Nonlinear Modeling in the Social Sciences, Sage University Papers, Thousand Oaks 1995.

[BU]   Berry, M.V. and C. Upstill, Catastrophe Optics: Morphologies of Caustics and their Diffraction Patterns; in: Progress in Optics XVIII, ed. E. Wolf, North Holland 1980.

[C1]   Casti, J., Alternative Realities: Mathematical Models of Nature and Man, John Wiley & Sons, New York 1989.

[C2]   Casti, J., Topological Methods for Social and Behavioural Systems, Int. Journal of General Systems *8*, 1982, 187–210.

[CH]   Castrigiano, D.P.L. and S. Hayes, Orbits of Lie Group Actions are Weakly Embedded, ArXiv, math.DG/0011241, 2000.

[Co]   Coste, M., Ensembles Semi-algebriques, Lecture Notes Math. *959*, Springer, Berlin 1982, 109–138.

[D]   Demazure, M., Bifurcations and Catastrophes, Springer, Berlin 2000.

[Die]   Dieudonné, J., Foundations of Modern Analysis, Academic Press, New York 1969.

[GG]   Golubitsky, M. and V. Guillemin, Stable Mappings and their Singularities, Springer, New York 1973.

[Gib]   Gibson C.G., Singular Points and Smooth Mappings, Pitman, London 1979.

[Gil]   Gilmore, R., Catastrophe Theory for Scientists and Engineers, John Wiley & Sons, New York 1981.

[GM]   Gromoll, D. and W. Meyer, On Differentiable Functions with Isolated Critical Points, Topology *8*, 1969, 361–369.

[Gol]   Golubitsky, M., An Introduction to Catastrophe Theory and Its Applications, Siam Review *20*, 1978, 352–387.

[Gu]   Guckenheimer, J., The Catastrophe Controversy, The Mathematical Intelligencer 1, Springer, New York 1978, 15–20.

[H]   Hofmaier, F., Non-equivalent germs remain non-equivalent when adding quadratic forms in new variables: Nonlinear Analysis: Series A Theory and Methods *53*, 2003, 951–956.

[HS]   Hirsch, M.W. and S. Smale, Differential Equations, Dynamical Systems, and Linear Algebra, Academic Press, New York 1974.

[J]   Jänich, K., Caustics and Catastrophes, Math. Ann. *209*, 1974, 161–180.

[KMS]   Kolář, I.P.W., P.W. Michor and J. Slovák, Natural Operations in Differential Geometry, Springer, Berlin 1993.

[L]   Lojasiewicz, S., Ensembles Semi-Analytique, IHES, 1965.

[LJH]   Ludwig, D., D.D. Jones and C.S. Holling, Qualitative Analysis of Insect Outbreak Systems: The Spruce Budworm and Forest, Journal of Animal Ecology *47*, 1978, 315–332.

[M]   Martinet, J., Singularities of Smooth Functions and Maps, Cambridge University Press, Cambridge 1982.

[M III] Mather, J., Stability of C^∞ mappings: III. Finitely Determined Map-Germs, Publ. Math. IHES *35*, 1968, 127–156.

[MW]   Marsden, J. and A. Weinstein, Calculus III, Springer, New York 1985.

[P]   Pradines, J., How to Define the Graph of a Singular Foliation, Cahiers de Top. et Géom. Diff. *26*(4), 1985, 339–380.

[Pe]   Peters, E.E., Chaos and Order in the Capital Markets, John Wiley, Chichester 1996.

[PS]   Poston, T. and I.N. Stewart, Catastrophy Theory and its Applications, Pitman, London 1978.

[R]   Rosser, J.B., Jr., From Catastrophe to Chaos: A General Theory of Economic Discontinuities, Kluver, Boston 1994.

[Sch]   Schäffler, S., Classification of Critical Stationary Points in Unconstrained Optimization, Siam J. Optimization, 1992, 1–6.

[Sei]   Seif F.J., Cusp Bifurcation in Pituitary Thyrotropin Secretion; in: Structural Stability in Physics. Proceedings of Two International Symposia on Applications of Catastrophe Theory and Topological Concepts in Physics, ed. W. Güttinger and H. Eikemeier, Springer, Berlin 1979.

[Sie]   Siersma, D., The Singularities of C^∞-Functions of Right-Codimension Smaller than or Equal to Eight, Indag. Math. *25*, 1973, 31–37.

[Sin]   Sinha D.K., editor: Catastrophe Theory and Applications, John Wiley & Sons, New York 1981.

[Sma]   Smale, S., The Mathematics of Time; in: Essays on Dynamical Systems, Economic Processes and Related Topics, Springer, New York 1980.

[SZ]   Sussmann, H.J. and R.S. Zahler, Catastrophe Theory as Applied to the Social and Biological Sciences: A Critique, Synthese *37*, 1978, 117–216.

[T1]   Thom, R., Stabilité Structurelle et Morphogénèse, Benjamin, Reading 1972.

[T2]   Thom, R., Semio Physics: A Sketch—Aristotelian Physics and Catastrophe Theory, Addison-Wesley, Redwood City 1990.

[Tou]   Tougeron, J.C., Idéaux de fonctions différentiables I, Ann. Inst. Fourier *18*, 1968, 177–240.

[Tro]   Trotman, D.J.A., The Classification of Elementary Catastrophes of Codimension ≤ 5, in [Z1].

[U]   Ursprung, H.W., Die elementare Katastrophentheorie: Eine Darstellung aus der Sicht der Ökonomie, Springer Lecture Notes in Economics and Math. Systems *195*, 1982.

[W]   Wassermann, G., Stability of Unfoldings, Springer Lecture Notes in Mathematics *393*, 1971.

[Z1]   Zeeman, E.C., Catastrophe Theory (Selected Papers 1972–1977), Addison-Wesley, Reading 1977.

[Z2]   Zeeman, E.C., Applications of Catastrophe Theory, Proc. Int. Conf. on Manifolds and Related Topics in Topology, University Tokyo Press 1975.

Notation Index

Subject Index

9 780813 341255